NEW QUERIES I

ANALECTA HUSSERLIANA

THE YEARBOOK OF PHENOMENOLOGICAL RESEARCH

VOLUME XXXVII

Editor-in-Chief:

ANNA-TERESA TYMIENIECKA

The World Institute for Advanced Phenomenological Research and Learning
Belmont, Massachusetts

PHENOMENOLOGY IN THE WORLD
FIFTY YEARS AFTER THE DEATH OF EDMUND HUSSERL

Book 1 THE TURNING POINTS OF THE NEW PHENOMENOLOGICAL ERA
Husserl Research — Drawing upon the Full Extent of His Development

Book 2 HUSSERLIAN PHENOMENOLOGY IN A NEW KEY
Intersubjectivity, Ethos, the Societal Sphere, Human Encounter, Pathos

Book 3 HUSSERL'S LEGACY IN PHENOMENOLOGICAL PHILOSOPHIES
New Approaches to Reason, Language, Hermeneutics, the Human Condition

Book 4 NEW QUERIES IN AESTHETICS AND METAPHYSICS
Time, Historicity, Art, Culture, Metaphysics, the Transnatural

The Editor acknowledges the assistance of Robert Wise in the technical preparation of these volumes.

NEW QUERIES IN AESTHETICS AND METAPHYSICS

Time, Historicity, Art, Culture, Metaphysics, the Transnatural

BOOK 4

Phenomenology in the World Fifty Years after the Death of Edmund Husserl

Edited by

ANNA-TERESA TYMIENIECKA

The World Phenomenology Institute

Published under the auspices of

The World Institute for Advanced Phenomenological Research and Learning

A-T. Tymieniecka, President

SPRINGER SCIENCE+BUSINESS MEDIA, B.V.

Library of Congress Cataloging-in-Publication Data

New queries in aesthetics and metaphysics : time, historicity, art, culture, metaphysics, the transnatural / edited by Anna-Teresa Tymieniecka.
p. cm. -- (Analecta Husserliana ; v. 37) (Phenomenology in the world fifty years after the death of Husserl ; bk. 4)
English, French, German, and Spanish.
Chiefly papers from the First World Congress of Phenomenology held in Santiago de Compostela, Spain, Sept. 26-Oct. 1, 1988.
"Published under the auspices of the World Institute for Advanced Phenomenological Research and Learning."
Includes bibliographical references.
ISBN 978-94-010-5501-7 ISBN 978-94-011-3394-4 (eBook)
DOI 10.1007/978-94-011-3394-4
1. Husserl, Edmund, 1859-1938--Congresses. 2. Phenomenology--Congresses. 3. Aesthetics, Modern--20th century--Congresses. 4. Metaphysics--History--20th century--Congresses. 5. History--Philosophy--History--20th century--Congresses. I. Tymieniecka, Anna-Teresa. II. World Congress of Phenomenology (1st : 1988 : Santiago de Compostela, Spain) III. Series. IV. Series: Phenomenology in the world fifty years after the death of Husserl ; bk. 4.
B3279.H94A129 vol. 37
[B829.5]
142'.7 s--dc20
[142'.7] 91-8232

ISBN 978-94-010-5501-7

Originally published by Kluwer Academic Publishers in 1991
Softcover reprint of the hardcover 1st edition 1991

TABLE OF CONTENTS

PART THREE
THE LIFE-SIGNIFICANCE OF LITERATURE AND ITS INTERPRETATION

PART FOUR
METAPHYSICAL ISSUES IN AESTHETICS

FOREWORD

This collection is the final volume of a four book survey of the state of phenomenology fifty years after the death of Edmund Husserl. Its publication represents a landmark in the comprehensive treatment of contemporary phenomenology in all its vastness and richness. The diversity of the issues raised here is dazzling, but the main themes of Husserl's thought are all either explicitly treated, or else they underlie the ingenious approaches found here. Time, historicity, intentionality, eidos, meaning, possibility/reality, and teleology are the main concerns of this collection devoted to studies in aesthetics, metaphysics and literary interpretation, written by such authors as, among others, R. Cobb-Stevens, C. Moreno Marquez, J. Swiecimski, Sitansu Ray and M. Kronegger.

These original studies of phenomenological aesthetics and literary theory by scholars from all parts of the world were gathered by the World Institute for Advanced Phenomenological Research and Learning during the year 1988/89 during its assessment of the phenomenological movement, fifty years after Husserl's death.

A-T. Tymieniecka (ed.), Analecta Husserliana, Vol. XXXVII, ix.

ANNA-TERESA TYMIENIECKA

WORLD-WIDE PHENOMENOLOGY FULFILLING HUSSERL'S PROJECT

An Introduction

What is the status of Husserl's phenomenology today? Does it play any significant role or is it relegated to strictly historical research? Has the phenomenology initiated by Husserl come to an end? There is hardly any orthodox Husserlian today. But what is or could be an orthodox Husserlian?

These questions come to mind when, even after fifty years of discussions among scholars since the death of this great master of phenomenology, we do not have a unified interpretation of his thought. Moreover, such a unifying interpretation is altogether impossible in view of Husserl's unfolding of his ever-expanding doctrine down to the very end of his life, and of his reaching ever-new perspectives. The possibility of a consensus about his thought recedes further and further as rival or competing interpretations have stimulated new phenomenologists and younger representatives to move in their own directions, often stimulated by non-Husserlian factors and nourished by new ideas. Lastly, the now vast field of research claiming allegiance to phenomenology is diversified into numerous sectors inspired by the developing thought of other classic phenomenologists and their followers.

As a matter of fact, it is often pointed out that phenomenology as a philosophical trend is not due to one single thinker but was somehow "in the air" at the beginning of this century. We trace its direct origins to Franz Brentano and Edmund Husserl who, as the disciple interpreting the master's intuitions in his own fashion, had elaborated the starting point and foundations of phenomenology as a *philosophia prima*. Yet, we acknowledge that the vigor, decisiveness, convincing force, dissemination, as well as its launching as a new philosophical approach by Husserl was supported, invigorated and carried out by colleagues and friends who gathered around Husserl, such as Moritz Geiger, Fritz Kaufmann, Adolph Reinach, A. Pfaender, Oscar Becker and Max Scheler. They joined Husserl in his convictions while he inspired and formed a group of students around him. Their work not only contributed initially to launching the main *porte parole* of this new

A-T. Tymieniecka (ed.), Analecta Husserliana, Vol. XXXVII, xi—xx.

way of thinking, the *Jahrbuch für Philosophie*, but their own original phenomenological research has inspired in the past and is now inspiring phenomenological investigations in various regions of philosophical questions that they respectively undertook to investigate.

In short, it is obvious that the powerful current of thought into which phenomenology gathered its momentum was the result of the meeting of several minds, meeting in a strong conviction and prompted by their personal inventive and talented efforts. It was truly a significant moment in the history of Occidental culture that gave rise to this trend as it is certainly also a significant situation of contemporary culture at large that phenomenology, after having formed a school of thought, did not fold its wings after one or two generations as did NeoKantianism but rather is being acutely heard within the world, not only Occidental or Oriental, but within the world wherever the present culture calls for genuine philosophical inspiration.

In view of this vast expanse of thought and research which go on in the present day in lines of innumerable diversifications, we naturally must ask whether there is still a trend of shared features that could fall under the common label of "phenomenology." I answer this question emphatically in the affirmative. It is precisely in pointing to some basic ideas of Husserl that they converge.

Don't we find, in fact, a pervading thread of the idea of intentionality, although *extended to new areas*? Is not the expansion of phenomenological inquiry due to the *discovery of the work of constitution in previously unsuspected areas*? In mentioning here just these two main tenets of classic phenomenology expanded into present-day thought, we cannot overlook the fulguration of thought provoked by inquiries into the later Husserl's intuitions and the subsequent *discoveries of historical, cultural and life elements entering into and affecting present experience.*

Recognizing, on the one hand, the essential contributions to the classic phenomenological foundation-laying phase of phenomenology by Husserl's associates, then and now a valid source of our investigation, and, on the other hand, the innovative philosophical work by the following generation, Sartre, Merleau-Ponty, Rombach and others not forgetting such mavericks as Heidegger and Ortega y Gasset which improved upon the pioneering ideas of the Husserl of his earlier and middle period, we cannot fail to acknowledge the central role which the work of Husserl plays within the entire phenomenologically oriented orbit and far beyond it. Indeed, the immense, painstaking, indefatigable and ever-improving effort of Husserl to find ever-deeper and more reliable foundations for the philosophical enterprise (as well as his constant *critical* re-thinking and perfecting of the approach and so-called "method" in order to perform this task and thus cover in this source-excavation an ever more far-reaching groundwork) stands out and maintains itself as an inepuisable reservoir for philosophical reflection in which all the above-mentioned work has either its core or its source. In fact, in his undertaking to re-think the entire philosophical enterprise as such and to recreate philosophy upon what he sought to be at least a satisfactorily legitimated basis, Husserl, through his already systematised and "authorized" work, and his courses, and later on in his spontaneous reflection (which did not find its way into a definitive corpus but was nevertheless sufficiently coherent with his previously established body of thought to be considered a continuation of it), uncovers perspectives upon the universe of man and projects their new philosophical thematisation that brings together all the attempts by philosophers (e.g., Merleau-Ponty, who drew upon this material and found there his own inspiration) who succeeded him with foundational intentions; it also gives a core of philosophical ideas and insights for the younger generation of philosophers today.

It is also true that the present-day culture — not only this or that specific culture but what we might call the cultural spirit of the world — shows a receptivity, a thirst for the ideas which only phenomenology appears able to offer. It is also true that the cultural climate of the last two decades fostered a new dynamism in those who are phenomenologically inspired, one even more vigorous than before. As its result, phenomenology today is completing an entire phase of its self-critical course, the *third phase* which I announced two decades ago (*Analecta Husserliana*, Vol. II, 1971).

As a matter of fact, because of the fundamentally self-critical character of phenomenological principles (cf. A-T. Tymieniecka "Phenomenology Reflects Upon Itself," I and II in *Analecta Husserliana*, Vols. II and III), there is today an enormous proliferation of thought in new and very diverse directions which, however, remain attached to the basic tenets of phenomenology. And this crucial significance of the self-critical principles of phenomenology applies in the strongest sense to Husserl himself who, as pointed out above, has not only sought to perfect his approaches and formulations but also in this self-critical effort expanded his range of positive, constructive insights in various circuits of reflection. In fact, since the Second World War, Husserl research and the influence of his thought have followed the progressive advance of Husserl himself as the various posthumous publications secured by Husserl disciples and directed by the enthusiastic Fr. H. L. Van Breda and released from the Husserl Archives at Leuven to the expectant philosophical world: With each major volume the perspectives upon Husserl's thought have changed and expanded. Now, as we read in the latest publications of his inedita (e.g. *Intersubjektivität, Ethische Vorlesungen*, . . .), Husserl's thought seems to have encompassed an entire cycle of philosophical reflection upon the human being within his life-world and even beyond it leading toward the divinity. It is from this complete cycle that the present-day generation of phenomenological scholars draws inspiration and enlightenment. For this and other major reasons which we will briefly treat below, the present four-volume collection not only gives us the essential panorama of what phenomenology is at the present moment (we could say a truly culminating moment of its fruitful progress) — a vigorous thought inspiring inventive minds around the world in all cultures, languages, nations, political orientations, and economic conditions — but further makes a point of getting a fix on this newly self-completing phase of the phenomenological development as such. We could say that the "third phase" of phenomenology, into which two decades ago phenomenology was entering, leaving the classic and post-classic phases, has reached its full growth and precisely this in still one more quite major turn in the (then) unforeseeable enrichment of all lines of Husserlian thought and within innumerable ramifications of these lines.

This collection is composed mainly from the papers submitted for the First World Congress of Phenomenology organized by The World Phenomenology Institute in September of the year 1988 in Santiago de Compostela, Spain, commemorating the fiftieth annivarsary of Husserl's death, as well as from selected work presented at other programs of the Institute which took place the same year and with the same intent in the US and England (cf. the report: "Phenomenology in the World Fifty Years after the Death of Edmund Husserl," Bk 1, p. xxi). These programs carried out on two continents, at two world congresses (the other being the XVIIth World Congress of Philosophy in Brighton, England, 1988) have been an exceptional occasion to bring together our collaborators dispersed in the world with many other phenomenologically inspired scholars attracted by these rare opportunities to come together and air their views, interests, concerns. This accounts for the truly world-wide sounding of what phenomenology is today; it allows the surprisingly extensive and colorful fulguration of interests, problems and formulations of ideas to appear.

It is not possible, in fact, to put the spectrum of philosophical issues in their original varied colorful richness which we have here into fixed philosophical categories; they are too full of ingenious new twists, aspects, insights, views, indications, hints. . . . Consequently, in their arrangement we will follow a rather standard differentiation by disciplines and themes.

Nevertheless, while declining to prematurely attempt a systematic, interpretative differentiation of this wealth of ideas which has emerged so profusely, we must indicate, first, their allegiance to phenomenology and to legitimizing it; second, we must trace the origin of this unexpected fecundity which phenomenology, now a century old, displays as on the first day.

The first reason for this new wave of renewal of the entire field lies in the first place in the above-cited availability of the entire cycle of Husserl's thought, renewing all in itself already or having germinal thought toward it. But it can be traced also to four other factors. We will endeavor to trace them while we present the main sectors of our anthology.

1. The present collection of essays marks in a striking way the special new phase in strictly Husserlian research. Although inroads into phenomenology drawn from the *integral Husserl* corpus have already been initiated in recent years, as witnessed in the latest volumes of *Analecta Husserliana* and elsewhere, it is in the present collection that we see it in a vast spread of ideas, themes and insights; this collection does, in this sense, inaugurate the new integral phase in Husserl research proper.

2. Yet we gain not only new vistas and new precisions about the thought of Husserl on the one hand, but also a deeper view into the great puzzles of phenomenology, by confronting Husserl's thought with other great phenomenological (and other) thinkers. Our second book groups these studies. It covers a great range of issues, bringing them into a new light. Also, in the strictly thematic essays, viewed literally or obliquely, the great classic issues remain openly and intrinsically the focus of concentration. Throughout these studies and reflections by the

new generation of scholars we find not only the work of Husserl and the classic phenomenologists but also the ruminated and digested presence of the classic interpretations of Husserl (e.g., of E. Fink, R. Ingarden, L. Landgrebe and his school). The later thought of M. Merleau-Ponty, H. G. Gadamer, E. Levinas, and Paul Ricoeur, phenomenlogically inspired albeit divergent in other aspects, are either directly treated or implicitly alluded to. We might say that in this vibrantly new fragrance of thought we feel the new generation of scholars breathing the air of their forerunners.

What makes this vaste expanse of thought phenomenological, or, what makes its allegiance to phenomenology, is, in the first place, the predominance of the direct concern with the great classic issues of Husserlianism: intentionality, evidence, consciousness, subject-object, intuition, constitution, reason, empathy, certainty, method, relation, transcendentalism, foundationalism, originality, time, horizon, historicity, intersubjectivity, life-world, etc. In the enormous variety of approaches, queries, insights, versatility of points of view, these dominant issues undergo an infinite adumbration in nuancement and refinement.

3. This richness and its spread is also due to the immersion of scholars in the debates going on in the philosophical streamlets of today — debates in which they participate and solidarize themselves vicariously — because it can be said that the entire span of the philosophical arena of today, whether positive or negative, constructive or decadent, is indebted to the vigorous Husserlian proclaiming of phenomenology and its unfolding. We distinguish Husserlian phenomenological concerns in all the streamlets of present-day philosophical thought. Whether it be structuralism, semiotics, dialogism, communicative action, existentialism in its various shades, deconstruction, etc., in spite of their emphatic disclaiming of any allegiance to phenomenology, each displays basic controversies or issues which can easily be shown to be related to or issuing from Husserl's inspiration. We may detect a Husserlian influence at the very heart. First of all we might say that Husserl's vigorous struggle against relativism and his quest for a *neutral framework* for the formulation and resolution of philosophical questions are visible in Habermas' efforts and those . . . of Foucault where we see a startling example of the old drive for a unitary framework; the drive also underlies the most recent *phenomenology of life* (Tymieniecka). The old Realism/Idealism issue is still vigorously debated having taken on

new forms, e.g., moving from transcendental idealism to the metaphysical "onto-"realism.

As already mentioned, the trends of structuralism, post-structuralism, deconstructionism, etc., can all be related to Husserl's main emphasis upon pure forms, absolute certainty, evidence of *eidoi*, etc. Had not, in the final account, his *critique of reason* in the hands of his followers and others in contemporary philosophy led to disastrous aporias? But it also stimulated the determined countering of the tendencies which lead to the total decadence of our culture, by seeking a major way out of them in a new attempt at rethinking the starting point and the context of phenomenology precisely in the *phenomenology of life* which takes all these aporias in its stride.

Phenomenology appears to have laid bare the bone of contention to be taken up by the main debates in the decadent philosophies of the present historical moment; it has brought forth the subjacent arteries of issues denouncing the mystification or twist or biased approaches and subsequent formulations. (They are led astray into dead-end streets or float upon spurious waves at the thinnest surfaces of this human universe of discourse). The decisive issues thus brought forth by phenomenology such as objectivity/subjectivity, individualism/intersubjectivity, cognition of reality/transcendental constitution, idealism/realism, horizon, analysis and passive synthesis, life/reason, structure/content, intellect/passions, cognition/action, individual/community, etc., constitute the centers of these streamlets and are reformulated according to the different starting points which the thinkers take, giving dynamism to the new debates in which these streamlets play.

Consequently, immersed in a much vaster network of philosophical discussions than the strictly phenomenologically encircled one, the present-day scholar in phenomenology is in his very own insights and formulations of questions influenced by the philosophies of today through those of their aspects congenial to phenomenology and yet different due to their own biases. Hence we witness even in strictly Husserlian research and everywhere beyond it a wealth of new ingenious twists and new intuitions with which the great issues of the core of the phenomenological patrimony are adumbrated and enriched. The almost infinite proliferation of perspectives upon the great classic themes is overwhelming and eludes any hasty categorization.

When we propose the picture of the phenomenological spirit within the entire world in which it is alive today, we cannot overlook the fact that when classic phenomenological ideas fall upon a ground quite different culturally from the one in which they emerged, these ideas undergo specific variations and nuancements. Since it is the human being within this life-world that is the center of the phenomenological concern, different types and modalities of the life and societal world also play their role in giving a special "flavor" to the work of the spirit, special enrichment.

This should not be understood, nevertheless, as indicating the dispersal of phenomenology today. Besides being differentiated into fields of study, some new vigorous self-critical attempts, instructed by the criticism of classic and post-classic phenomenological inquiry, bring forth a new interpretation of the phenomenological project in the reformulation of the philosophical enterprise as such *in toto* (unlike the attempts of those of the post-Husserlian period who took up some major innovative task but did not bring it to a conclusion that alone could allow a judgement as to the validity of the total effect, e.g., Ingarden, Merleau-Ponty, etc.). We find this reformulation within the present collection as a low but vigorous profile of this vast spread of thought, making its way through it and taking on substance.

Yet the most remarkable thing which I have been emphasizing over and over again is that scholars from the West and East, from the North and South, from all the continents, social milieux, and political tendencies meet at conferences of The World Phenomenology Institute and find in our core themes, the *phenomenology of human life and of the human condition* a unique ground for intimate communication through and beyond all the divergencies which they otherwise bear.

In fact, after we see the wheel of critical reflection upon the various phases of phenomenology turn its full cycle, we find at the pole opposite to Husserlian intentionality as the sovereign function of the human being, the passions; intentionality's constitutive/cognitive mode of operation is dethroned from its primordial position by the *creative act* of man and his creative function; the intentional network of functioning is challenged by the *creative orchestration*; and the life-world with the absolutism of transcendental consciousness at its center is, in its position of pole of reference, dismissed to a secondary command, receding to the subjacent life with its pre-human, pre-subject/object division, to *the unity-of-everything-there-is-alive* (Tymieniecka and the work of The World Phenomenology Institute expounded in the forum of the *Analecta Husserliana* series).

4. But, as we all know, Husserl's intent that phenomenology should function as a *philosophia prima* with respect to all fields of scholarship, all fields of knowledge, has been fully realized. Indeed, from its incipient stage, phenomenology not only encompassed all the philosophical disciplines such as philosophy of mind, logic, aesthetics, ethics, ontology, anthropology, etc., but already was applied to jurisprudence (Reinach), social science and economics (Scheler), sociology (Schutz), religion (Otto), art (Geiger), biology (Conrad-Martius), etc. The early phenomenological schools of psychology and psychiatry have burgeoned (Binswanger, Bujtendinck, Boss, Straus, Minkowski), and their works are classics by now. But this first wave of the influx of phenomenology into the sciences of man has intensified and spread in the period after the Second World War and now, toward the end of the century in its tenth decade, it can be said without exaggeration that there is hardly any human science or art theory which does not bear directly or by proxy a mark of phenomenological inspiration in its incredibly varied and rich spectrum of ideas, insights, bents, illuminations, etc.

Seeking to systematise the fruitful exchange between phenomenology and the sphere of knowledge, the sciences and the arts, our continuing research program carried out under the heading of "The Interdisciplinary Phenomenology of Man and of the Human Condition" coordinated in the systematic progression the world-wide research into what is called "phenomenological praxeology" by The World Phenomenology Institute for the past two decades. (cf. *Analecta Husserliana*, Vols. 1—32 and *Phenomenological Inquiry*, Vols. 3—14). Phenomenology has proven itself to be enlightening beyond the strict humanities, extending to biology, all branches of sociology, technical studies and architecture, and the phenomenology of life has much to contribute to ecology and environmental studies.

In summary, phenomenology in all its variants is present beyond the scholarly sphere in all realms of educated life, on every continent, wherever the local culture seeks some serious and innovative philosophical inspiration.

We conclude this survey by stating that, after a long period of reception, criticism, dissemination of germinal ideas, and progressive discovery of deeply seated intuitions, phenomenology in the world of scholarship, science, art, thought and culture has come of age. What would be the most appropriate historical moment to bring it into the open? This is the conclusion that the reader, aware of the philosophico-scientific and cultural sphere of the present-day wide, wide world, will come to make after study of our four volumes.

Clockwise from the lower left: Angela Ales Bello, Marie-Rose Barral, Robert Wise, Anna-Teresa Tymieniecka and Carmen Balzer during a reception.

The First World Phenomenology Congress, Santiago

A session of the conference

PART ONE

TIME, HISTORICITY, CULTURE

RICHARD COBB-STEVENS

HUSSERL AND HISTORICISM: FIFTY YEARS LATER

The term 'historicism' was first used extensively in late nineteenth century debates among political economists who were concerned about the relativistic implications of making economic theory excessively dependent on economic history. Ernst Troeltsch later defined historicism more broadly as the tendency to regard the conceptual systems of both the natural and human sciences as world views whose presuppositions are determined by contingent historical transformations. He also observed that historicism is one of the two fundamental discoveries of modernity, the other being the naturalism that follows from the reductive interpretation of nature as an ensemble of quantifiable entities and forces.[1] Vico's maxim that we know best what we have made or created ("the true and the made are convertible") is commonly regarded as the founding principle of historicism. Vico claimed that we can achieve greater certitude in historical studies than in the natural sciences, because history is a product of familiar human passions and adaptive strategies, whereas the physical realm is not our making. He added, however, that we should not presuppose, in the manner of Hobbes, that human nature has remained static and immutable throughout the ages. Hobbes ascribed to primitive peoples attitudes, feelings, and rational decisions which in fact could only have emerged under subsequent historical conditions. Such anachronistic speculation may best be avoided by paying closer attention to the historical evidence of modifications in our self-understanding, which may be discerned in the evolution of primitive myths and fables, and in the etymological genealogy of the guiding metaphors for our mental processes. Finally, Vico observed that even in our investigations of the physical world we know best what we have made comprehensible by reason of constructed hypotheses and arranged experimental situations.[2]

In "Philosophy as a Rigorous Science," Husserl begins his discussion of historicism by citing Dilthey's perceptive warning: "The formation of an historical consciousness destroys more thoroughly, than does surveying the disagreement of systems, a belief in the universal validity of any of the philosophies that have undertaken to express in a

A-T. Tymieniecka (ed.), Analecta Husserliana, Vol. XXXVII, 3—16.

compelling manner the coherence of the world by an ensemble of concepts."[3] Husserl then predicts that the historicist spirit will eventually undermine even the objectivity of the sciences, for these too are cultural formations whose content is predetermined by given historical relationships: "The ideas of truth, theory and science would then, like all ideas lose their absolute validity."[4] The final step, he further suggests, will be to consider historicizing rationality itself as a contingent cultural formation. These comments have proved to be remarkably prophetic. In recent years there has occurred a proliferation of movements which have in common the conviction that rationality itself is an artifact of history. Often referred to as "post-modernist," these movements are in fact the logical outcome of the specifically modern tendency to account for reason in terms of something other than itself.

Within the analytic tradition, the drift towards historicism has its remote origins in the works of Bacon and Hobbes who first defined knowledge as power and reason as reckoning. It might seem at first that nothing could be more alien to the historicist spirit than their confidence in the power of calculative rationality. However, Hobbes' celebrated definition of reason explicitly links rationality to convention: ". . . Reason . . . is nothing but reckoning, that is adding and subtracting of the consequences of general names agreed upon for the marking and signifying of our thoughts."[5] This passage is preceded by an explicit criticism of Aristotle's theory that the cognitive intellect takes on the shape, or "becomes somehow" the forms of the things that it knows. The new science demonstrates, Hobbes claims, that perception is reducible to the movements of elements within a formless nature. All of the forms that we seem to discern in things (colors, savors, etc.) are illusory impressions provoked by the body's reaction to the pressure on its organs by motions of matter. It follows that the "looks" of things are appearances, not in the sense of modes of presentation, but in the sense of mere appearances. Their appearance to us is "fancy."[6]

For some time, the dramatic success of the new sciences obscured the relativistic implications of this position. The ancient conviction that human beings are knowers of essences survived in the expectation that the new sciences would eventually reveal the hidden structures of reality. The power of grasping essences was thus tacitly transferred from cognitive intuitions to calculative procedures. Yet Hobbes' critique of Aristotle had in fact undermined the connection not only between ordinary names and the species-looks of things, but also between mathematical models and the hidden structures of things. He explicitly rejected Aristotle's general thesis that a word's definition is founded upon an intuitive grasp of the essence represented by the word: "For considering that new names are daily made, and old ones laid aside; that diverse nations use different names, and how impossible it is either to observe similitude, or make any comparison betwixt a name and a thing, how can any man imagine that the names of things were imposed from their natures?"[7] He then masked the historicist implications of this thesis by making an exception for the definition of the Aristotelian term 'materia prima', noting that this expression ". . . is not of vain use; for it signifies a conception of body without the consideration of any form or other accident except only magnitude or extension, and aptness to receive form and other accident."[8] However, he offered no explanation of why this conception of body should be any less vain, and why extension should be any less fanciful a property than color or savor? If signifiers are labels whose meanings cannot be correlated with the essential structures of things, then why should mathematical expressions be privileged over ordinary words? It seems clear in retrospect that Hobbes' implicit justification was pragmatic. His point was really that Galileo's approach works better, not that its mathematical language is more truthful.

Twentieth century analytic philosophers have generally subscribed to the modern preference for calculative rationality, and have been suspicious of the pre-modern emphasis on intuitive rationality. After Frege's critique of psychologism, the project of analytic philosophy was to defend rationality against relativism by detaching logic and semantics from dependence on subjective intuitions. Notational systems and calculative techniques borrowed from mathematics were imported into logic and semantics. Concepts were redefined as functions, expressions of generality were recast in the notation of quantifiers and variables, and the interplay of sense and reference was construed in terms of substitutivity of identical extensions. It has gradually become clear, however, that logical analysis alone cannot establish a connection between the senses of ordinary or mathematical expressions and the overt or hidden structures of reality. Having endorsed the modern interpretation of intuition as exclusively sensuous and empirical, analytic philosophers could not appeal to the ancient thesis that the species-looks of things are given to intellectual insight. The only solution, therefore, was to reaffirm Hobbes' nominalism, and to settle for the vague notion that linguistic competence is the product of routinized practice in the use of words. Advocates of this pragmatic approach to semantics no longer regard reference as a link between words and world, but rather as a successful move in a language-game. They point out that linguistic systems are the products of long-forgotten contingent conventions, and openly acknowledge the historicism entailed by this position. Indeed, proponents of the new pragmatism claim that we must rid ourselves of nostalgia for pre-modern certitudes by admitting once and for all that rationality is contingent in its origins and in its historical transformations. As Quine puts it, there is simply no right way of describing or explaining reality. The totality of sentences that we hold to be true is "a web of belief," comprised of deeply entrenched categorial habits towards the center, and more unstable linguistic usages at the periphery. Jolts registered on the periphery provoke minor realignments. The considerations which incline us to choose from among various "cultural posits" (e.g., the gods of Homer, ordinary physical objects, the particles of physics) are not always rational, but ". . . where they are rational, they are pragmatic."[9] We give preference to the particles of physics over other cultural posits, because the resolution of all things into ensembles of matter in motion renders the universe amenable to quantitative analysis and thus enables us to deal with it more efficiently. Regarded as the mirror of our mathematical constructs, the world is more familiar and more malleable. In short, Vico's historicism is ultimately more powerful than Hobbes' naturalism. The true and the made are convertible.

Within the hermeneutic tradition, historicism no doubt also has its remote origins in Vico's maxim. More proximately, however, it seems to be derived from Kant's theory of concept formation and from Nietzche's perspectival account of truth. Having accepted the modern premise that sensory affections caused by hidden and unknowable things-in themselves provide the only link between mind and nature, Kant's strategy was to appeal to the mind's mysterious but consistent application of rules (i.e., conceptual categories) to the flow of sensory data, in order to account for the objectivity of our knowledge. However, he never adequately explains how empirical concepts are acquired in the first place, or how the understanding manages to select just the right rules to fit specific sensory affections. His vague references to the mysterious workings of the transcendental imagination concealed within the depths of the human soul do little more than restate the problem.[10] On the grounds that our perceptions are guided by culturally determined interpretative horizons, post-Kantian philosophers tend to look to the repository of language for a solution to these issues. The conceptual constructions which we impose upon the sensory manifold are said to be relative to the history of interpretations encoded within our language. Various languages carve up the spectrum of meaning differently. Hence, we acquire our concepts not from the unmediated intuition of the species-looks of things, but by being introduced to the institutionalized praxis of a language having a contingent history and an inaccessible pre-history.

There is only a small step from this position to Nietzsche's thesis that meanings are illusory cultural formations generated by the will to power. If there is no essential link between meanings and the structures of reality, and if concepts have their origins in an always prior and hence ultimately ungrounded linguistic matrix, then there is no reason to suppose that the patterns of meaning which we seem to detect in the world are anything more than useful interpretations effected by an instinctual will for survival. Like Hobbes, Nietzsche defines science as the "... transformation of nature into concepts for the purposes of governing nature."[11] What we call truths are those well-established perspectives, interpretations, or fictions without which we could not survive.[12] His perspectivalism casts a long shadow over the hermeneutic tradition. Even those who speak in defence of reason are hesitant about making any universal claims. For example, followers of Heidegger are hard pressed to defend his hermeneutic phenomenology against the criticism that a philosophy which stresses the radical historicity of human existence cannot coherently claim to have discerned the essential structures of our mode of being in the world. Consistency requires that hermeneutics should acknowledge its own dependence on conditions of historical existence and therefore degrade its philosophical claims to the level of contingent interpretations. It would seem that a philosophy which gives primacy to history must ultimately abandon the notion that human beings are knowers of essences. Deconstructionists accordingly reject the notion of objective reference entirely, and speak only of signifiers signifying other signifiers. The world is a text susceptible of multiple interpretations. Or rather, there is no text at all. There are only interpretations. In short, theories are Apollinian dreams, powerful fictions, necessary illusions. The world is a work of art, and its apparent rationality is an illusory facade masking nature's ultimately irrational spontaneity.

These are highly inflated claims. It suffices to peruse the pages of

Herodotus in order to see that pre-modern thinkers were hardly oblivious to the diversity of languages and cultures. Moreover, contemporary techniques of historical research have yielded as much evidence of continuity as of differences among diverse cultures and traditions. Husserl nevertheless took the danger of historicism seriously, for he recognized how difficult it is to strike the mean between the extremes of relativism and rationalism. His early works tend to stress the incoherence of the historicist notion that radically alien conceptual systems might be as viable as our own. In *Logical Investigations*, for example, he points out that the historicist thesis trades on a tacit acknowledgment of sufficient communality among alternate conceptual systems to justify their recognition as alternate interpretations of the same phenomena.[13] His later works deal more extensively and more positively with the relationship between philosophy and its own history. "There is no doubt," he observes in *Cartesian Meditations*, "that we must engross ourselves in historical considerations if we are to be able to understand ourselves as philosophers and understand what philosophy is to become through us."[14] In *The Crisis of European Sciences and Transcendental Phenomenology*, he adopts a specifically historical approach to the development of modern philosophy, stressing how Galileo's mathematization of nature transformed the traditional philosophical understanding of the powers and limits of reason.

This method yields a perceptive analysis of the link between Hobbes' nominalism and Kant's constructivism. Husserl first calls attention to the mixture of rationalism and relativism implicit in Hobbes' exaltation of the power of reckoning. This revolutionary account of reason seemed to promise that a more developed physics would eventually explain not only the processes of nature but also the workings of the soul. Hence, Hobbes did not regard his naturalistic account of the psyche as a humiliation of reason.[15] It required a genius such as Hume, says Husserl, to take the naturalism initiated by Hobbes and further developed by Locke to its logical conclusions. Hume realized that if intuition cannot break out of the circle of impressions and ideas, there is no justification for supposing that reckoning can yield any less fanciful results. Berkeley had already observed that inferences drawn from impressions do not justify conclusions about the structure of reality. The fundamental categories requisite for a mathematicized version of nature (e.g., identity, contiguity, materiality, extension, "world," "extra-mental reality") must somehow be derivable from information provided by the manifold of impressions. On Hobbes' account, however, sensory impressions yield only illusions. It follows therefore that scientific theories too are productions of fancy. This realization is the key to Hume's skepticism: "Hume goes on to the end. All categories of objectivity — the scientific ones through which an objective extra-psychic world is thought in scientific life, and the pre-scientific ones through which it is thought in everyday life — are fictions."[16]

Husserl next points out that Kant did not fully appreciate the enigma that had engendered Hume's skepticism. Hume's radicalism had led him to shake the foundations of that ". . . dogmatic mathematicizing objectivism, so inspiring to people of the time, which ascribed to the world itself a mathematical-rational in-itself (which we copy, so to speak, better and better in our more or less perfect theories.)"[17] Unfortunately, Kant's critique of reason failed to attain the radical level achieved by Hume. Concerned exclusively by the threat posed by Hume's skepticism to the scientific mode of objectivity, Kant did not take seriously enough Hume's claim that the world of everyday life must also be related to the reach of rationality. On Husserl's view, therefore, Kant made the following methodological error. He set about to establish that the same faculties which express themselves in mathematical thinking also function ceaselessly in the rationalizing of sense data. He thus hoped to justify the account of nature proposed by Newtonian physics. However, his preoccupation with the transcendental conditions for the appearance of scientific objects led him to neglect the more fundamental problem of justifying the objectivity of the everyday pre-scientific knowledge on which scientific knowledge is dependent. Husserl recognizes, of course, that Kant repudiated Hobbes' derivation of reason from natural drives. Indeed, he praises Kant for stressing the ontological difference between transcendental and empirical dimensions. What Kant failed to discern, however, was the root error of modern naturalism, i.e., its rejection of the traditional notion that the intellect grasps the species-looks of things through intellectual intuition. Husserl elsewhere adds that, despite his passionate and unrelenting radicalism, Hume made the mistake of ascribing to the processes of mind the same sort of "transcendent" reality that he had reduced to the status of fiction. In other words, he took it for granted that mental processes have the same ontological status as the transcedent things that we would reach, if we could somehow break out of the circle of immanence.[18] It is this incoherent presupposition that governs the tendency of naturalistic philosophy to reduce intentional directedness to a relationship of the empirical type that obtains between things within the causal network of nature.

The latter argument turns the tables on historicism by calling attention to the fact that the modern interpretation of the empirical realm fails to recognize itself as an interpretation. It is always inappropriate, Husserl suggests, to attempt to think out the relationship between mind and the rest of reality by first ascribing a priority to empirical conditions, as though the notion of the empirical domain were itself uncontaminated by a history of interpretations. The very thinkers who appeal to the abyss-like character of the history of concepts as proof of the irremediably hermeneutic character of all human knowing often fail to think in a hermeneutical fashion about history itself. It never occurs to them to relate the decision to consider the problem of mind historically to the problem of how the sense of history is related to the work of mind. As a result, they tacitly postulate a mythical observer who possesses an interpretation-free understanding of the term 'empirical' and who therefore can conclude, in an unproblematic fashion, that a history of interpretations has empirical priority over any philosophical thought about history. In short, historicist hermeneutics is naive about the conditions of thought, for it forgets that the sense of the empirical dimension has a history.

Husserl's positive response to the crisis of reason provoked by its reduction (in Hobbes) to reckoning and by its limitation (in Kant) to the phenomenal and mathematicized realm of scientific objectivity was to attempt to restore confidence in the rationality of intuition. As early as 1894, he rejects the modern reductive interpretaion of intuition of the reception of sensory data, and firmly asserts that our intuitive consciousness reaches out beyond sensory affections and grasps the intended object itself: "... an intuition is a 'setting before' in an authentic sense, where the object is actually put before us in such a manner that the object is itself the topic of psychical activity."[19] His theory of categorial intuition in *Logical Investigations* essentially reaffirms Aristotle's description of the complementarity of sensuous and intellectual intuition, and endorses Aristotle's account of the derivation of sortal predicates from the species-looks of things. Husserl describes particular and categorial intuition as having a reciprocal founding/founded relationship. What presents itself to our intelligence when we perceive a particular thing is the look that it shares with a family of things. We grasp the look both as a form whose surplus sense exceeds the particularity of the instance, and as the condition for the manifestation of the particular as such. There is no insight into the species-looks of things that is not based on a prior seeing of the form and its instantiation together, and there is no seeing of particulars except as instances of some form.[20]

Husserl also distinguishes between the vague grasp of family resemblances and the thematic intuition of essences. He rejects the notion that essences are ideal forms having some separate ontological status. Like Aristotle, he insists that discernment of essences entails a work of discrimination that takes the initial looks of things as its point of departure, but then calls upon sophisticated sorting procedures, investigation of variants, and imaginative consideration of limit-cases.[21] The kind of certainty that we should assign to the results of this procedure varies in proportion to the type of access involved, and to the rigor of critical analysis and scope of imaginative variation. Husserl argues that the relative immediacy of access to the basic structures of rationality provided by our tacit awareness of intelligent performances makes for a stronger certainty than is available in other areas of research.[22] However, given the oblique and unthematic character of such apperceptions, and given the distorting influence of prevalent philosophical theories and models, reflective descriptions are often vague and confused. In the course of developing and refining his phenomenological method, Husserl discovered again and again that the method often brought to light presuppositions that had structured his original questions. Hence, he developed a method of "regressive questioning," which combines reflective intuition and critical interrogation of presuppositions.[23] As a result, his discoveries often took the form of a retroactive realization that the original question motivating some inquiry had been guided by an unsuitable anticipatory grasp of the topic. He thus came to recognize that the phenomenological method was a powerful instrument for achieving progress in truth by regressive clarification of the historicity of philosophy itself.

Eugen Fink claims that Husserl never adequately understood the implications of this properly hermeneutic discovery. Fink observes that it is always impossible to achieve full thematic clarification of all the concepts which shape our thinking. He distinguishes between operative and thematic concepts. Operative concepts are those ideas, models, or schemata which serve as instruments for fixing the sense of thematic concepts, while themselves remaining opaque and unclarified. The ensemble of such operative components makes for a tacit conceptual network, in function of which the thematic conceptual apparatus is constructed, ordered and explicated. Attuned to the history of concepts and the weight of words in their development, the philosopher should strive to achieve an ever more extensive appreciation of operative concepts, even though total elucidation will always be impossible.[24] Jacques Derrida carries this style of criticism a step further by noting that Husserl's ideal of total clarification testifies to the fact that he clearly underestimated the importance of this issue. Derrida observes that our entire philosophical tradition relies upon original metaphors (the sun, the cave, seeing, the return, foundation) whose metaphorical character has been forgotten as a result of constant usage. Hence, Husserl ought to have extended the work of regressive clarification to the suppressed metaphorical and even mythological themes operative within the conceptual network of Western philosophy. These themes are so fundamental to our thought patterns that we cannot find a platform free from their influence, from which we might pass judgment upon their validity.[25] This critique returns us to the fundamental issue of whether or not philosophical reflection is capable of distantiation from its own history. If phenomenology cannot recognize all of its own presuppositions, then it would seem that its descriptions are as perspectivally conditioned as any other mode of thought. Without claiming access to a detached, non-historical viewpoint, how can Husserl maintain that philosophical reflection is not rendered relativistic by reason of its own operative conditions?

David Carr calls attention to the ambiguity of the later Husserl's comments on the lifeworld. In the *Crisis*, Husserl claims that it is possible to uncover the universal structures of the lifeworld that have been obscured by the sedimentations of Galileo's mathematization of nature.[26] In *Cartesian Meditations*, however, he refers to multiple lifeworlds and thus hints that it might be impossible to retrieve a unique and common world that simply presents itself to us in a manner free of cultural and historical interpretations.[27] Carr suggests that Husserl may have finally realized that all we can do is attempt to struggle against distorting prejudices and strive to come as close as possible to the ideal of a nonrelative truth.[28] Yet even in those later passages which stress the horizon-structure of the most primitive pre-predicative experiences, Husserl seems to think it possible for phenomenological method to "explicate" the contents of all such horizons: "When we follow this procedure . . . we have the truth then, not as falsely absolutized, but rather, in each case, as within its horizons — which do not remain overlooked or veiled from sight, but are systematically explicated."[29] Husserl seems therefore to have maintained to the end that philosophical thinking achieves an orientation that differs from what we ordinarily understand by a perspective or a point of view. Philosophy is not a point of view alongside others, for it so relates to the transcending power of mind that it situates itself not within some limited perspective, but within the space of freedom generated by the discernment of our engagement in this or that perspective.

A clue to the reason for Husserl's tenacity on this issue may be found in his comments on the perspectival character of perception. He criticizes the Kantian distinction between the thing-in-itself and the thing as it appears to us, on the grounds that this comparison suggests that human perception delivers images or signs of things, rather than the things themselves. Moreover, he contends that physical things are such that perspectival presentation belongs to their mode of being. Perceptual objects are necessarily given in successive profiles rather than all at once. He thus defines perceptual givenness as incomplete without measuring its incompleteness against a postulated non-finite mode of knowing. The experience of alternate perspectives requires the positing of an "empty X" as a placeholder for further inquiry, but it does not require that the "X" be taken as a thing-in-itself known by an *intuitus originarius*. However, Husserl does retain the regulative idea of an infinity of profiles. The idea of an endless continuum of perspectives is entailed by the discovery that it is always possible to look at something from a new angle. Kant had described the idea of the "object = X" as an a intention without intuition. According to Husserl, however, all intentions are geared to intuition. Since the idea of an infinity of profiles is not itself an infinity, the idea can be given adequately: ". . . seeing intellectually that this infinity of necessity cannot be given does not exclude, but rather requires, the intellectual seen givenness of the idea of this infinity."[30] Note that Husserl does not here claim that truth requires the having of an infinity of perspectives. Indeed, he stresses that inexhaustibility is an essential characteristic of a thing's mode of being, and hence of its truthfulness. He does assert, however, that there are non-perspectival intuitions. The intuitive grasp of perception's essentially perspectival character is not itself another perspective on the same level as the perspectival presentations of a thing. To grasp the principle governing the endless continuum of profiles is not to enjoy another profile but to identify an invariant structure of appearing.

While it is true that philosophical metaphors and models may obscure our reflective intuition of such structures of mind, it is inappropriate to think of such models as generating distorting perspectives in the sense of illusions. This is true both of ordinary perception and of reflective intuition. The elaboration of strategic distinctions and the critique of distorting prejudices help us to see better, and thus reveal the partial or confused character of earlier insights. This is no reason, however, to describe the achievement of clearer and more comprehensive insights as the spontaneous generation of new interpretative constructions. Husserl therefore rejects historicism's inflated perspectivalism, on the grounds that it obliterates differences in modes of truthfulness. To enjoy the truth is not to be in possession of some more powerful or life-enhancing illusion, but to have discerned the essential from the adventitious in some domain. Truth comes in different forms: there are the clarities attained by mathematical reason, the necessities discovered in moral praxis, the complementary structures uncovered by interpretation and explanation, the perspectives articulated in judgments of perception, and the illuminating distinctions achieved by philosophical reflection. It is just as inappropriate to blur these nuances by asserting that all knowing in perspectival, or that another interpretation in always possible, as it is to look for mathematical certitudes in the ethical and political domains.

Boston College

NOTES

[1] Ernst Troeltsch, *Der Historismus und seine Probleme* (Tübingen, 1922); "Die Krisis des Historismus," *Die Neue Rundschau*, XXXIII (1922), 572–590. See also Maurice Mandelbaum, "Historicism," in *The Enyclopedia of Philosophy*, ed. Paul Edwards (New York: Macmillan, 1967), IV, pp. 23–5.

[2] Giambattista Vico, *Opere complete*, eds. B. Croce, G. Gentile and F. Nicolini (Bari, 1914–41), I, pp. 136ff.

[3] Wilhelm Dilthey, *Weltanschauung, Philosophie und Religion in Darstellung* (Berlin: Reichel, 1911), p. 6.

[4] Edmund Husserl, "Philosophy as a Rigorous Science," trans. Quentin Lauer, in *Phenomenology and the Crisis of Philosophy* (New York: Harper and Row, 1965), p. 125.

[5] Thomas Hobbes, *Leviathan, English Works*, ed. William Molesworth (Aalen: Scientia Verlag), III, v.

[6] *Ibid.*, III, i.

[7] Hobbes, *Concerning Body*, EW, I, p. 16.

[8] *Ibid.*, I, p. 118.

[9] Willard Van Orman Quine, "Two Dogmas of Empiricism," in *From a Logical Point of View* (Cambridge: Harvard Univ. Press, 1953), pp. 42—6.

[10] Immanuel Kant, *Critique of Pure Reason*, trans. Norman Kemp Smith (Toronto: Macmillan, 1965), A 141 (B 181).

[11] Friedrich Nietzsche, *Werke*, ed. K. Schlecta (Munich: 1954—6), iii, p. 440.

[12] *Ibid.*, iii, p. 844.

[13] Husserl, *Logical Investigations*, trans. J. N. Findlay (London: Routledge and Kegan Paul, 1970), I, Prolegomena, sec. 40. David Carr calls attention to the similarity between Husserl's argument and Donald Davidson's thesis that an alternate conceptual framework must have enough in common with our familiar framework to be translatable into our language. It follows that recognition of an alternate scheme as alternate establishes that it is not radically alternate. See Donald Davidson, "On the Very Idea of a Conceptual Scheme," *Proceedings and Addresses of the American Philosophical Association*, XLVII (1974), pp. 5—20; David Carr, "Phenomenology and Relativism," in *Interpreting Husserl: Critical and Comperative Studies* (The Hague: Nijhoff, 1987), p. 27.

[14] Husserl, *Cartesian Meditations*, trans. Dorion Cairns (The Hague: Nijhoff, 1960), p. 391.

[15] Husserl, *The Crisis of European Sciences and Transcendental Phenomenology*, trans. David Carr (Evanston: Northwestern Univ. Press, 1970), sec. 11.

[16] *Ibid.*, sec. 23.

[17] *Ibid.*, sec. 24.

[18] Husserl, *The Idea of Phenomenology*, trans. William P. Alston and George Nakhnikian (The Hague: Nijhoff, 1964), lect. 1, p. 16.

[19] Husserl, "Psychological Studies in the Elements of Logic," trans. Dallas Willard, *The Personalist* LVIII (1977), 304.

[20] *Logical Investigations*, II, vi, sec, 40—52. See Aristotle, *Metaphysics*, Z, 103lb 6—7; 20—21.

[21] Husserl, *Experience and Judgment*, trans. James S. Churchill and Karl Americks (Evanston: Northwestern Univ. Press, 1973), sec. 82. See Aristotle, *Posterior Analytics*, 74a 32—35; 89a 33—7; 100a 12—b 3.

[22] Husserl, *Ideas Pertaining to a Pure Phenomenology and to a Phenomenological Philosophy*. Book I. trans. Frederick Kersten (The Hague: Nijhoff, 1952), sec. 42, 44, 49.

[23] Husserl, "The Origin of Geometry," in *The Crisis*, appendix vi, pp. 353—78.

[24] Eugen Fink, "Operative Begriffe in Husserls Phänomenologie," *Zeitschrift für philosophische Forschung*, II (1957), 321—7.

[25] Jacques Derrida, *Edmund Husserl's Origin of Geometry: An Introduction*, trans. John P. Leavey (Stony Brook: Nicholas Hays, 1978), pp. 139—41. See also Derrida, "La mythologie blanche: la métaphore dans le texte philosophie," *Poétique*, V (1971), 1—52.

[26] *The Crisis*, appendix vii, pp. 379—83.
[27] *Cartesian Meditations*, p. 133.
[28] Carr, *Interpreting Husserl*, pp. 12, 90—3.
[29] Husserl, *Formal and Transcendental Logic*, trans. Dorion Cairns (The Hague: Nijhoff, 1969), sec. 105.
[30] Husserl, *Ideas* I, sec. 143.

ELLA BUCENIECE

THE TELEOLOGY OF THE HISTORICAL BEING IN HARTMANN AND HUSSERL*

"There have been thousands of purposes so far, because there have been thousands of people. What is lacking, however, is bridles for thousands of heads — one purpose is lacking. Do tell me, my brethren — in the absence of one purpose for Mankind — should we not say that there is no Mankind as yet . . . ." These are the words of Zarathustra, as Nietzche tells us. No one before him has put it so succinctly, no one has proposed to seek the unity of mankind and of culture through a common purpose or aim.

True, Herder, who was one of the first to try to understand the development of history and culture, struck upon the notion of teleology; yet he did not develop this notion beyond the boundaries of its traditional interpretations. The time had not come yet.

The Nietzschean question was remembered much later. Such, according to Theodor Adorno, is the structure of the question in history — duration and oblivion.

What I propose to do is to call to mind the two lines of thought, prominent in the twentieth century philosophy, which tackle the abovementioned problem. One of these is that of Nicolai Hartmann (who, incidentally — and I am glad of the opportunity to remind you of this — was born in Riga, the city I am from). The other is that propounded by Edmund Husserl.

Admittedly, this may not seem to be the very best line of comparison. It may, perhaps, more appropriate to compare Hartmann's philosophy with that of Roman Ingarden, as Anna-Teresa Tymieniecka does in her book *Essence et Existence. Etude a propos de la philosophie de Roman Ingarden et Nicolai Hartmann* (Paris: 1957). Likewise, Husserl's teleological conception may, perhaps, be best elucidated by comparing it with the teaching of Merleau-Ponty, as is done by Bernard P. Dauenhauer in "Teleology of Consciousness: Husserl & Merleau-Ponty" (*Analecta Husserliana*, Vol. IX). Indeed, it is not very easy to compare Husserl and Hartmann, and this is so for at least two reasons. First, it is difficult to establish Hartmann's place within the phenomenological tradition (here I subscribe to the evaluation of Hartmann in Herbert

A-T. Tymieniecka (ed.), Analecta Husserliana, Vol. XXXVII, 17–22.

Spiegelberg's *The Phenomenological Movement: A Historical Introduction*, 2 vols. [The Hague: Nijhoff, 1969—1971] and the assessment made by J. Smith in his book *The Disputation over Teleological Thinking* [Mainz: 1969]). The second reason is this: Hartmann does not make a positive contribution to the discussion of the teleology of historical being. He divides the world into strata of being, allotting the supreme place to the being of Man, the finality of which is the central feature of the world. Thus is an impressive structure of the world built; yet it is precisely a structure, and not a process, not a history. It should be mentioned, plainly, that I base my deliberations here mainly upon two works: Hartmann's *Teleological Thinking* and Husserl's *The Crisis of the European Sciences and Transcendental Phenomenology*.

The main feature of Hartmann's book is its critical ethos. In a categorically precise and substantiated way he offers a critique of the historical forms of teleology, beginning with the ancient Greeks, and discusses the functioning of the teleological principle on various levels of consciousness: the naive, the scientific, the speculative-metaphysical. Hartmann caps his critical analysis of the teleological principle with a declaration that final determinism is a fiasco. We should not be dismayed by the critical attitude taken by Hartmann. Criticism has acquired in the twentieth century (to be more exact, since Kant's famous Critiques), a specific ontological sense. It is not so much the result of criticism, the so-called "positive programme," that is of importance. Rather, the very process of critical assessment is what holds essential value. This aspect of criticism is well brought out in Adorno's *Metacriticism of the Theory of Cognition*. Criticism of Husserl's ontology of consciousness constitutes for Adorno the basis for the formation of the ontology of Nothing which he proposes. This is made possible because his critique, to borrow a phrase from Paul Ricoeur, is of such a nature, that it is capable of forming a new realm of objectivity and intelligibility. (P. Ricoeur, *Die Interpretation* [Frankfurt am Main: 1969], p. 442)

Hartmann establishes teleology as a new realm of intelligiblity by way of making a comparison between causal and final determinism. He considers finality to be an inverted causality, as the reversal of the direction of the time-process: the dependence of something that is prior upon something that takes place later. (*Das Teleologische Denken*, p. 3). Dependence on a telos is such a type of determination, which flows in the direction opposite to the one that characterizes the flow of the time-process. The subsequent stages of any process are determined not by the states that it has gone through, but by its final stage. This final nexus, or the last stage, is the purpose, the aim, the telos — both the end and the fulfillment of the process. Such reasoning adds to the various stages their end, and to the given the result, and thus the singular, holistic, all-embracing picture of process is achieved.

Causal reasoning moves backwards in a search for a chain of endless causes, and hence any progress in the development of the causal scheme inevitably leads to pluarlity and eludes singularity. (*Das Teleologische Denken*, p. 4) This distinction between the two forms of determination — the causal and the teleologically meaningful — is of great help in the understanding of cultural phenomena in general, and of literature in particular. In order to grasp the full meaning of outstanding events, it is not enough just to know the causes that brought them about. Boris Pasternak, for example, when writing of revolution, has this to say: "The events of such magnitude require no dramatic proofs. I will believe them anyway. There is no use to look for the causes of cyclopic events. There are no such causes. It is only household quarrels that have their genesis . . . . Everything that is really great is without beginning. Like the Universe . . ." (B. Pasternak, *Doctor Zhivago* [Novij Mir, 1988] No. 2, p. 100).

For Hartmann — and here he differs from Husserl — finality and causality are not two different positions of consciousness. They represent different realities: causality as a category of the processes of nature is distinct from finality, which is a category of the mind. For Husserl, the categories of understanding are simultaneously categories of being as well.

So, similar trains of thought are observable in these two philosophies, but Husserl unlike Hartmann, postulates concrete designations for the cognitive forms. He relates causal determinism to objective science, while final determinism he relates to transcendental philosophy and to phenomenology in particular.

Husserl sees that one of the reasons why "philosophy could not develop into an actual science" is the impossibility of substantiating the final aim (*Endzweck*), and thus of holding that the historical task is natural and necessary. According to Husserl, the aim, the final end, has been treated in philosophy from a position of natural science, naturalistically, has been perceived as being outside the subject, as an aim for the subject, but not as the subject itself. In the objective sciences, as well as in philosophy that is modeled along the same lines, aims are being set continously, of course. But the final aim cannot be substantiated there, for there a continuous transformation (*Verwandlung*) takes place wherein the final aim always becomes a means, a prerequisite for other aims. Having an infinity of objective aims, the sciences becomes a negative infinity, an absolutely arbitrary setting of aims, and in the long run, this grows into aimlessness. According to Husserl, the objective sciences cannot achieve true insight into the task of history. (*Die Krisis* . . . , p. 144). The objective sciences occasion the appearance of new aims and substantiate them with necessity (*Bedürfnis*). But from these ever-changing, occasional necessities which constitute Man's natural life, one cannot proceed to the common aim, the integral meaning of activity. At the same time, Husserl is convinced, that in all actions, both scientific and extra-scientific, which are historical actions, there emerges, in the long run, a strong typicality (*feste Typik*); this is not a mere generalization of the already completed activity, but "essential typicality" that is a pure a priori pre-determination of everything that is performed. Hence, the necessity that phenomenology advance to "the postulate of a universal science of a new kind — to subjectivity given prior to the appearance of the world." (*Die Krisis* . . . , p. 150). In the setting of its purposes (*Zwecksetzung*), the universal science of a new kind should not orient itself to things, objects. Instead, it should strive to achieve the horizon of the human world, or the life horizon. Therefore, transcendental subjectivity, being based in the life world as a source of all meaning, emerges as the final aim, an aim that makes the setting of all other aims unnecessary and pointless. This final aim does not complete a development, but opens up the horizon, appearing thus as a "final aim — task."

Hartmann relates finality to Man; he speaks of the meaning of finality to Man. Finality, according to Hartmann, belongs neither to a spiritless being, nor to some other objective spirit; it constitutes the centre of Man's existence. (*Das teleologische Denken*, p. 80) Man's essence, or personal spirit is, in its turn, formed by the being of the soul and of the spirit. Hartmann has a very peculiar way of arguing for the specific place of Man, the place he occupies in the structure of Being: "the highest ontic mode is always the most dependent and the most exposed one." *Ibid*., p. 80) Yet, the personal spirit, which is the most dependent being of all, attains "independence" due to its aim-setting activity. These two functions, the setting of aims and the selection of means for their attainment, allow him to become ontically the highest being. The personal spirit should use this capacity — the aim-setting capacity — as a means to rule over spiritless being.

However, in opposition to Hartmann's claim that Man thus asserts his own existence freely, I doubt if this could properly be called freedom — this is just the choice of means and the discharge of the function of ruling. This is actually backward movement within the structure of Being — a return to its lower stages. Moving forward would entail the step of an autonomous personality towards another personality — steps towards genuine humanity.

For Husserl, the purpose, or aim, is not just the setting and solving of some final tasks, nor does it mean a pure consideration of aims and purposes in the world; it is an endless change of orientation in the subject, one set by history. As G. Hoyos puts it: "The setting of the mind towards telos need not be only a mistaken or a speculative aim-setting." (G. Hoyos: "Zum Teleogiebegriff in der Phänomenologie Husserls" in *Perspektiven transcendentale — phenomenologischen Forschung*, (The Hague: 1972, p. 68).

The adequacy and apodicticity of a task set in the present guarantees regard for its prime basis (*Urstiftung*) as a "teleological beginning" (*teleologische Anfang*). The prime basis, as the givenness of an aim, at the same time constitutes a subsquent substantiation (*Nachstiftung*), and in this transformation (*Abwandlung*) historical goal-setting is shaped (*neuversuchte Erzielungen*), a chain embracing all history is made. The prime basis as a teleological beginning contains the ultimate substantiation (*Endstiftung*) set for the whole historical process.

The idea of a close connection between pre-substantiation and ultimate substantiation is further developed by Paul Ricoeur in his conception of a dialectic of archeology and teleology. An *arche* is possessed only by an agent who has a telos. (P. Ricoeur, *Die Interpretation . . .*" [Frankfurt am Main: 1969], p. 219.)

"History," Husserl writes in *Passive Synthesis* (p. 219), "means the constituting of ever higher meaning by stages, which is permitted by an immanent teleology."

Thus, the way toward a unified Humanity, toward a global culture, lies through teleology, through the formation of a common purpose. Jurij Lotman, a Soviet specialist in the semiotics of culture, is convicted that we stand now on the threshold of a global culture. He considers that global culture-building will be the task of the coming century. In my opinion, the philosophical apprehension of history is advancing on the same path previously covered by those who investigated nature: division, analysis, and hierarchy-setting in the initial stages, and the comprehension of the essential unity of Nature, and of the categorical necessity of its preservation later. The same processes are at work with regard to history: objective prognostication of the future in order to foster its realization must come from the reasonable consideration both of the past and of the future. Moreover, we should not forget the unity of Mankind, we should not overlook or neglect a single individual, and we should keep in mind the human meaning of the process.

The phenomenological interpretation of teleology may be of great help here. One can repeat with approval the words of G. Hoyos: "Phenomenological reflection first and foremost shows the relativeness of all truths and values within the horizon of the endless historical development; here is not the absoluteness of relevant historical epochs, but the ever-present telos within the horizon of each culture, being constituted by the subject (agent) who attains autonomy in the sense of the highest absoluteness of universality. This is why it seems possible with the help of goal-setting to engage in an intersubjective general praxis for the attainment of true humanity." (G. Hoyos, *op. cit.*, p. 92). I will end with the words of Antoine de Saint Exupery: "Do not strive to forsee the future; let it manifest itself."

Institute of Philosophy and Law
Academy of Sciences, Latvia

NOTE

* Translated by Jānis Vējš.

OLIVER W. HOLMES

HISTORICAL TIME, MIND AND CRITICAL PHILOSOPHY OF HISTORY

> Time presupposes a view of time.
>
> Maurice Merleau-Ponty

> History is the intellectual form in which a civilization renders account of its past.
>
> Johan Huizinga

> Time is a discovery which is only made by thinking. We create it as an idea and do not begin until much later to suspect that we ourselves are Time, inasmuch as we live.
>
> Oswald Spengler*

Critical philosophy of history comprises a field of study much concerned with the problem of time and mind. The intimate relation between philosophy and history has given rise to a series of issues concerning the "territory of time." The theme, the time and mind, has been treated from a variety of perspectives. Historically, it first received treatment from philosophers as part of the problem of the determination of the forms of thought, conceived in a way which continues to be regarded as the distinctively philosophical attitude to the question. Meanwhile, philosophies of history which interpret the historical process have reflected the centuries-old debate centering around the crucial issue: how do we know the past? At the turn of the twentieth century challenges to Newtonian physics, particularly to the notion of time as being "absolute," contributed to the lively controversies concerning the nature of historical time and the epistemological status of historical knowledge. The controversy emerged in the 1880's and became an active issue during the early decades of the twentieth century. Central to the issue has been the view that the point at which philosophy and history meet provides a comprehensive scheme within which historical data may be systematically analyzed and comprehended: the individual mind.

Since the late eighteenth and early nineteenth centuries there have been numerous studies of the philosophy of history which have characterized time as "lived experiences." Time as "event" was associated

A-T. Tymieniecka (ed.), Analecta Husserliana, Vol. XXXVII, 23—54.

objectively with individuals who live events and time; thus understood, it became subjective. To understand time as it is experienced, modern critical philosophy of history has analyzed those features of human life which are expressive of the passage or the succession of time. These analyses have been primarily concerned with the problem of historical time and our knowledge of the past and, consequently, have emphasized conceptual and epistemological characterizations of history.

Certain of the primary concerns underlying the epistemological status of history are: What are the conditions under which historical knowledge becomes possible? How can we gain a valid knowledge of the past while we, as interpreters of the past, are experiencing the "passing moments" in the broader passage of historical time? What are the methods to be employed by the historian-interpreter to transcend the limitations of his or her subjective imagination thereby achieving an objective perspective on history? How does the historian's philosophical construction of historical "reality" develop from the "data" of immediate, "experienced" reality? The latter question provided the basis for theoretical challenges to the view of "history as a science," a view which claimed to account for a past event "as it actually happened." The purpose of history, according to the narrow, fact-oriented successors of Ranke, is to reproduce the past factually. This "positivistic" approach did not clarify how any "factual reproduction" of past historical reality actually shaped the contents of reality.

The constitutive power of the human mind with respect to nature, in formulating general laws of nature, had been acknowledged in the natural and social sciences. These laws of nature were considered to embody a coherence outside of the mind. The view that the human mind has the same constitutive power with respect to history has not been acknowledged by the more positivistic observers of historical events. For historical empiricism, history is the straightforward reproduction of past events and any discrepancy between "history proper" and the event reveals a shorter version of "facts" that may be enlarged quantitatively. The radical empiricist proposition that ultimately history is unknowable, meaning that our perception of the past as distinct from the past itself changes constantly, limits historical knowledge. The acceptance of this position points clearly to an epistemological break between the past and present.

This paper opposes the empiricist viewpoint, and establishes the Kantian a priori basis of historical knowledge. The paper approaches the problem of time and mind, as an issue, the understanding of which combines the philosophical and the historical attitudes in the form of critical philosophy of history. Critical philosophy of history thus refers to the position that characterizes the world we know and act in as a product of mind. Through this characterization of the world, this paper will identify the epistemological status of historical knowledge. It will discuss all these concerns as they are treated in the theories of Croce, Collingwood, Ortega, and to a lesser extent Dilthey, Husserl, and Heidegger. These thinkers are selected because they are modern examples of reflective thinkers who approach history from philosophy and because their theories of history as a source of human knowledge have epitomized the tendency to connect the concepts of time and mind. Together they have challenged positivistic approaches to history and contributed an important aspect to the modern concept of history: the tenet that there is a connectedness and a meaning in human history which comes from a continuous principle in individuals themselves.

Critical philosophy of history provided a framework within which the boundaries of historical knowledge may be determined. It emphasized the logical, conceptual, and epistemological characterization of history, and analyzed the concepts and assumptions of historical knowledge and inquired into the arguments and procedures of historians. To understand the boundaries it established for "positivistic" history, certain epistemological principles concerning the general nature of knowledge which inevitably arise must be plumbed.

Historical empiricists' efforts to reproduce a factual account of historical reality were based on fundamental epistemological principles. In observing and in analyzing empirical reality, an empirical theory of perception rests on the interaction between our sensory organs and physical objects. The experience of the world of physical objects is marked by this interaction. The individual perceives several surrounding physical objects. As every idea is derived from some impression which is similar to it, the impressions must either be sensations derived from perceptible vision, or some internal impression arising from these sensations. Our internal impressions are rational, a priori principles and, accordingly, are the conditions under which one conceives the ordered system of the natural, physical world. One of the fundamental problems of perception is whether physical objects are directly perceived. The contention that one directly perceives sense data, instead of physical objects, makes an interpretation of these data difficult. For the so-called "laws of nature" and all concepts of objects are specifications of a priori principles. The human mind and hence human actions too become data of history.

In the interest of studying human behavior, however, some late seventeenth century political theorists did apply the word "science" to the idea of a science of politics. For them, the study of politics proceeded from the assumption that the same cause will always produce the same effect, one of the fundamental postulates of science. Hobbes, Locke, and Spinoza, thinkers who worked out the various aspects of human nature, developed their respective political theories as components of general philosophy and operated from this very assumption. Some other thinkers who advanced the idea of a science of politics had more the "art of statecraft" in mind than a theoretical doctrine. In putting forward their position, they connected this idea of scientific politics with the teachings of the Greeks and Romans and with historical study. One such thinker was the pamphleteer Henry Neville who, in 1681, contented that

> whosoever sets himself to study politics, must do it by reading history; and observing in it the several turns and revolutions of government; and then the cause of such change will be so visible and obvious, that we need not impute theft to any man that finds it out: it being as lawful and as easy for any person . . . to read Thucydides, Polybius, Livy or Plutarch; and if he do so with attentiveness, he shall be sure to find the same things there that they have found.

Political wisdom was to be found, he continued, in reading "Aristotle, Plato, or Cicero, or any of those great artists ancient or modern, who teach that great science of the governing and increasing great states and cities." A help in "foreseeing" political "danger" to any governing body is the politics, or art of governing, "a science to be learned and studied by counsellors and statesmen be they never so great, or else mankind will have a very sad condition under them, and they themselves a very perplexed and turbulent life, and probably a very destructive and precipitious end of it."[1] To understand human actions therefore required studying social and political institutions, which were external to the mind.

There are significant continuities, however, between traditional views concerning the subject of historical knowledge and the modern notion of it as a predominantly critical discipline, and we should consider these in our discussion. The diverse forms of philosophical attitudes to time and mind, as manifested in critical philosophy of history, emerged in the eighteenth century. A general discussion, therefore, of the eighteenth century background that will serve to explain how philosophical history emerged, will inform our understanding of its attitude toward time and mind.

In the second half of the eighteenth century, a new and radically different attitude toward the classical world and to history developed. Plato and Aristotle already taught that human experience and the historical past are essential categories for any form of human understanding. In beginning with the conception of the individual as a social or political animal, they insisted that, since the individual is social, it is natural for him or her to live in society. The way of life in which individuals live isolated from other individuals can in consequence only be developed in society. Indeed, to Greek thinkers, a philosophy of history would have been considered a contradiction in terms. To them history consisted of political history and was, accordingly, the proper study of historians and statesmen. Classical theorists, however, were by no means at one in their opinion on the kind of interpretation to which historical phenomena are susceptible.[2]

In contrast to classical theorists, eighteenth-century thinkers postulated that the individual is a "natural" being who naturally forms an association or society — which, in turn, was viewed to be artificial. The eighteenth century was the opening of the age when the foundation for the modern concept of history was prepared. During this period, history became thoroughly secularized and was revitalized in the course of various attempts to unify Western society. History had come to resemble, for some, the natural sciences in its seeking of truth and explanation, but to differ from them in that the truths and explanations it seeks are different: particular not universal truths. Finally, history came to be understood as the history of the individual.

In 1765, Voltaire initiated one of the earliest usages of the category "philosophy of history." Although the political theorist Jean Bodin, in referring to "all the historical writings" of Philo Judaeus of Alexandria, had used the word "*philosophistoricus*" approximately one hundred years earlier, it was Voltaire who received credit for having employed the term 'philosophy of history' in its contemporary secularized sense.[3] In this understanding the philosophy of history has the character an attempt to provide a rational explanation of the course of historical events. This field of endeavor proceeds from the assumption that laws or principles regulate past events and attempts to discern a meaning and logical order in the historical process. Voltaire's *Essay on Manners* was published in 1756. His *Philosophy of History* essay, dedicated to Catherine II and first published in 1765 under the pseudonym, "the late Abbé Bazin," was prefixed to the *Essay on Manners* as an introduction in the 1769 edition. The organizing principle of the work was a design to trace the course of human civilization, the origin and manifestations of manners as a national phenomenon, and the history of the human mind and of human society from Charlemagne to Louis XIII. Without returning to the traditional conceptions of history found in the writings of the Hebrew prophets, St. Augustine, and Orosius, Voltaire chose to criticize Bossuet's *Discourse on Universal History*, a study which regarded history as the "handiwork of God." Voltaire's de-Christianized history and rationalism, typical of the contemporary philosophy of history and resembling the skeptical empiricism of his contemporary Hume, refuted the belief in a "Providence" through which "God governs human affairs," the notion of the "transcendent and miraculous operation of Providence" in theological interpretations of history.

Voltaire emphasized the will of man and human reason, not the will of God and Divine Providence, as the fundamental elements for historical analysis.[4] To Voltaire, a "philosophical history" signified "the lessons which history teaches the *philosophe*," and a *philosophe* was known to be committed to the active life of the reformer. This signification perhaps explains why he dedicated the essay to Catherine II and what he meant by the pronouncement: "One must write history as a philosopher." Voltaire reasoned that insofar as history provides for a human race what experience provides for the individual, and insofar as philosophy often provides "practical wisdom," the category "philosophy of history" aptly connotes a review of a series of past events in combination with their appropriate moral or lesson, which should serve as "mirrors for magistrates."[5] An extended application of the concept in its contemporary sense was Herder's *Ideas Toward A Philosophy of the History of Mankind*. For Herder and the post-Kantian idealists of the nineteenth century, the "philosophy of history" is fundamentally concerned with the question: Is it at all possible to discern in the history of mankind any plan, meaning, or consummation? A fundamental thesis of this brand of speculative philosophy of history was the claim that, despite contingencies, a providential purpose was unfolding unconsciously in the actions of individuals and nations.[6] The shift in emphasis from a transcedent world to the world of everyday experience placed the individual and the individual's mind at the center of historical inquiry. Human agency thus replaced Divine Providence as the important factor in interpreting and reconstructing past events.

Speculative philosophies of history have in common the objective of giving a comprehensive account of the historical process that may be viewed to have meaning and the claim to possess a comprehensive and articulated set of concepts and hypotheses about the principal lines of historical development as a whole. In rejecting theological explanations of the natural and social worlds, and in de-emphasizing theories of mathematics and the physical sciences, certain eighteenth-century theorists turned from "nature" to history; to understand human nature, they turned from the theologians' "Voice of God" and from Rousseau's "nature's voice" to the "voice of time" of Herder, for whom the voices of time spoke in a multiplicity of languages. "The muse of Time," Herder observed, "History, herself sings with a hundred voices, speaks with a hundred tongues."[7] The concept of history thus presupposes an ordering principle of successive moments. Through a rational account of the past, enlightenment thinkers began to associate time and mind more directly. The identifying characteristic of their idea of history, therefore, was philosophical history which outlined the general history of humanity to demonstrate the rational coherence of the course of human events. As Divine Providence was removed from history, the province of history became human reality.

This emphasis on rationalistic theories was accompanied by the growth and development of modern science. The concept of a rational universe was thus a presupposition inherent in the very conception of the philosophy of history. The objective of historical knowledge was not so much to go beyond experience as to explain experience. However, since not all experience was sense-experience, not all philosophical statements about the structure of experience were "sensible verifiable," and even less "sensible verifiable" were the logical explorations of the conditions presupposed by the possibility of experience. The important role played in this process by the growth of modern historical thought should also be acknowledged. For although there were theorists who were inspired by the example the physical sciences had set, there were others who insisted that the scientific paradigm was inappropriate to history and who called for a different procedure in historical analysis. The latter position emphasized the differences rather than the similarities between the events of natural phenomena and the processes of human thought and action. The efforts of those who took this point of view, appeared to provide distinctive methodological features that embodied specific claims about the actual character and scope of historical thought. Subject matter (the individual), in effect, determines method. History is primarily the study of past human actions, events, and situations. Philosophy of history is the study of the study of these issues. The philosophy of history thus becomes a study which distinguishes history from the physical sciences in that, through history, the "muse of time," individual minds speak to one another. As magical and religious forces were increasingly excluded from the analysis of empirical reality, history, as the history of the individual, was henceforth understood as providing the knowledge of human actions and thoughts.

After the eighteenth and nineteenth centuries, there were historical studies that remained under the influence of the natural sciences. Some more positivistic observers insisted that the task of historical research and analysis was the reconstruction of the past from the sources which, if certain falsehoods were expunged, would tell the story of the past by themselves. The steady accummulation of "pure" sources would in due course overcome whatever subjectivity existed in the individual historian. Philosophy, for some, was to be excluded from the field of history, since philosophers were given to imposing an arbitrary pattern that was alien to historical reality. By extracting "facts" from the historian's account of the past, philosophers had undertaken to provide a comprehensive view of events which should explain why and how events occur as they do.

At the turn of this century the procedural suggestions advanced by those theorists who rejected the assumption of similarities between the natural and the social sciences re-emerged. Such thinkers as Dilthey, Croce, Collingwood, Heidegger, and Ortega challenged the claim that the study of human and social phenomena is fundamentally contiguous with the scientific examination of natural phenomena and insisted that the historian's approach to his or her subject matter, by its very nature, diverges from the approaches adopted in the scientific domains of inquiry. Previously, history provided the "facts" or "examples" for philosophy; in the twentieth century, history performs the function of synthesizing "facts" and philosophy. Philosophy and history are closely connected, though philosophy cannot determine the contents of history by mere speculation. It is as dependent on history as it is on science.

But it is not merely the epistemological concern which brings philosophy and history together. It is the fundamental problem of philosophy, vis-a-vis a theory of reality, itself: namely, it is the question "what is the individual?" which compels the philosopher to study history. For the subjective nature of man can be approached only through history. There is no new beginning in human life. One's thoughts and every action presupposes and contains an earlier background of thought and action. What proves to be true for individuals applies equally to cultural, historical, and social phenomena. As an alternative to presupposing that the historical person and the historian have identical "natures," this approach identifies one mind with another as it moves from "nature" to history.

In an effort to overcome Hume's skepticism, Kant attempted, among other things, to establish both the validity of knowledge and the impossibility of knowledge without sensory experience. According to the form of empiricism opposed by Kant, observation of reality consists solely in receiving sense impressions. For Kant, observation of reality entails something other than and more than merely the reception of sense impressions. When Kant wrote his *Critique of Pure Reason*, he described the methods employed by the natural sciences, maintaining that empirical and rational principles should be well coordinated, particularly since a priori principles are the conditions under which we form a conception of the ordered system of the natural world. The principles of constraint, therefore, for Kant, are reciprocal.[8]

The idea of time derived from the succession of our perceptions — impressions as well as ideas and impressions of reflection as well as of sensation — affords us an instance of the a priori and the historical character of time. For, as it is from the disposition of visible and tangible objects that we receive the idea of space, it is also from the succession of ideas and impressions that we form the idea of time. Though there may be real succession in objects, we must have successive perceptions of these objects to have a conception of time. Through these physical phenomena, the idea of time gradually makes its appearance to the mind; time is always discoverable by the perception of a succession of changeable objects. Kant's formulation of the "metaphysical exposition" concerning the "concept of time" conveys our position best when he remarks:

Time is not an empirical concept that has been derived from any experience. For neither coexistence nor succession would ever come within our perception, if the representation of time were not presupposed as underlying them as *a priori*. Only on the presupposition of time can we represent to ourselves a number of things as existing at one and the same time (simultaneously) or at different times (successively).

Time is a necessary representation that underlies all intuitions.[9]

Through his a priori formulation of time Kant thus fostered the Enlightenment's emphasis on reason as the unifying principle of human and historical knowledge, and specified the limits of the historical reality it could unify.

In his *Lectures on the Philosophy of History*, Hegel already identified the problem further as he proceeded beyond Kant to synthesize empirical and rational principles:

History combines in our language the objective as well as the subjective side. It means both the *HISTORIA rerum gestarum* and the *res gestae* themselves, both the event and the narration of event. (It means both *Geschehen* and *Geschichte*). This connection of the two meanings must be regarded as significant and not merely accidental. We must hold that the narration of history and historical deeds and events appear at the same time; a common inner principle brings them forth together.[10]

Despite his prominent position as a speculative philosopher of history, Hegel's notion of a "common inner principle" — the mind — becomes an important principle in our analysis of time and mind and the problem of historical knowledge. His programmatic pronouncement that a common inner principle links events and narration, proves to be useful in our discussion in that the course of events and narration of the course of events are conjoined as being "subjective" and "objective." Both are synthesized by a "common inner principle," and the conclusion that we derive from such a relation is that a radical separation in subject matter and in the method of philosophy of history would be inappropriate. From this standpoint, thus, the word "history" refers both to what has "happened" in the past, meaning an ("objective") historical process, and to the historian's account of it ("subjective"). Human actions and creations are the subject matter, and the mind of the historian-interpreter shapes the subject matter, giving it either an analytical or narrative treatment.

For the thinkers selected in this paper, the historical present is not a mere mathematical point on the time line.[11] There are methodological problems inherent in reconstructing the past, metaphorically, in order to enhance historical knowledge. Bergson and the early Husserl described our conception of time in spatial images and in metaphors, as an intellectual construct which deceptively represents real concrete time which is grasped intuitively by "internal or inner consciousness," and which "does not require to be seen, but to be lived."[12] If we perceive, metaphorically, the passing of time as movement along such a line, we are prompted to ask "What moves?" and are disposed to respond, as Husserl argues, "events continue moving into the past"; one tends to neglect the fact that the word "move" becomes a metaphor to account for events which do not "move" or "change." From this viewpoint, those who temporalize events, conceiving of it as an order within which events occupy different places, are construing events. "The now as the actual now," Husserl observed, "is the givenness of the actual present of the temporal position. As a phenomenon moves into the past, the now acquires the character of a past now. It remains the same now, however. Only in relation to the momentarily actual and temporally new now does it stand forth as past."[13] "A now is always and essentially the edge-point [*Randpunkt*] of an interval of time."[14] "The act of perception sinks back in time in the same way as the perceived in the appearance, and in reflection each phase of the perception must be given identically the same temporal position as the perceived."[15] "If we also denote," he continues, " 'internal consciousness' as 'perception,' then we have here in fact strict simultaneity of the perception and the perceiver."[16] As we reflect on Husserl's "reflection of the past," and pursue other paths that might take us away from the "closed sphere" of "internal consciousness," and as we turn to someone like Collingwood for counsel, we will discern positions on time which are similar in approach but which differ in emphasis.

An individual experiences life through two collateral historical times: that is, one consisting of what occurs to his or her body through human behavior, and an other consisting of what occurs in and to his or her mind. The events in the first are external events in the physical world, those in the second are internal events in the mental world. This antithesis of outer and inner worlds has, at best, been construed as a metaphor. Underlying this metaphorical representation of the bifurcation of an individual's life-experiences rests a profound and philosophical assumption — that of two different kinds of historical existence or status. What exists or happens in the past may have the status of physical existence, or it may have the status of mental existence; some existing is physical existing, other existing is mental existing.[17] An essential feature of what has historical mental existence is that it occurs in time, as constituted by the mind and the actions of individuals, and as formulated in the mind of the historian. What has historical physical existence is composed of individual behavior and physical objects, or else is a function of individual actions and physical objects; what has mental existence consists of "internal" consciousness, or else is a function of consciousness. Collingwood's formulation of the "inside-outside" theory of historical events characterizes this position well:

> The historian, investigating any event in the past, makes a distinction between what may be called the outside and the inside of an event. By the outside of the event I mean everything belonging to it which can be described in terms of bodies and their movements: the passage of Caesar, accompanied by certain men, across a river called the Rubicon at one date, or the spilling of his blood on the floor of the senate-house at another.
>
> By the inside of the event I mean that in it which can only be described in terms of thought: Caesar's defiance of Republican law, or the clash of constitutional policy between himself and his assassins. The historian is never concerned with either of these to the exclusion of the other. . . . His work may begin by discovering the outside of an event, but it can never end there; he must always remember that the event was an action, and that his main task is to think himself into this action to discern the thought of his agent.[18]

The historian's rendering of past human acts demonstrates the limits imposed on historical time in the mind of the philosopher-historian. Collingwood supplanted the postivistic view of history with one which postulated thought as the fundamental concept of historical inquiry. According to this view, to comprehend the actual nature of past events it becomes necessary to penetrate the "inside" of the events and to discern the content of the thoughts of the historical agents under investigation. The historian's task entails rethinking these thoughts. This process involves reconstructing both the time and situation in which historical agents were located and the manner in which they envisaged it. History narrates the actions of individuals in society. In the world of physical objects, things change, in the social world of individuals, events happen. References to historical reality, or to historical time, therefore, are defined in terms of human motives, actions, and reactions and therefore as unique events.

The historical accounts that emphasize the internal consciousness of historical agents have been the subject of theoretical controversy in recent years, various criticisms being leveled at it. Chief among them is the contention that the historian approaches his or her work as an artist.[19] The theorists critical of historical writing claim that the historian, in an effort to "make sense" of the "facts" known at the moment of writing, constructs a particular image of a past world which has been arranged in such a manner that the image has order, unity, clarity, and thoroughness. Whether the "facts" of history are characterized as "poetic" or as "fictive," the order and unity attributed to the events observed, according to this claim, are afforded by the methodological constructive imagination of the historian.[20] For the thinkers represented in this paper, the narrative, as a mode of historical explanation, constitutes one of the basic assumptions under which historians proceed to analyze and to interpret such source materials as human documents, written and spoken, archeological artifacts and human actions. These documents provide the facts which are available to the historian.

What are historical facts? "It used to be said," E. H. Carr has remarked, "that facts speak for themselves. This is, of course, untrue. The facts speak only when the historian calls on them: it is he who decides to which facts to give the floor, and in what order or context . . . the facts of history never come to us 'pure' . . . they are always refracted through the mind of the recorder."[21] To Ortega, "all knowledge of facts," as "isolated" data, "is, to be precise, incomprehensable and can be justified only when used in the service of a theory." Our capacity to comprehend the connectedness in the external world is served best "in the synthesis of facts," for "the unity of facts" is "found not in themselves" but is found "in the mind of an individual."[22] Historical reasoning thus, for Ortega, has a specific form of narration. "In short, the reasoning, the *reason*, that sheds light here consists in a narration. Facing pure physico-mathematical reason there is, then, a narrative reason. In order to comprehend anything human, be it personal, or collective, it is necessary to tell its history. . . . Life only returns a small degree of transparency in the presence of *historical reason*."[23]

Comparable observations informed Croce's concept that history is intuitive:

For since history is history of the spirit, and since spirit is value, and indeed the only value that is possible to conceive, that history is clearly always history of values; and since the spirit becomes transparent to itself as thought in the consciousness of the historian, the value that rules the writing of history is the value of thought. . . . For doubtless imagination is indispensable to the historian: empty criticism, empty narrative, the concept without intuition or imagination, are altogether sterile. . . . Without this imaginative reconstruction or integration it is not possible to write history, or to read it, or to understand it. But this sort of imagination, which is really quite indispensable to the historian, is the imagination in and for thought, the concreteness of thought, which is never an abstract concept, but always a relation and a judgement, not indetermination but determination.[24]

And finally: "Since a fact is historical is so far as it is thought, and since nothing exists outside thought, there can be no sense whatever in the question, What are *historical facts* and what are *non-historical facts*?"[25]

Collingwood carried Croce's concept a step further considering all history as the history of thought. For Collingwood, however, re-thinking the thought of another person is never "a passive surrender to the spell of another mind"; re-thinking is mentally active.

The history of thought, and therefore all history, is the re-enactment of past thought in the historian's own mind. . . . It is thus the historian's picture of the past, the product of his own a priori imagination, that has to justify the sources used in its construction. . . . It follows that historical knowledge can only grow out of historical knowledge; in other words, that historical thinking is an original and fundamental activity of the human mind, or, as Descartes might have said, that the idea of the past is an "innate" idea.

Historical thinking is that activity of the imagination by which we endeavour to provide this innate idea with detailed content. And this we do by using the present as evidence for its own past. Every present has a past of its own, and any imaginative reconstruction of the past aims at reconstructing the past of this present, the present in which the act of imagination is going on, as here and now perceived.[26]

We may also add Croce's contention that "it is evident that only an interest in the life of the present can move one to investigate past fact. Therefore this past fact does not answer to a past interest, but to a present interest, in so far as it is unified with an interest of the present life."[27]

For Collingwood, the present may explain the historian's perception of a given "picture of the past." This perspective, however, "is not an argument for historical skepticism." Rather, it

is only the discovery of a second dimension of historical thought, the history of history: The discovery that the historian himself, together with the here-and-now which forms the total body of evidence available to him, is a part of the process he is studying, has his own place in that process, and can see it only from the point of view which at this present moment he occupies within it.

But neither the raw material of historical knowledge, the detail of the here-and-now as given him in perception, nor the various endowments that serve him as aids to interpreting this evidence, can give the historian his criterion of historical truth. That criterion is the idea of history itself: the idea of an imaginary picture of the past. That idea is, in Cartesian language, innate; in Kantian language, *a priori*. . . . Like other ideas of the same sort, it is one to which no fact of experience exactly corresponds.[28]

Collingwood's notion that the historian's idea of an imaginary picture of the past involves a Kantian a priori, as he acknowledged, for the idea is not an inference from that which is observed. Though he remarked elsewhere, "History is . . . , as a specific form of experience, identical with "perception,"[29] he expressly affirmed the imaginary picture of the past. Historical "data, on the one hand," as Collingwood outlines it for us, "and principles of interpretation, on the other hand, are the true elements of all historical thought. . . . They exist together or not at all. . . . Anything whatever may serve him as data, if he can find out how to interpret it. The historian's data are the entire present."[30]

History, as a discipline of study, has staked a claim in both the social sciences and the humanities. Theories of knowledge, based on this claim, have been rendered difficult since they involve several disciplines of learning: history, philosophy, psychology, logic, mathematics, and the physical sciences. The difficulty increases when we consider that such alignments of disciplines, which may often result in confusion between the different points of view, have been constant sources of methodological problems in the humanities and the social sciences. By its methods and subject matter history, as a discipline in the humanities and in the social sciences, could be distinguished from other arts of sciences. But the central aspiration to afford a total explanatory account of the past has been less sympathetically regarded, as evidenced by the critical reception given by positivistic historians and philosophers alike to our problem. More generally, it has been objected that historians, and in some instances, philosophers, have tended to rely upon unexamined a priori assumptions, regarding both the methods they employ and the materials they select. It is to such considerations that contemporary theoretical attention has frequently been directed.[31]

A social scientific theory of knowledge itself may be conceived of in two different ways and provide a further source of potential confusion. In the first instance, in accepting as knowledge whatever science recognized as such, we may ask: how have we acquired this knowledge and how best can our analysis synthesize it into premises and inferences? — the familiar model of the natural sciences.

In the second case, we may adopt the rationalist, Cartesian standpoint whereby we seek to divide what passes for knowledge into more certain and less certain propositions. These two modes of inquiry are not so distinct as they might appear, for, since the forms of inference involved are not demonstrative, our premises clearly will have more certainty than our conclusions. This tendency, however, merely makes it the more difficult to avoid confusion between the "factual" and "imaginary" modes of inquiry. Descartes' sharp distinction between mind and matter presented the argument that there could be an inquiry into the internal world which would be completely distinct from inquiries into the external world. The realm of philosophy came to be thought of as paralleling the realm of physics: the sciences of man contrasting with the sciences of nature. Ortega argued that by extending Cartesian rationalism from the realm of nature to history, nineteenth-century idealism had affirmed Descartes' general principle of the primacy of mind over matter with the result of the individual's being locked into his or her own ego under the false assumption that a priori reason possessed the key to cognizable reality. "For philosophical idealism since Descartes," he observed, "man is reduced to *une chose qui pense*, *res cogitans*, thought or ideas. The world has no reality of its own, it is only an ideated world."[32]

Yet nineteenth century idealism derived the mind empirically through the individual. Collingwood discussed this view through his disagreement with Descartes. His philosophy of mind rejected the Cartesian emphasis of mind over matter in positing that we cannot study the mind, in itself, but only "man as mind."[33] In speaking of "man as mind" and of man as being constituted by "*whatever he is conscious of being*," Collingwood took issue with Descartes' notion that acts of consciousness are self-revealing. For Collingwood, the individual has to perform additional acts in order to become aware of any of his or her acts of consciousness. Thus no act of consciousness becomes analogous to perceiving entities or to any other form of sensation.[34] Collingwood repudiated the proposition that one becomes aware of one's acts of consciousness through introspection. One cannot know, he argued, one's own mind by turning the inner eye, as it were, on its procedures, because introspection can solely bring to the mind something of which one has already become aware.[35] The individual with which Collingwood becomes concerned, in his account of the history of civilization, is a " 'Man' of the mind" and, since "man's mind is made of *thought*," "man's mind is *whatever he is conscious of being*."[36]

One of the fundamental problems in modern historical knowledge, as in philosophy, has been the distinction between "objective" or "factual" and "subjective" or "imaginary" realities. Ortega and Collingwood addressed this problem repeatedly in their philosophical writings. For instance, we assume that perception can cause knowledge although it may accredit mistakes if we are careless in our logic. To confirm this point we may add that all human knowledge of matters of fact is in part determined by perception. However, a principle of this kind is clearly one which can only be known by inference. General principles of inference, in the social world, whether inductive or deductive, are very much connected to a human point of view. In the social sciences, perception is not impartial, but proceeds from a center; our perceptual world is a perspective view of the common social world. From a human point of view, that which is near time and space generally gives rise to a more vivid and distinct memory of the object perceived than that which is distant. From an historical perspective, what is distant is often considered to be more clear and concrete than what is near by way of contemporary events.

As a philosopher interested in interpreting events of the past, Collingwood cannot accept the claim that "one part of time follows after another" because "events are not related to one another in this way." Historical events, however, may be construed as relating to each other with the implication that human agents create meaning by connecting these events into some purposive activity. The task of the historian, therefore, as an interpreter of the "past" — and as forecaster of the "future" if he or she may be so inclined — is to discern the "pattern" in human behavior that reveals how events unfold and are subsequently "related to one another." Through individual and collective memories, and through inferences in logic, we derive knowledge of the past. To avoid the extreme positions of either denying the reality of the past and future, or in assigning the status of "real" to the past and "unreal" to the future, the subjective dimensions of our experience of time need to be acknowledged and emphasized: "We do not remember *the* past, but only *our* past; and we do not expect *the* future but only *our* future."[37]

These statements were based on the recognition that historical time differs from time in nature. Clearly, any natural phenomenon occurs in time. Nevertheless, references to historical reality or to historical time, often have been defined in terms of human motives, actions, and reactions. For this reason historical events have often been characterized an unique events, in contrast to the laws of nature where similar elements meet under similar circumstances. The history of the individual therefore means more than mere change in time. Historical time becomes meaningful through human actions. In connecting historical time to human actions, critical philosophy of history sought to disclose the mind's activity as a tangible process. As with Dilthey, Croce, and Collingwood, Ortega in his philosophical orientation was affiliated with the tradition which, by way of contrast with classical British empiricism, portrayed the human mind in active terms; in other words, the Neo-Kantian tradition treated mental activity as creative and self-determining, displaying a constructive role in human experience rather than passively responding, in mechanical fashion, to the prompting of external stimuli. Kant had already established that all knowledge is a function of the human mind. The mind, by means of its own a priori forms, structures the entire domain of knowledge. This formal extension of the domain of mental activity appealed to Ortega. The particular importance that Ortega ascribed to history and his contention that its character had been radically misinterpreted by empiricist philosophers, reflected the presupposition underlying his portrayal of the individual as an autonomous historical agent. He viewed the empirical, or positivistic, objective of formulating a theory of human nature in accordance with principles drawn from the physical sciences as providing faulty findings. In adopting this position one assumes that human behavior is subject to invariant, universal regularities, and that in this area of analysis explanations necessarily consist in subsuming what occurs beneath "a general law of nature." Against such a view Ortega argued that human "nature" involves a continual process of reflective self-transformation, which shows individuals to be continually modifying and reformulating their experiences. In this connection, the notion of a fixed human character, conforming to immutable principles which are valid for all individuals during all historical periods, was unacceptable to him.

To explain reality, Ortega continued, the natural sciences, in their concern with the existence of objects in natural phenomena, aim at discovering the general concepts or the natural laws under which these objects may be subsumed. In order to understand the relationship between human life and reality, therefore, the individual must escape from "the terrorism of the laboratory." Through his various readings of Kant, Ortega learned that the autonomous mind must liberate the self from the enslavement of our empirical existence by nature. To avert compounding the ontological distinction he felt should be made between a philosophy of physical nature and a philosophy of human nature, Ortega maintained that "man has no nature." The individual must free himself or herself from the shackles of the natural scientists for the latter are too involved with general, abstract, "concepts of nature," and "man has no nature." In this sense, the generalized interpretations of the natural scientists exclude the most essential element of human reality: the life of the individual.

Ortega rejected positivistic positions that presupposed the adoption of the external standpoint of the "observer"; this standpoint was wholly inappropriate to the context of the human and the social world. For the phenomena with which we are concerned in the social world have to be conceived under the category of purposive human agency, and to apprehend an event as an action in contradistinction to a mere natural occurrence is to perceive it as the expression of human thought and volition. A perception of this sort contains the embodiment at the level of observable physical reality of that developing rationality which is the individual's distinguishable attribute. Thus, the individual as a knowing subject produces physical and historical realities and thereby the mind constitutes itself as "history."[38]

If man has no nature, Ortega pronounced that "man has history." "Man lives" and, as a "living being," relates to other living beings such that vital, operative factors are in motion: the "living experiences of man." History, therefore, "*sensu stricto*" constitutes "human history."[39] For history "is the reality of man. He has no other. Through history he has made himself such as he is."[40] To Ortega human life is "not a thing" but "a drama"; human life "is a gerund, not a participle: a *faciendum*, not a *factum*."[41] Finally, as expressed differently by Collingwood: "what nature is to things, history, as *res gestae*, is to man. . . . Man finds that he has no nature other than what he has done himself."[42] For Collingwood, history is the study of "*res gestae*: actions of human beings done in the past"; and, because it studies human activity, it is the study of purposive activity.[43]

From the perspective of purposive human activity, interpreting what individuals do becomes a matter of eliciting the processes of thought that inform their discernible behavior and are intrinsic to its correct identification and analysis. Such interpretations are the metier of historical, not scientific, understanding. For the historian, unlike the natural scientist, does not seek to explain human events by correlating them with other events in view of some well-attested uniformity or general law; rather the historian aims to uncover the inner rationale of human events — which can only be done by "re-enacting" within one's own mind the thoughts of the person, or persons, with whose activities one is concerned. Historical knowledge differs irreducibly from scientific knowledge because the former studies human conceptions, intentions, desires, and feelings. Only by such an interpretation may historical inquiry acquire an acceptable understanding of individuals as diverse and ever changing.

Individual lives are historical. Human life has a beginning and an end and, on entering the world, the individuals enter a social world that is given to each in conjunction with the historical process of his or her time dimension. History has neither a new beginning nor a final ending. It can only be conceived of as the realization of the potentialities of man in historic time. In retracing the individual's struggle to control nature and in reviving the thoughts and conscious life of the past, historical study offers any contemporary generation the challenge of making its own vital decisions on the basis of a critical knowledge of the full scope of former human experiences and achievements. History is understood here in its dual meaning as a factual world of phenomena and as a discipline concerned with only its relevance to life in the present. "Everyone brings his own mind to the study of history," Collingwood reminded us,

> and approaches it from the point of view which is characteristic of himself and his generation; naturally, therefore, one age, one man, sees in a particular historical event things which another does not, and *vice versa.* The attempt to eliminate this "subjective element" from history is always insincere — it means keeping your own point of view while asking other people to give up theirs — and always unsuccessful. If it succeeded, history itself would vanish.[44]

Starting from the premise that consciousness, in the sense of cognitive awareness, always should be consciousness of something, modern theorists have perplexed themselves with such questions as what consciousness is in itself and how it is related to the thing, or the facts, which are its objects. It does not appear to be identical with its objects, yet, as indicated earlier, neither does perception. Historical consciousness constitutes a concern with the past as being relevant to our understanding and apprehending of the human condition. Its key attribute consists of our awareness of ourselves as "being-in-time." It implies that the temporal ordering of events refers to the very nature of human existence. Historical consciousness thus enables the individual, a civilization, and the historian, to expand their experiences beyond their actual historic time frame to absorb the results of events in which they had no part. The study of history thereby opens the door to participation in the completeness of human history.

This participation in human history has to be understood as being direct and active. Where the natural scientist perceives phenomena which have a reality independent from the observer, history has the character of being real in the consciousness of the historian. The word "history" means both history as actuality — the *res gestae* of Collingwood and Ortega — and written history; "history" as what has happened in the past and "history" as the historian's account of what has happened in the past. Clearly, any attempt to distinguish between historical "reality" and historical "accounts of reality" implies the interpreter *knows* what constitutes historical reality. The past is present only insofar as it is relived by the historian through sympathy and understanding. The central problems of an historical epistemology and methodology revolve around the understanding that an objective knowledge of the past can only be attained through the subjective experiences of the investigator. In this connection, Dilthey maintained that: "The human mind understands best what it has created." And so Croce said that "All history is contemporary history . . . an act of thought." Ortega y Gasset echoes that "History is . . . a science of the present. . . . The past is not there, at the date when it happened, but here, in me." Collingwood declares that "All history is the history of thought." This is what Heidegger means when he claims that Being lives in a temporal existence and is thereby "historically situated." Temporality signifies a constitutive trait of "Being" and "is also the condition which makes historicality possible."[45]

When the historian's subjective experiences prompt his reliving of the past, the study of history offers no escape from historicity and all the limitations of the historic individual apply as well to the study of history. Indeed, without subjectivity we could neither synthesize nor penetrate history. Interpreters of history will not always gain answers to the problems of history, but should be able to formulate questions with which to approach the past. "The beginning of historical research," Collingwood observed, "is therefore not the collection or contemplation of crude facts as yet uninterpreted, but the asking of a question which sets off looking for facts which may help one to answer it."[46] The underlying concept of the individual ultimately determines the critical character of historical interpretations, which results from both a reading of the past, and simultaneous self-analysis by the observer. In this critical revival of the past, the contemporary generation achieves a higher historical consciousness of its own being. Internal consciousness of temporal changes within the individual coupled with awareness of the external manifestations of change in the social world of human reality dispose one to see how Ortega's concept of the generation constitutes a concept of historical methodology:

> The concept of the generations, converted into a method of historical investigation, consists in nothing more than projecting that structure across the past. Anything other than this is to renounce discovering the authetic reality of human life in every period of time — that is the mission of history. The method of the generations permits us to see that life from within itself, in its actuality. History is virtually to convert that which has already passed into that which is present. For this reason — and not only metaphorically — history is to relive the past. And as living is nothing else but actuality and the present, we have to transmigrate from the actuality and the present which are ours to those of the past, looking at them not from without, not as living experiences which have been, but as those which persist in being.[47]

Historical study, thus, is a humanistic enterprise. Currently, history can be our approach to knowledge of the individual and of humanity and through history we acquire the wealth and wisdom of past cultures. The critical awareness of the potentialities of the individual enables us to act in our own historical time with deeper insights and with an active commitment. The humanism of historical consciousness stems from the realization that our lives constitute time. For the "past is past," Ortega remarked, "not because it happened to others but because it forms part of our present, of what we are in the form of having been: in short, because it is *our* past. Life as reality is absolute presence: one cannot say that *there* is anything if it is not present, of this moment. If, then, *there* is a past, it must be as something present, something active in us *now*."[48] In this connection, Ortega was in accord with Croce's notion that "we know that history is in all of us and that its sources are in our breasts."[49] According to this notion, the recordings of "chronicle's dead data" of the past fail to register the reality of history and human life which is the "vital," living action of the present, and is one with the past and the future. History characterizes *what we are*, instead of something that we possess. Through history we learn *who we are* by examining *what we have done.* In view of this position, that we are what we have done, history shapes itself in and through ourselves.

Historical time thus carves the essence of individual configurations. For Ortega "historical reason" perceives and registers the tangible facts of historical reality through the demonstrative process of how the present originated from the past in order to engender the future. In conjunction with the historicization of the individual, "historical reason" became the medium that circumscribed and eventually supplanted "physico-mathematical reason" as the unitary principle of ultimate reality: human life, and thereby "historical reason" was now viewed as possessing, at once, the principle of diversity and unity. Hence, history was rationalized and reason was historicized, as with the individual:

> . . . Man alienated from himself encounters himself as reality, as history. And, for the first time, he sees himself compelled to occupy himself with his past, not from curiosity nor in order to find normative examples, but because he *has no* other thing. Things are never done seriously but when, truly, they are absolutely necessary. For this reason, the present hour is the time for history to re-establish itself as historical reason.
>
> Until now history has been contrary to reason. In Greece the two terms "reason" and "history" were opposed. And until now, in effect, hardly anyone has been concerned to search in history for its rational substance. . . . Hence the expression "historical reason" must be understood in all the rigor of the term. Not an extrahistorical reason which appears to fulfill itself in history but, literally, *a substantive reason constituted by what has happened to man*, the revelation of reality transcending man's theories and which is himself, the self underlying his theories.[50]

Through this discussion the ground upon which historical knowledge has been based for Ortega becomes apparent. For speculative philosophies of history, the "owl of Minerva" began its flight at dusk; for Ortega, and certain twentieth-century philosophers of history, "historical reason" proclaims the dawn of human history.

At present the problem of historical knowledge and the philosophy of history, as well as the tradition to which the thinkers under discussion belong, has a revived interest. Croce, Collingwood, and Ortega, and others who associated with this tradition, expressed sympathy with those who claimed that the interpretation of human actions raised difficulties for standard empiricist accounts of explanation of the human and social world. For the most part, they were wholly committed to universal human objectives and placed a very high value on the uniqueness of the individual. They were concerned with the philosophical implications of historical knowledge and made attempts to develop systematically the problem of relating philosophical presuppositions the problem of historical knowledge. They sought to relate their specialty, philosophy and the philosophy of history, to historical knowledge. "History", Ortega observed,

> is perfect continuity. Every idea of mine comes from another idea of mine or from the idea of some other man. [Historical thought therefore] proceeds with respect to human phenomena — philosophy, law, society, arts, and letters, language, religion — the same thing began with the natural sciences, as established in the works of Kepler and Galileo, when they proceeded to bring together emprically the simple facts of material phenomena.[51]

Ortega's contribution of this area of study has been to emphasize what he considered to be the epistemological implications of historical knowledge. By such phenomenologists as Husserl, Heidegger, and Scheler, Ortega was challenged to a continual emphasis on a strict and accurate observation of the various elements of human experience. Ortega was convinced that isolated empirical "facts," apart from any understanding of their interior meaning, their inner connection to human life and the meaning of the social and the historical process, were limited if not futile. The results of Ortega's studies in the philosophy of history and in historical knowledge led him to the perspectivist viewpoint that "the position of the observer influences the results of thought."[52] For him, therefore, the perspective of the observer must always be considered but a partial and limited perspective.

Collingwood carried this perspective further and incorporated psychological categories. He is known for his position that the kind of history one writes, or the way one thinks about history, is ultimately a reflection of the kind of person one is. Clearly, the reverse is also true and may be applied to a thinker like Ortega. The "experience of life," Ortega reflected, "presents us with what I call 'aspects.' . . . Perhaps this may not be a reality, but only as an aspect [of reality]. For that reason, I call it accordingly. But as an aspect it is real and influences us — whether we like it or not."[53] In studying the past, one immerses oneself, as an historian, in the ambience of the period. "The events of history," Collingwood reminded us, "are things which the historian looks, not at, but through, to discern the thought within them."[54] The historian may know the personalities of a historical period through an accumulation of biographic detail, letters, documents, biographies, as well as through personal acquaintances of figures. The historian's approach and attitudes toward the past, when honed successfully, can result in his or her ability to comprehend and communicate with the past on its own terms. The historical approach thus serves as a communicative link between the past and the present and the historian, in making such communication possible, serves as a mediator between the worlds of the past and the present. The historical lines linking the past and present also link time and mind. To attain this end, Ortega eventually grounded historical knowledge in a thorough knowledge of the individual. "History," he wrote, "is the systematic science of that radical reality which is my life. ... There is no *actio in distans.* The past is not there, at the date when it happened, but here, in me. The past is I — by which I understand, my life."[55]

"Autobiography," Dilthey wrote, "is the literary expression of the individual's reflection on his life. When this reflection is transferred to the understanding of another's existence it emerges in the form of a biography."[56] The central concern of the historian in recapturing relevant biographic details of someone's life is one of eliciting the meaning of what has been done, and this, it may be argued, can only be attained insofar as we show these details to be intelligible in terms of reasons, explanations, or perceptions attributable to the historical agent in question. Thus to understand an action, or past actions, implies comprehending it as an expression or process of meaning whereby action has been connected conceptually to some underlying pattern of thought or intention, rather than as a completely contingent relation. Moreover, such an underlying pattern presupposes a background structure of established rules or practices of the sort that may prove to be integral to human life and experience as we apprehend them. For to claim, as did Collingwood, that such an account of historical understanding requires the re-enactment of past thinking merely directs attention to the method, admittedly familiar among historians, of imaginatively projecting oneself into another's position for the purpose of arriving at an explanatory hypothesis. Ortega, as well, claimed that,

> man invents for himself a program of life, a static form of being, that responds satisfactorily to the difficulties planned for him by circumstance. He essays this figure of life, attempts to realize this imaginary character he has resolved to be. He embarks on this essay illusioned and creates, completely, the experience of it. ... But on experimenting with it, the limits of this vital program make apparent its insufficiencies. It does not resolve all the difficulties, and it creates new ones.[57]

The claim was to demonstrate that the explanations historians actually provide are similar in their structural analysis to those put forward in other areas of investigation in the social sciences; this demonstrates that an explanation of past historical actions, by reference to motivational factors, is possible in presupposing the validity of pertinent psychological generalizations. What then, we may finally ask, is history for? "My answer is," Collingwood wrote,

that history is "for" human self-knowledge. It is generally thought to be of importance to man that he should know himself: where knowing himself means knowing not his merely personal peculiarities, the things that distinguish him from other men, but his nature as man. Knowing yourself means knowing, first, what it is to be a man; secondly, knowing what it is to be the man *you* are and nobody else is. Knowing yourself means knowing what you can do; and since nobody knows what he can do until he tries, the only clue to what man can do is what man has done. The value of history, then, is that it teaches us what man has done and thus what man is.[58]

Taking these considerations into account, philosophers, historians, and philosophers of history may learn from psychologists and psychoanalysts that constructive comparisons can be made between understanding what individuals do and understanding what they say. Through this connection the philosopher of history may reconstruct historical time as a form of human understanding and, if so inclined, as the process through which self-analysis may be attempted. The interpretation of human actions raises difficulties for standard empiricist accounts of explanation of the human and social world. History thus may be the approach of choice to knowledge of the individual and of humanity; through history, the epistemological break is not with the past but with empirical accounts of the past.

The historical act of recreating past human creations and of projecting oneself into the thoughts of others, makes salient critical philosophy of history's synthetic function of connecting time and mind through the process of internal time consciousness. "Very deep is the well in the past,"[59] Thomas Mann wrote at the begining of his novel, *Joseph and His Brothers*. In bringing human connectedness and meaning to historical time and mind, critical philosophy of history has provided the synthesis that serves to prevent the deep well of the past from becoming bottomless.

Wesleyan University

NOTES

* Maurice Merleau-Ponty, *Phenomenology of Perception*, trans. Colin Smith (London: Routledge and Kegan Paul, 1962), p. 411; Johan Huizinga, "A Definition of the Concept of History," in *Philosophy and History: Essays Presented to Ernst Cassirer*, ed. Raymond Klibansky and H. J. Paton (Oxford: The Clarendon Press, 1936), p. 9; Oswald Spengler, *The Decline of the West*, ed. Helmut Wagner, trans. Charles F. Atkinson (New York: The Modern Library, 1962), p. 77.

1 Henry Neville, *Plato Redvivus or, A Dialogue Concerning Government* in *Two English Republic Tracts*, ed. Caroline Robbins (Cambridge: Cambridge Univ. Press, 1969), pp. 68, 83, 111.

2 See Aristotle's *Politics* where he observes, in a critical discussion of Plato, that "we are bound to pay some regard to the long past and the passage of the years. . . ." — *The Politics of Aristotle* trans. with an introd. by Ernest Barker (London: Oxford University Press, 1946), p. 52. *Introduction to Aristotle*, second ed. with an introd. by Richard McKeon (Chicago: University of Chicago Press, 1973), pp. 681—82. Compare elsewhere, in the *Politics*, where Aristotle makes an important distinction between the functions of the historian and the poet. The difference between history and poetry is not a difference in diction, between prose and verse (for "you might put the work of Herodotus into verse, and it would still be a species of history") but a difference between writing about what has happened and what might have happened. "From what we have said," he observed, "it will be seen that the poet's function is to describe, not the thing that has happened, but a kind of thing that might happen, i.e. what is possible as being probable or necessary. The distinction between historian and poet is not in the one writing prose . . . it consists really in this, that the one describes the thing that has been, and the other a kind of thing that might be. Hence poetry is something more philosophic and of graver import than history, since its statements are of the nature rather of universals, whereas those of history are singulars." *Introduction of Aristotle, op. cit.*, pp. 681—82. Herodotus, who was called "The Father of History" by Cicero and others, states that the purpose of his "research" or "inquiry" was "to preserve the memory of the past by putting on record the astonishing achievements both of our own and of other peoples; and more particularly, to show how they came into conflict." Herodotus, *The Histories*, trans. Aubrey de Selincourt, introd. and notes A. R. Burn (Bungay, Suffolk: Penguin Books, 1983), p. 41.

3 Voltaire, *Essai sur les moeurs et l'esprit des nations, et sur les principaux faits de l'histoire depuis Charlemagne jusqu'à Louis XIII* (Paris: Editions Garnier Frères, 1963), 2 vols., introd. René Pomeau; Jean Bodin, *Methodus, ad Facilem Historiarum Cognitionem*, (Amsterdam: Neudruck Der Ausgabe, 1650; Darmstadt: edition Scientia Verlag Aalen, 1967), pp. 342ff; see also *Method For the Easy Comprehension of History*, trans. Beatrice Reynolds (New York: Columbia University Press, 1945), pp. 319—333, 368. For his discussion of the biblical historical writings of Philo Judaeus and Josephus Flavius, see especially pp. 319—359. It is in Chapter X where Bodin writes "Philo the Jew, who can be called a 'philosophic historian' . . .", p. 368. In categorizing historical knowledge, Bodin brought juristic knowledge to the formation of a philosophy of history. He also drew from the tradition in the literature of Biblical and Classical antiquity. "Of history," he writes, "that is, the true narration of things, there

are three kinds: human, natural, and divine. The first concerns man; the second, nature; the third, the Father of Nature. One depicts the acts of man while leading his life in the midst of society. The second reveals causes hidden in nature and explains their development from earliest beginnings. The last records the strength and power of Almighty God and of the immortal souls, set apart from all else." (*Method*, p. 15; *Methodus*, p. 8) In 1853, Henri Baudrillart considered Bodin's *Methodus* as "already the first considerable essay on the philosophy of history," and identified Bodin as the "true father of the philosophy of history" by way of his "philosophic interpretation" and "synthesis of history and law." See Henri Baudrillant, *Jean Bodin et Son Temps: Tableau des Théories Politiques et des Idées Economiques au Seizième Siècle* (New York: Burt Franklin, 1969), p. 145. See also pp. 146—167.

[4] Voltaire, *Essai sur les moeurs*, I: 3, 55; 196—97, 202—203. Arnold J. Toynbee, perhaps the best-known example of an historian who turned to philosophy, an exponent of the traditional theology of history, characterized history as "a vision of God's creation on the move, from God its source to God its goal." Toynbee makes this statement opening a chapter entitled "The Attractiveness of the Facts of History." See Arnold J. Toynbee, *A Study of History*, 10 vols. (New York: Oxford University Press, 1954), 10: 3—41.

[5] Voltaire, *Essai sur les moeurs, op. cit.*, 2: 800—802.

[6] Johann G. von Herder, *Outlines of a Philosophy of History of Man*, trans. T. Churchill (New York: Bergman, 1800). A journal "devoted exclusively to the interests of Speculative Philosophy" was initiated as late as 1867 for English reading audiences. See editor's note "To the Reader," and on "The Speculative" in *The Journal of Speculative Philosophy*, Vol. I, num. 1 (1867): 1—6.

[7] Herder, *op. cit.*, p. 458.

[8] David Hume, *A Treatise of Human Nature*, 2 vols., introd. by A. D. Lindsay (London: Everyman's Library, 1911), I: 73—96; Immanuel Kant, *Critique of Pure Reason*, trans. Norman Kemp Smith (New York: Macmillan, 1963), p. 20, and Immanuel Kant, *Prolegomena to Any Future Metaphysics*, ed. Lewis White Beck (New York: Bobbs-Merrill Co., 1950), pp. 122—23.

[9] Immanuel Kant, *Critique of Pure Reason, op. cit.*, p. 48.

[10] G. W. F. Hegel, *Reason in History*, trans. Robert S. Hartman (Indianapolis: Bobbs-Merrill, 1953), p. 75. For an important discussion of historical realities as external manifestations of underlying ideas and the notion that "every human individuality is an idea rooted in actuality," see Wilhelm von Humboldt's essay, "On the Historian's Task," *History and Theory*, VI (1967): 57—71. For two interesting interpretations of Hegel, see E. Voegelin, "On Hegel — A Study in Sorcery," in *The Study of Time*, Vol. I, ed. J. T. Fraser *et al.*, 1972) pp. 418—451; and W. Mays, "Temporality and Time in Hegel and Marx," in *The Study of Time*, Vol. II, ed. J. T. Fraser and N. Lawrence (New York: Springer-Verlag, 1975), pp. 98—113.

[11] Henri Bergson, *Time and Free Will: An Essay on the Immediate Data of Consciousness*, trans. F. L. Pogson (New York: Harper Paperback, 1960), pp. 119—121.

[12] *Ibid.*, pp. 80, 181—83; 191.

[13] Edmund Husserl, *The Phenomenology of Internal Time-Consciousness*, ed. by Martin Heidegger and trans. by James S. Churchill with an introd. by Calvin O. Schrag (Bloomington: Indiana Univ. Press, 1964), p. 90.

[14] *Ibid.*, p. 95.

[15] *Ibid.*, p. 97.

[16] *Ibid.*, p. 148.
[17] For a discussion of the immediate experiences upon which our knowledge of time is based, see Russell's, "On the Experience of Time," in *The Monist*, Vol. **25** (1915): 212—233; particularly where he defines time which arises through the relations among objects as "physical time," and time which arises through the relations of subject and object as "mental time." Russell's theory of time requires a definition, of what "one (momentary) total experience" signifies, that does not presuppose the notion of time. That is to say: "One (momentary) total experience" consists of "a group of experiences such that the objects of any two to them are experienced together, and anything experienced together with all members of the group is a member of the group." The "specious present" of a momentary total experience constitutes "the period of time within which an object must lie in order to be a sense-datum in that experience" (*Ibid.*, p. 218).
[18] R. G. Collingwood, *The Idea of History* (New York: Oxford University Press, 1956), p. 213.
[19] Arthur Schopenhaur, *The World as Will and Representation*, trans. E. F. J. Payne (New York: Dover, 1966) II. 403—410, 439—442.
[20] Some of the most stimulating and controversial discussions of the general issues involved in the methodological discussions of the general issues involved in the methodological debate over the analysis of historical narratives occur in certain recent studies of historiography, literary theory, and philosophy of history. Noteworthy examples include: Paul Ricoeur, *Time and Narrative*, 2 vols., trans. Kathleen McLaughlin and David Pellauer (Chicago: The Univ. of Chicago Press, 1984), I: 91—230; and David Carr, *Time, Narrative, and History* (Bloomington/Indianpolis: Indiana Univ. Press, 1986), pp. 100—121; Hayden White, *Metahistory: The Historical Imagination in Nineteenth-Century Europe* (Baltimore: Johns Hopkins U. Press, 1973); and White's "The Structure of Historical Narrative," *Clio*, Vol. **1**, num. 3 (1972): 5—20; Louis Mink, "The Autonomy of Historical Understanding," *History and Theory*, Vol. **5** (1965): 24—47; and Mink's "History and Fiction as Modes of Comprehension," *New Literary History*, Vol. **1** (1970): 541—58; Peter Gay, *Style in History* (New York: Basic Books, 1974); Maurice Mandelbaum, *The Anatomy of Historical Knowledge* (Baltimore: Johns Hopkins U. Press, 1977); W. H. Dray, "On the Nature and Role of Narrative in Historiography," *History and Theory*, Vol. **10** (1971) 153—71; and Leon Goldstein, *Historical Knowing* (Austin: University of Texas Press, 1976).
[21] Edward Hallett Carr, *What is History?* (New York: Random House, 1967), pp. 9, 24.
[22] José Ortega y Gasset, *Obras Completas*, 11 vols. (Madrid: Revista de Occidente, 1969) I: 316—317.
[23] Ortega, *Obras, op. cit.*, 6: 40.
[24] Benedetto Croce, *History: Its Theory and Practice*, trans. Douglas Ainslie (New York: Russell and Russell, 1920), pp. 36, 39, 108.
[25] Collingwood, *op. cit.*, pp. 215, 245, 247.
[26] *Ibid.*, p. 248.
[27] Croce, *op. cit.*, p. 12.
[28] R. G. Collingwood, *op. cit.*, p. 248; see also H. Wildon Carr, " 'Time' and 'History' in Contemporary Philosophy; With Special Reference to Bergson and Croce," in *Proceedings of the British Academy* (1917—1918): 331—349.
[29] R. G. Collingwood, *Speculum Mentis* (Oxford: The Clarendon Press, 1924), p. 205.

30 R. G. Collingwood, "The Philosophy of History" in *The Historical Association Leaflet*, num. 79 (1930): 14; R. G. Collingwood, "The Nature and Aims of a Philosophy of History," in *Proceedings of the Aristotelian Society*, Vol. **25** (1925): 163—73.

31 A major claim for history and historiography as a science, if not at least as a profession, rests on the attitude of historians toward documents and documentation. As Marc Bloch observes: "A science, however, is not to be defined entirely in terms of its object. Its limits can be fixed quite as well by the character of its appropriate methods." See Marc Bloch, *The Historian's Craft*, trans. Peter Putnam (New York: Random House, 1953), p. 47. In this regard see note 1 above. See also R. G. Collingwood, "Are History and Science Different Kinds of Knowledge?", *Mind*, Vol. **31** (1922): 443—451; R. G. Collingwood, "The Limits of Historical Knowledge," *Journal of Philosophical Studies*, Vol. **III** (1928): 213—222; C. O. Weber, "The Reality of Time and the Autonomy of History," *The Monist*, Vol. **37** (1927): 521—540; and Peter Winch, *The Idea of a Social Science and its Relation to Philosophy* (London: Routledge & Kegan Paul, 1958).

32 Ortega, *Obras, op. cit.*, 7: 116.

33 R. G. Collingwood, *The New Leviathan* (Oxford: The Clarendon Press, 1924), p. 7.

34 *Ibid.*, pp. 30—31, 60; R. G. Collingwood, *The Principles of Art* (New York: Oxford Univ. Press, 1958), p. 222.

35 R. G. Collingwood, *An Essay on Metaphysics* (Oxford: The Clarendon Press, 1948).

36 Collingwood, *New Leviathan, op. cit.*, pp. 6—7.

37 Collingwood, "Some Perplexities About Time: With An Attempted Solution," in *Proceedings of the Aristotelian Society*, Vol. **26** (1926): 142—43, 146. Collingwood's views on the "perplexities" of time compare favorably with St. Augustine's position: "Some such different times [past, present, future] do exist in the mind, but nowhere else that I can see. The present of past things is the memory; the present of present things is direct perception; and the present of future things is expectation. If we speak in these terms, I can see three times and I admit that they do exist." (St. Augustine, *Confessions*, trans. R. S. Pine-Coffin (Harmondsworth: Penguin Books, 1961), Book XI, Section 20, p. 269; also see Section 28, p. 277. Unlike their speculative predecessors, most contemporary contributors to historical knowledge and to the philosophy of history take it to be a secondary form of inquiry with the aim, not of trying to elucidate and assess the human past itself, but rather of trying to elucidate and assess the ways in which historians typically describe or comprehend that past. In other words, the philosophy of history is now treated as having to do with such matters as the presuppositions underlying historical narratives, the categories implicit in historical judgment and explanation, and the modes of argument whereby historical conclusions are supported or established. See also Collingwood, *The Idea of History, op. cit.*, and White, *Metahistory, op. cit.*

38 Ortega, *Obras, op. cit.*, 6: 24.

39 *Ibid.*, 6: 32, 41.

40 *Ibid.*, 6: 32—33.

41 *Ibid.*, 6: 41. Hermann Cohen, Ortega's teacher at Marburg, in his *Die Religion der Vernunft aus den Quellen des Judentums* (Wiesbaden: Fourier Verlag, 1978), 366—369 provides an interesting comparison here through the observation that historical thinking in the Christian Church is the result of prophetism, as expounded in the prophetical

historism of Joachim of Floris: "The Concept of history is a product of prophetism. . . . What Greek intellectualism could not produce, prophetism has achieved. In Greek consciousness, (*historien*) is equivalent to inquiry, narration, and knowledge. To the Greeks history remains something we can know because it is a matter of 'fact' [*factum*], that is, of the past" (pp. 305—308, 366—369).

[42] Collingwood, *The Idea of History op. cit.*, p. 9.

[43] Collingwood, "The Philosophy of History," *op. cit.*, p. 15. In an historical period where we have become conscious of the ideological implications inherent in any form of inquiry, we have also become aware of another rule of historical inquiry: the perceptions of the observer-interpreter of events are equally as important as the perceptions of historical agents in the past. See Karl Mannheim, *From Karl Mannheim*, ed. with introd. by Kurt H. Wolff (New York: Oxford University Press, 1971), p. 97; and Karl Mannheim, *Ideology and Utopia* (New York: Harcourt Brace and World, 1936), pp. 133—271.

[44] R. G. Collingwood, *An Autobiography*, intro. by Stephen Toulmin (Oxford: Oxford University Press, 1978), p. 110; R. G. Collingwood, *Religion and Philosophy* (London: Macmillan and Co., 1916), pp. 49—51; Ortega, *Obras, op. cit.*, 7: 148—50, 152—53, 415—16. In emphasizing the subjective dimensions of our experience of time, Collingwood maintains that the "past as *past*" and the "future as *future*" are not existent *qua* real in the present but are "purely ideal." That is, "the past as living in the present and the future as germinating in the present are wholly real and indeed are just the present itself. . . . That which is ideal is for a mind, and has no other being except to be an object of mind. . . . Hence, if there were no mind, there would at any given moment be no past and no future; there would only be a present in which the past survived transformed and in which the future was present in germ. The past *as past* and the future *as future*, in contradistinction from their fusion in the present, have being for mind and only so. We do call the past, *as such*, into being by recollecting and by thinking historically; but we do this by disentangling it out of the present in which actually exists, transformed, and re-transforming it in thought into what it was. Hence time, as succession of past, present and future, really has its being *totum simul* for the thought of a spectator. . . ." ("Some Perplexities About Time: With An Attempted Solution," *op. cit.*, pp. 149—50)

[45] Ortega, *Obras, op. cit.*, 6: 39, 9: 360; Croce, *History: Its Theory and Practice, op. cit.*, pp. 60—61; Benedetto Croce, *History as the Story of Liberty*, trans. by Sylvia Sprigge (New York: Meridian Books, 1955); Collingwood, *The Idea of History, op. cit.*, pp. 215—220, and *An Autobiography, op. cit.*, pp. 107—19; Wilhelm Dilthey, *Patterns and Meaning in History: Thoughts on History and Society*, ed. with an intro. by H. P. Rickman (New York: Harper & Brothers, 1962), p. 125; Martin Heidegger, *Being and Time*, trans. by John Macquarrie and Edward Robinson (New York: Harper & Row, 1962), p. 41. For an interesting interpretation of Heidegger, see Jacques Derrida, "A Note to a Footnote in *Being and Time*," in *Phenomenology in Perspective*, ed. F. J. Smith (The Hague: Martinus Nijhoff, 1970), pp. 54—93.

[46] Collingwood, "The Philosophy of History," *op. cit.*, p. 14.

[47] Ortega, *Obras, op. cit.*, 5: 40.

[48] *Ibid.*, 5: 40, 55; 6: 33.

[49] *Ibid.*, 7: 178—79.

[50] *Ibid.*, 6: 49—50.
[51] *Ibid.*, 6: 167, 184.
[52] *Ibid.*, 7: 398, 405.
[53] *Ibid.*, 9: 26—27; 6: 44.
[54] Collingwood, *The Idea of History, op. cit.*, p. 214.
[55] Ortega, *Obras, op. cit.*, 6: 40—41.
[56] Dilthey, *Patterns and Meaning in History, op. cit.*, p. 89.
[57] Ortega, *Obras, op. cit.*, 6: 41.
[58] Collingwood, *The Idea of History, op. cit.*, p. 10.
[59] Thomas Mann, *Joseph and His Brothers*, trans. H. T. Lowe-Potter (New York: Alfred A. Knopf, 1945), p. 3.

MARIA BIELAWKA

DOES MAN CO-CREATE TIME?

The paper presents my attempt to answer the question of whether man is a co-creator of time. It is based on Edmund Husserl's discovery of the very deep, "wonderful," or, as Husserl sometimes says, "marvelous" part of man — inner time-constituting consciousness and its absolute Ego.

A positive answer to the question assumes acceptance of an anthropological conception in which at least one of man's spheres remains *beyond time*, beyond its annihilating power. A negative answer, on the other hand, is linked to experiencing time as the only reality whose process of change annihilates the whole of man.

The most outstanding Polish disciple of Husserl, Roman Ingarden, very convincingly gives two possible anthropological solutions which have been manifest in the history of philosophy. He describes, especially in his well-known book *Man and Value*, two fundamental experiences of time within overselves. These experiences give us the ground for the two completely opposing philosophical solutions mentioned above. The first experience, states Ingarden, teaches us that time is only a phenomenon, of which we, in particular our true Ego, are independent. In *Man and Value*, we read:

> Neither the very occurrence of change in my psychic structure and in my body, even deep and multi-faceted ones, nor even the consciousness of such changes having taken place hinders me in the least in this feeling of being myself through the course of my entire life. . . . I do no take time into account: I do not bother with it. No abstract *knowledge* about the constant passing of events in the world surrounding me, and in me myself, can in any way change the feeling that time is entirely without significance for me myself.[1]

In this way of feeling, time is not intuited as reality, it "has the kind of appearance-mode that enables me to meander over an actuality which is, as it were, all-pervasively present."[2] Such an experience occurs, first of all, Ingarden continues, when we perform unconstrained, free deeds in border situations, especially in the face of death. Thus we so radically strengthen our inner power that we become independent of time. At such moments we discover within ourselves traces of "a being which is

A-T. Tymieniecka (ed.), Analecta Husserliana, Vol. XXXVII, 55—64.

not subject to the passage of time," our "permanent ego,"[3] "extraordinary in its primariness and uniqueness."[4]

The second experience, on the other hand, is completely different. It reveals to us, according to Ingarden, the radically devastating role of time and great fragility of the whole of real being. We read in *Man and Value*:

For the moment the possibility of non-existence of what is actual, as well as the fact of the destruction of real objects, unveils itself to us directly; the *essence* of time also shows itself, as it seems — an essence which constrains everything temporal to being in the present, and at the same time does not allow what is present to *endure*, since it ousts it from being into the past and non-being, through a continually new present.[5]

Time does not spare our deep ego either. The ego is unable to preserve identity and is constituted in various time perspectives, and thus may be conceived of only as "a present ego which continually originates itself anew."[6]

In Ingarden's phenomenology there is no final solution to the problem under discussion. On the basis of his philosophy it is impossible to determine whether ego is independent of time or whether time dominates us completely. One can say that Ingarden's concept of man can be situated between the intuition of immortality and the brutal reality of our passing.[7]

In the case of Husserl's philosophy we deal with a different situation. Contrary to Ingarden, Husserl categorically chooses the first solution. Why? First of all Husserl consistently tries to realize the old idea of "first philosophy" — old, but in modern times not explored before Descartes. The main aim of "first philosophy" is to approach definitively and absolutely the fundamental philosophical questions posed by man. Thus Husserl, entering the domain of absolute, apodictic givenness, finds within himself the sphere which in his texts is often referred to as *timeless, non-temporal.*

Some of Husserl's claims are as follows: "Therefore sensation, if we understand this as consciousness (not the immanent, enduring red, sound, etc., i.e., the sensed), and also retention, remembering, perceiving, etc., is untemporal, viz. it is nothing in immanent time."[8] "*Each cogito with all its elements comes into being or passes away in the stream of experiences. But the pure Ego is not created and does not pass away. . . .*"[9] "*The Ego is unatemporal.* Naturally, it would make no sense to treat it as temporal being. The Ego is timeless, it is a pole of the mode of my activity directed *to* the temporality, it is the subject which assumes a given attitude *towards* time."[10] "The Ego in its original primacy is *nothing in time*. . . ."[11] ". . . as a matter of fact, primitive time is not real time, it is only the preliminary stage of time as form of coexistence."[12] "I am actual and concrete as the enduring present, that is my concrete being."[13] "Looking back into epoché leads us to the original constant stream — in a way to the 'nunc stans', standing 'present'."[14]

It is important to say, and I agree here with the opinion of G. H. Shin,[15] that it is possible to detect a distinction between all that is constituted as a temporal object and the primal, non-temporal constituting consciousness, for example, in the period of Husserl's lectures on time delivered in the winter semester of 1904—5 at Göttingen. At that time, however, Husserl did not possess any special technical term with which to designate the absolute sphere, a term which could express its non-temporality. Such a term — the "living present" — appears in the twenties. Later, especially in manuscripts from the thirties, it is replaced by the Latin phrase "nunc stans," probably because of the ambiguity of the term "living present" and the possibility of using it not only to denote the primal consciousness with its Ego, but also a monad.[16]

Commentators who attempt to give an account of Husserl's notions indicate and describe — as it seems to them in agreement with Husserl — the structure of inner consciousness. It is not clear to them, however, whether the absolute sphere of man is really discovered in the immediate, apodictically certain intellectual experience that Husserl should use in accordance with his famous "principle of all principles" or whether it is, instead, an abstract philosophical construction. Robert Sokolowski, for example, openly asks: "Are these structures (of inner consciousness — M. B.) truly intuited and registered, as Husserl claims, or are they constructs?"[17]

Of course, it is extremely difficult to understand and interpret the untemporal consciousness mainly because of paradoxes within the primal domain. Those paradoxes were pointed out by Husserl himself and by commentators. We can divide them into two basic groups. The first, the "epistemological" class, refers to the manner of experiencing of the inner time-constituting consciousness. The second class deals with structural, paradoxes, especially the paradox of the permanent-flowing nature of inner consciousness and, directly connected with it, the mysterious relationship between non-temporality and time and between the constituting and all that is constituted. We are particularly interested in the second group.[18]

What does the extremely important structural paradox of inner consciousness mean?

We must keep in mind that Husserl often emphasizes that the primal consciousness and its Ego is the standing, persistent activity of the present. But, at the same time, he says that absolute subjectivity itself is the "stream" — the primary change. Klaus Held, Iso Kern, and Robert Sokolowski notice this paradoxical situation within the framework of the primitive subjectivity. Iso Kern points out among other things that in Husserl's lectures on time there is a paradoxical phrase which refers to the primal consciousness, namely "a modification or a continuum of changes which is no change."[19] For example, according to Thomas Seebohm, Klaus Held stresses that we may say about the primitive subjectivity discovered by Husserl only that "it is," and that, on the other hand, "a stream dominates within it." How in this case, wonders Held, can we imagine a unity of the "stream" and the persistent being "in" ("within") *nunc stans*?[20] "How," Sokolowski asks, as well, "can we resolve the paradox of a present that is both permanent and flowing?"[21]

I would like to present another interpretation of this problem. It seems to me that the *untemporal constancy* is the most important moment of the primal constituting time consciousness. Inner consciousness *does not flow*, but *it is*; it remains in its pure all-embracing present. The term "flow" or "stream," usually placed by Husserl in quotation marks, denotes only the means or the mechanism of our process of constituting all that is temporal. In this sense, the absolute "flow" *is not* — contrary, for example, to Sokolowski — "pictured as a regular stream of consciousness" and *it does not* "resemble our ordinary stream of moods, feelings, and acts, except that it is more elemental."[22] Furthermore, it is not — as Krzysztof Michalski, for example, seems to believe — time par excellence, emerging (rising) *living time*, which is "untemporal" only in the sense that it has not hardened, like lava, into ready temporal identities — the inner objects.[23]

The living present of our consciousness itself, the "nunc stans" sphere that Husserl reached with the help of apodictic intellectual intuition, *is not time, never*. It is a really permanent "source" of time, an all-present "origin," "space," where time springs up, as opposed to the inner units flowing in time.

This does not mean, however, that there are no modifications inside the frame of the inner time-constituting consciousness. The existence of time would not be possible without any changes. But there are no *temporal* changes, there are only modifications of *quality*. Let us consider this situation carefully.

In our inquiries we find it helpful to apply a simple scheme. We are fully aware that the scheme can only help us to bring the domain of the constitution of time to light; it cannot replace our private intellectual experience. Husserl's well-known diagram of time is of no use here: this first of all, because it suggests the punctual, momentary being of inner consciousness while we are interested in pointing out its all-embracing present; furthermore, the structure of his diagram is, in my opinion, more complicated than reality itself.

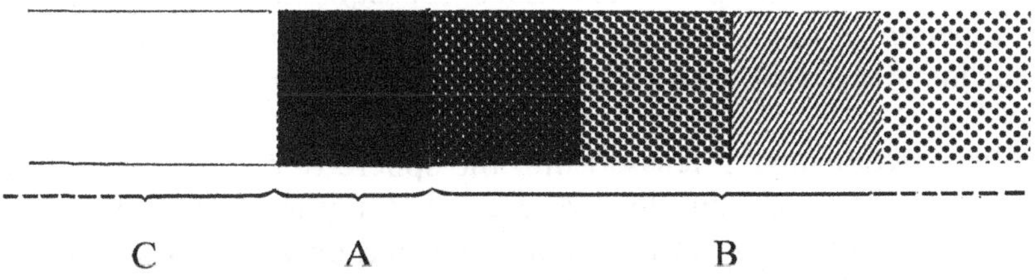

The scheme presented above takes into account most of all the passive aspect of constitution. On the higher, active level enormous "intertwinings" of intentionality, which are the forms of expression of the ego's active life, are built. The problem of active constitution, however, is beyond the scope of the present paper.

Segment A stands for our immediate apprehension[24] of what is present — that part of our mind which in the Eleventh Book of St. Augustine's *Confessions* is called the attention of man.[25] Segment B represents retention — primary memory with its modifying "phases"; and Segment C represents protention, i.e., primary empty expectation. The protentional-perceiving-retentional structure is well known to Husserlians, and there is no need to discuss all its aspects. The special importance of two of them, however, should be stressed.

First of all, we must stress that the structure of the absolute consciousness is *itself* the *present unity* and only in an abstract way can one distinguish within A from B and C. Husserl writes: "Thus we have in the present day the abstractively obtained *core of the authentic present* — a phase distinguished within the stream, which does not include anything which has passed or is to come, but contains the *pure* present."[26] The "core" and the two horizons which surround it, retention and protention, exist *side by side, together*; they belong to the same unique present which Husserl sometimes calls the "full present."[27] Husserl often pays attention to the fact that it is wrong to say that these three elements are simultaneous, because the word "simultaneous" suggests temporality, and they are completely timeless. In the manuscript already quoted we read further on: "The double meaning of the present — the constituted present and the constituting subjectivity that is the original living — needs more adequate terminology."[28] In fact, Husserl radically differentiates between the momentary present as a part of the constituted identity of the immanent object (the said punctual momentary present of all that is given to us "in person" and which, due to retentional modifications, subsequently develops into the phenomenon of time), and the full, all-embracing present, of which St. Augustine was already aware when he spoke of the immortal human mind.

When accompanying Husserl into the sphere of our deep conscious life, we have to see on the evidence that nothing but our *present expectation* waits for what is coming, and that only due to our *present apprehension* do we receive all that is present, all that after a momentary stay within the sphere of immediate experiencing disappears from real being, and is changed and saved for us owing to our *present retentional modifications*. Everything that happens within the frame of the "absolute" life takes place in the *enduring, persistent present*. We cannot force our way out of this present to see, for example, through our past expectation "objects" that are actually given.

The problem then appears of how the illusion of consciousness which flows in time and which never stays in the "full present" comes into being.[29]

In my opinion the solution of this problem lies in the realm of "temporalization" (*Zeitigung*) and the changes of quality which occur there. This is the second aspect I would like to focus on. Let me illustrate it with an example from our everyday visual experience on its passive level.

When one is experiencing a transcendent object belonging to the world, sensations must appear in the "field of perception." There are, in

Husserl's words, "felt" color points which comes to us, spots given to our mind which, as we can say with Ingarden, resemble the points put on the canvas by impressionists. These "fields" or complexes of sensations (a simple sensation can be detached only theoretically) continually modify themselves; they flow, as Bergson put it. This "flowing of quality" is shown inside frame A in our scheme. Out of this frame, outside of the immediate non-intentional apprehension, sensations are being "extracted" in their momentary stages (as "momentary pieces"), and these pass on to the inside of retention (the inside of frame B in the scheme). Then they are carried regularly away from the pure present center; they become already past, the further past, and so on. Retention *saves them for us*; they are held and kept alive in it, since the extracted qualitative sensation states do not really exist any longer; they disappear from actual being as soon as new qualitative states replace them. Thus, retention and the identifying function of passive synthesis[30] are responsible for the *phenomenon of flowing fields of sensations* that spread before us.

The temporal extension of states of primal apprehension occurs analogically. Simple qualitative states of apprehension, obviously with complexes of sensations appropriate to them, seem to be "pulled out" from the sphere of the pure present (sphere A in the scheme). They appear as the content of retention (within frame B) and here, being saved due to our mind, they continuously elapse. The temporal extension applies to all other modes of conscious activity as well, for example, to the intentional recognition ("constitution") of transcendent objects, which is performed on the basis of the fields of sensations and the eidetic knowledge we possess. The problem of active constitution is, however, too broad to be discussed now.

Such a mechanism of "temporalization," including two kinds of qualitative changes (modifications of pure qualities in the sphere of the pure present, and changes in the sphere of retention qualitative in the sense that something that occurs here, maintains the same nature, but every now and then becomes not only past, but also clearer or more dim) leads us to the *phenomenon of a division of consciousness*, in which consciousness *itself* turns into a temporal entity *for itself*. This secondary, derivative being of consciousness, "temporalized in the monadic time of the world," is sometimes called by Husserl transcendental consciousness in the first sense of the word, while, of the primary inner consciousness with its absolute Ego, he says that it is the transcendental reality in the second meaning of the term, a non-temporal "bearer of all temporalizations and times, of all identities of being, of all the worlds."[31] Thus, the timeless consciousness discovered by Husserl in deep intellectual intuition is the basis of the constitution of all that is temporal — it is the constitution of the material and psychic world, of the world of man's products, of values, and even of the monads' transcendental temporal consciousness.

In my opinion, by constitution Husserl means neither 1) the creation of reality, nor 2) sense giving, but 3) the co-creation of reality through the agency of the structure of our mind here discussed. I suggest the term "co-creation," because Husserl, who reaches the very deepest sphere in man, does not consider himself a creator — of either the structure of his consciousness, or of impressions (sensations), or of the eidetic knowledge, which plays such as important part in the constitutive processes. Therefore he has to accept, *with the necessity of reason*, the existence of the absolutely creative consciousness, of the divine being,[32] in the reality of which we partake through our *timeless, constituting time consciousness.*

Jagiellonian University

NOTES

[1] R. Ingarden, *Man and Value* (*Książeczka o człowieku*), A. Szylewicz, trans. (Munich, Vienna: Philosophia Verlag, 1983), pp. 34 and 35. I have discussed "The Mystery of Time in Roman Ingarden's Philosophy" at the International Conference on "The Philosophy of Roman Ingarden," organized in Cracow on 25—27, October, 1985.

[2] *Ibidem*, p. 38.

[3] *Ibidem*, p. 48.

[4] R. Ingarden, *Spór o istnienie świata* (Controversy over the Existence of the World), Vol. II (Warsaw: 1961), p. 517.

[5] *Man and Value, op. cit.*, pp. 38—39.

[6] *Ibidem*, p. 39.

[7] There is a close analysis of the problem in my paper "The Mystery of Time in Roman Ingarden's Philosophy." See *Analecta Husserliana*, Vol. XXXIII, pp. 109—117.

[8] E. Husserl, *Zur Phänomenologie des inneren Zeitbewusstseins*, R. Boehm, ed. Husserliana X (The Hague: Nijhoff, 1966), pp. 333—334 (R. Sokolowski's English translation from his book *Husserlian Meditations* [Evanston: 1974], p. 133). This text was written in 1908 or 1909. In the same place Husserl adds: "These are highly important things, perhaps the most important in all of philosophy" (Sokolowski's translation, *ibidem*, p. 135).

[9] *Ideen zu einer reinen Phänomenologie und phänomenologischen Philosophie*, Vol. II,

M. Biemel ed., Husserliana IV (The Hague: Nijhoff, 1952), p. 103. Volume II of "*Ideas* . . ." is dated about 1912.

[10] Manuscript E III 2, 1920 or 1921, p. 50, quoted here from Klaus Held's book *Lebendige Gegenwart* (The Hague: Nijhoff, 1966), p. 117.

[11] Manuscript C 10, 1931, p. 21, quoted here from *ibidem*.

[12] Manuscript C 7 I, 1932, p. 17, quoted here from *ibidem*, p. 116.

[13] *Ibidem*, p. 5, quoted here from: R. Sokolowski, *Husserlian Meditations, op. cit.*, p. 160 (Sokolowski's translation).

[14] *Ibidem*, p. 30, quoted here from: A. Diemer, *Edmund Husserl. Versuch einer systematischen Darstellung seiner Phänomenologie* (Meisenheim am Glan: 1956), p. 32.

[15] G. H. Shin, *Die Struktur des inneren Zeitbewusstseins. Eine Studie über den Begriff der Protention in den veröffentlichten Schriften Edmund Husserls* (Bern: 1978), p. 88.

[16] On this subject see also the opinion of Klaus Held in *Lebendige Gegenwart, op. cit.*, p. 124.

[17] R. Sokolowski, *Husserlian Meditations, op. cit.*, p. 162.

[18] I examine the problem of the epistemological paradox in my paper "The Discovery of Inner Time-Constituting Consciousness" presented at the conference marking the fiftieth anniversary of Husserl's death organized by the Polish Philosophical Society in Cracow, April 29—30, 1988. In that paper I attempt to show a possible way of explaining the paradox through the acceptance of the thesis that Husserl's opening up of the domain of inner consciousnesss was possible because of a momentary, apodictically certain *transcendental or critical self-reflection*. See *Studia Filozoficzne* (Philosophical Studies), Nrs. 2—3 (291—292), 1990, pp. 130—138.

[19] I. Kern, *Idee und Methode der Philosophie* (Berlin, New York: 1975), p. 41.

[20] K. Held, *Lebendige Gegenwart, op. cit.*, p. 125 and 126.

[21] R. Sokolowski, *Husserlian Meditations, op. cit.*, p. 162.

[22] *Ibidem*, p. 166.

[23] See K. Michalşki, *Logika i czas. Próba analizy Husserlowskiej teorii sensu* (Logic and Time. An Attempt at the Analysis of Husserl's Theory of Sense), (Warsaw: 1988), pp. 202—204.

[24] It is important not to identify primary apprehension (real perception, authentic perceiving) with apprehension as an immanent temporal act. See, for example, Husserl's text from *Vorlesungen zur Phänomenologie des inneren Zeitbewusstseins*, M. Heidegger, ed. (Halle a. S.: Max Niemeyer, 1928), p. 444: "Thus, we must understand apprehension in a double sense: the one which is immanently constituted, and the one which belongs to immanent constitution, to the phases of the original stream itself. This is the primal apprehension (*Urauffassung*) which is no longer constituted." (Sokolowski's translation from his book *The Formation of Husserl's Concept of Constitution* [The Hague: Martinus Nijhoff, 1964], p. 86).

[25] G. Eigler's book *Metaphysische Voraussetzungen in Husserls Zeitanalysen* (Meisenheim/Glan: 1961), for example, contains a detailed comparison between St. Augustine and Husserl's views on the problem of time.

[26] Manuscript C 3 I, 1930, p. 9, quoted here from: A Diemer, *Edmund Husserl . . .*, *op. cit.*, p. 44.

[27] See manuscript C 3 III, 1931, the cover, quoted here from: *ibidem*.

[28] Manuscript C 3 III, p. 35, quoted here from: K. Held, *Lebendige Gegenwart*, p. 125.

[29] Roman Ingarden, among others, maintains this illusion, when he writes that theories

such as Husserl's "make time constitute itself in the primary consciousness of time, and then, *nota bene*, consciousness is found itself in time." Ingarden also stresses, that Husserl in his 1916 discussions spoke about great difficulties related to it, especially about the danger of a vicious circle — see R. Ingarden, *Spór o istnienie świata*, Vol. II, part 2 (3rd edition, Warsaw: 1987), pp. 145 and 146.

[30] The special kind of conscious life, the primal intuition, which shows us that sensations elapsed in retention keep alive the sameness of their nature despite the fact that they become the "phenomenal being of the past," makes this passive synthesis possible.

[31] E. Husserl, *Zur Phänomenologie der Intersubjektivität. Texte aus dem Nachlass.* Dritter Teil. 1929—1935, I. Kern, ed. (The Hague: Nijhoff), p. 587.

[32] See, for example, the note from §51 of the first volume of *Ideas*. On the problem of God in Husserl's thought, who is conceived not only as the absolute aim for all reality, but also as the absolute creator of facticity, see also: A. Diemer, *Edmund Husserl . . .*, *op. cit.*, pp. 376—378 or S. Strasser, "Das Gottesproblem in der Spätphilosophie Edmund Husserls," in: *Philosophisches Jahrbuch des Görres Gesellschaft*, Vol. 67 (1959), pp. 130—142.

ALCIRA B. BONILLA

THE REACTIVATION OF THE PAST AS AN ETHICAL DEMAND ON THE PHENOMENOLOGIST

There is no better tribute to a deceased thinker than to emphasize the current importance and fruitfulness of his work, and to discuss both the literal meaning of his texts, and what the texts suggest or of what they are silent on but state in one way or another. Thus we recover not only the philosopher's image, but also his "*shadow*."[1] This "shadow" constitutes a shelter and a mission for all of us who have decided *to think in critical dialogue* with phenomenology.

In a now famous text in his *Dimensions de la conscience historique*,[2] Raymond Aron said that *history* is the unceasing work of "reconstitution, by and for the living, of the life of the dead." Man turns his attention to history, understood as the universal current of events that we can embrace each from our limited perspectives, trying to find the solution to an essential question: "*What can I hope*?".

Without doubt, Husserl has been one of those contemporary philosophers who has made possible the *theoretical framework* and given us the *categories* needed for answering the old question of the meaning of history and, above all, of the *conditions of the possibility* of the existence of a history with meaning given by us and for us. In this tribute, obviously, we will discuss only some aspects of this very large subject.

Our starting point are the *three degrees of historicity* (indistinctness, *Geschichtlichkeit, Historizität*) which appear in *Beilage* XXVI of *Die Krisis* . . . .[3] Here each degree of historicity is considered at the same time to be *a degree of humanity* (called in the text *das Menschentum, die Menschheit*). From other Husserl's texts about the degrees of *historie*, or rather, of *historiography*,[4] we could assert that the levels or degrees recognized in this text refer likewise to *degrees of knowledge* of historicity and the historic. Curiously, in *Beilage* XXVI, there is an *inverse relation* between the extent of the explanation that Husserl gives of each degree and its intrasystematic significance. It is because of this that we are going to center our attention on the interpretation of what is stated about the third degree of historicity.

The *original generative historicity (ursprüngliche generative Histori-*

A-T. Tymieniecka (ed.), Analecta Husserliana, Vol. XXXVII, 65–71.

zität) appears as the lowest degree. But *lowest* here also means *basic, founding*; it is something which belongs inherently to humanity, which really constitutes it into humanity as such. In fact, *ursprüngliche generative Historizität* is the name of the unity of spiritual life, given that the whole community of human persons shapes constantly a cultural world through individual and, at the same time, joint work. Husserl underlines over and over again that this historicity comes from human acting itself; the objective cultural world in its complete display of present and past and in its future projections depends on this doing and working of man. Here, as Landgrebe understands, "history consists of the connection of happenings from which arise human acting."[5] Thus we are faced here with the most spacious field of *praxis*.

What turns human existence and the correlative human surrounding world (*menschliche Umwelt*), understood as a surrounding world of things (*Sachenumwelt*) and as a personal surrounding world (*personale Umwelt*), into something *intrinsically* historic is man's constant *acting in conformity with a meaning obtained* from the inherited cultural and spiritual results and forms, that is to say, within a tradition whose transformation he is capable of undertaking.

In short, *three features* characterize original generative historicity: human *acting* at its origin, its happening in *time*, just as much the case with material as spiritual forms of acting, and the *meaning* constantly given to action and, at the same time, shaping it (*Sinn als Gestalt*). Although Husserl's description of this first level seems to emphasize the two first features, the last paragraph of this description reinforces the idea that he formulates further on: historical life is always a goal-oriented life (*Zweckleben*) and, therefore, this essential feature appears in every historical life, no matter how rich or primitive this may be.

The *ethical mark* of that which is exposed proves evident: "Menschsein ein Teleologischsein," says Husserl in the *Crisis*,[6] and, precisely because of that, "und Sein-sollen ist." The teleology, the demand of a life goal, means that historicity — mankind as a historic organic unity — is always set "in motion": "Das Ganze eine Einheit der Sinnhaftigkeit, in Seingebung und Sinn selbst als Gestalt, die nicht als ein Zwecksinn von den Menschen im voraus entworfen ist, sie sind nicht Funktionäre einer Zweckidee, die dieses totale Menshentum bzw. diese Umwelt gewollt und verwirklicht haben."[7]

With the arrival of *philosophy* (*philosophy at its first level*, that is to say, rational knowledge of the world and the man) and the subsequent shape taken by European culture, there appeared the *second degree of historicity*.

Philosophy, or rather, the particular philosophers establish a new kind of goal-oriented life (*Zweckleben*), which gives mankind a new shape. Thanks to philosophers, mankind assents to a first level of truth and so becomes the subject and object of a wonderful conversion: "L'idea di filosofia è un punto di vista che garantisce la teleologia della ragione nella storia, anche se, per ipotesi, non fosse possibile ritrovarci in presenza di un *telos* in qualque modo gia attuato o avviato ad attuazione, nella storia stessa."[8]

The third degree of historicity is formulated in the text in the following manner: "Umwandlung der Philosophie in Phänomenologie, mit dem wissenschaftlichen Bewusstsein von der Menschheit in ihrer Historizität und der Funktion, sie in eine Menschheit zu verwandeln, die sich bewusst von der Philosophie als Phänomenologie leiten lässt."[9]

In order not to lead this exposition out of its nuclear problematic, we are going to eschew the subject of the scientific nature or scientificality of phenomenology. On the one hand, this question may be discussed from a viewpoint external to phenomenology itself, but on the other hand, it may be asserted that Husserl remained loyal to his first ideal of scientificality, although in his last writings he tries to get beyond the traditional *dóxa-epistéme* antinomy with praise for pre-predicative experience and the outline of a new ideal of scientificality. At any rate, when Husserl speaks in this texts about ". . . dem wissenschaftlichen Bewusstsein von der Menschheit in ihrer Historizität," we must refer to the phenomenological consciousness of and in its own historicity which mankind has attained through the philosopher or the philosophical community. It is to this that we advance now.

After accomplishing the transcendental operations of the reduction, the philosopher, returning to the *leistende Subjektivität*, discovers the temporal and historic nature of the latter, and at the same time, its nature as the source of all human works and actions.

With his *abjuration* of the linear representation of time traditional in philosophy, and by his analyses of *consciousness of the "now"* (*Jetzt*) as the living-present, Husserl has made possible the understanding of mankind's building of history and making itself within it in the strict sense of its *positing* itself as a *telos*. Perhaps it has been Landgrebe who has best grasped these implications of Husserl's analyses of temporality and, at the same time, the possible ways to correct or complete the shortcomings of the Husserlian hypotheses. Some of these ways have already been covered, as in the case of Heidegger's research into temporality in *Sein und Zeit.*

In Landgrebe's opinion, which we share, the "wissenschaftlichen Bewusstsein von der Menschheit in ihrer Historizität" bases itself on the discovery and showing of *retentional consciousness.* In fact, every instant in which we experience that which is "now conscious" and which slips on into what has "just been," is still retained for a short lapse. This means that what has "just been" is found between the present and the past as such (the past as what has "already been"), and, at the same time, what has "just been" makes possible the consciousness of just this difference, since it is the consciousness of what is not any longer. The present is no longer the "now point," but is consciousness of what is now present, and, at the same time, retained consciousness of what has "just been." As Landgrebe says, time is not a form, somehow existent, that makes it possible to discern the order of succession; rather, and on the contrary, we can trace this form of time to its own roots in the process of its slipping into what has "just been." Thus, we reach the explanation of an activity that is an occurrence proper.[10]

Besides the consciousness of what has "just been," the immediate expectation of what will be "at once," that is to say, *protention,* corresponds to "consciousness of the now." With regard to this expectation, protention behaves in the same way as retention with regard to memory. At this stage, Landgrebe will criticize two aspects of the Husserlian analysis: 1) Retention and protention are placed by Husserl on the same level, which makes a teleological explanation of history difficult (and — we would add — the same problem arises for the theory of intentionality); for the achievement of such a teleological explanation of history, it is necessary to show how time temporalizes itself from the future. 2) To retention and memory Husserl devotes much more space in his works than to protention. This results in a gap in Husserlian thought, because there is lacking a theoretical nexus which allows understanding of the later Husserl's researches into man as teleological being. The lack of careful examination of the relationships between retention and protention also makes impossible the complete development of a theory of historicity and the historic, as well as of a historiographical methodology in agreement with the requirements of phenomenology as part of the history of philosophy.

In any case, we think that Husserl's works together with those of other phenomenologists, even though they may not in all cases be "orthodox" Husserlians (the first "unorthodox" one was Husserl himself), can today complete a phenomenological philosophy of history and — we dare to say — a transcendental phenomenological theory of history.

On the need for teleological acting and doing (which involves the formulation of a goal and the selection of means for reaching it), is founded all individual or communal resort to the past, historiography included. The need and the selective nature of every memory follows from this. So here all connects with Aron's declaration quoted at the beginning: History is the unceasing work of the "reconstitution, by and for the living, of the life of the dead. It is us, who are still alive, who create history anew in our memory, in accordance with our particular or cultural finalities or those typical of our time. The image of a historic time as causally occurring, complete and continuous, proves to be an illusion. Historic time orders itself in accord with a historic finality freely decided on by each man or each group, which telos decides the continuity of history and its sequences. It is because of all this, that in the *Crisis* history is defined as "die lebendige Bewegung des Miteinander und Ineinander von ursprünglicher Sinnbildung und Sinnsedimentierung";[11] only from the living present we can understand the past and get it out of the dark.

The philosopher's work is inserted into temporality and its dialectic. He must show the *truth of the present*, that is, its origin and *télos*, through an account of its sedimentations and, what is obvious, through an endless effort to make evident the rationality latent there for the achievement of higher and higher stages of *self-examination* (*Selbstbesinnung*) and *self-responsibility* (*Selbstverantwortung*) for the whole of mankind. And this points to the second consequence of the transformation of philosophy into phenomenology: the phenomenological philosopher, the philosophical community, has the function of transforming mankind, leading it from philosophy.

In this sense,[12] Husserl speaks in the *Crisis* of the philosopher as the *Funktionär des Menschheit*. However the responsibilities of the remaining members of society, of the remainder of human beings, have not been annulled thereby. Let us be clear about this; we ought to say that a higher degree of responsibility for the process of the humanization of the world attaches to the philosopher.

The radical nature of the phenomenological viewpoint does not abolish the historic nature of philosophy, but gives to it its full sense. The reactivation of the past, especially of the philosophical past, is an unavoidable moment of the *Umwandlung* from philosophy into phenomenology because each genuine philosopher, except at the initiation of philosophy in Greece, presents himself as effect and heir, as one connected with a tradition.

Man — we have suggested this before — lives among traditions and the whole cultural world in all its forms is present for him because of tradition. A deliberate ambiguity in the valuation of the term "*tradition*" makes possible, for Husserl, the outline of the dialectic that the philosopher must engage in with it. This is the way in which we try to understand the significance of the philosophical *Generativität.*

As philosopher inheritors of the past, we receive from it the original intention of philosophy, its concepts, problems, and methods. But, in philosophical reflection on the history of philosophy we can discern clearly between a *geistige Generativität*, the formative and creative power which emerges from the historic philosophies, and a *Traditionalität*, understood as the totality of conceptual sedimentations passively transmitted and accepted.

No scientific theory and no philosophical system can claim to be an exhaustive interpreter of reality. The philosophical past, exposed to the consideration of the modern philosopher "in der Weise einer *historisch-erinnernden Vergegenwärtigung*," must now be reached and critically accepted through retrospective reflections of a critical-historiographical nature. The philosopher, then, must carry out a real "phenomenological reduction"[13] of former philosophies, looking at them as moments in the realization of the philosophy-idea. Only by returning to the original intention which inspires each philosopher and his steps to its fulfillment, can the modern inheritor of the philosophical tradition obtain a certain enlightenment on the dialectic between discovery and encovery which stands out as a constant of the knowledge transmitted by means of the scientific-philosophical tradition; by these approaches he can also position himself to fulfill the guiding *telos-idea* of philosophy.

The question of language is not beyond this retrospective inquiry. This is evident in paragraph 9 of the *Crisis* and also, in a longer way, in *Die Ursprung der Geometrie.* To a great extent, mankind is a linguistic community. The original act of producing any cultural object and what is produced become understandable for other men and actively co-realizable through the communication established by the spoken or written word. But, in thinking deprived of access to the original activity and not feeling any need or interest in reactivating it, the reception of evidence shared through language can be merely passive. The passive association of words, whose sense becomes more and more diffuse, substitutes for the originally intuitive life which generates all cultural formations as much as the language that expresses them. This results in obliviousness, falsification, and the darkening of the original sense of words — to such an extent that it makes impossible its enlightening use in the exposure of some scientific or philosophical language. The return to the sources of science and philosophy in order to elucidate their meaning, that is to say, the return to the sources of thinking itself, means likewise a return to the sources of the linguistic forms wherein that thought has been expressing itself — and, we add, even conforming itself — as well as a revision of the development of its traditions.

University of Buenos Aires
University of Mar del Plata

NOTES

[1] The noun "*shadow*" in quotation marks refers to the title and intentions of that great homage to Husserl that Maurice Merleau-Ponty wrote in his article "Le philosophe et son ombre" in *Edmund Husserl, 1859—1959*, L. van Breda and J. Taminiaux, eds. (The Hague: M. Nijhoff), pp. 195—220.

[2] R. Aron, *Dimensions de la conscience historique* (Paris: Plon, 1961).

[3] E. Husserl, *Die Krisis der europäische wissenschaften* . . . , (hereafter, *Krisis*) Husserliana VI, pp. 502—503.

[4] We remember, for example, *Krisis*, p. 72, where Husserl shows two ways of dealing with history: from outside, from fact (*von Aussen, vom Faktum*), or from inside (*von Innen*). Following with this difference, Husserl discerns in *Beilage* III an inner historiography (*innere Historie*), which founds the useful historiography of facts (see *Krisis*, p. 386). This is not the place to pay attention to the methodological framework of such an *innere Historie*.

[5] L. Landgrebe, *Fenomenología e historia* (Caracas: Monte Avila, 1975), p. 222.

[6] Husserl, *op. cit.*, p. 275.

[7] *Ibid.*, p. 502.

[8] Antonio Ponsetto, *Razionalità storica e fenomenologia husserlianna* (Lecce: Milella, 1984), p. 152.

[9] Husserl, *op. cit.*, p. 503.

[10] See, Landgrebe, *op. cit.*, p. 225.

[11] Husserl, *op. cit.*, p. 380.

[12] *Ibid.*, p. 15.

[13] E. Husserl, *Ens te Philosophie*, Husserliana VII, p. 387.

CARLOS MINGUEZ

HARTMANN: THE HISTORICITY OF CULTURAL DATA

INTRODUCTION

1. The evolution of phenomenology has progressively evidenced the fact of the historical situation, which was "reduced" — in a first and more basic stage — within its conception of philosophy as a strict science. Husserl's posthumous work legitimizes the role of history in the formation of thought and, as Montero Moliner (1987, p. 51) points out, leads investigation towards "the search for the genuine phenomena which are the bases for those concrete lived worlds realized from the perspective which our position in the twentieth century gives us," as well as to the recording of the phenomenological analyses which the great philosophers of all times are able to master, and that recent phenomenological bibliography has been slowly making evident.

2. Both directions of investigation have been studied by Nicolai Hartmann (1882—1950). Recording the phenomenological moments of the great philosophers is an exercise often repeated in his research, inasmuch as the "problem situation" (*Problemstellung*) has to be disentangled from the analyses and interpretations commonly given in the history of thought; the great philosophers bring us close to genuine problems and phenomena with brilliant mastery. He searched for the genuine phenomenon which are the bases of concrete lived worlds, because the "concrete lived world" summarizes the cultural ambit, the spirit of an age, which history is to be understood by. The determination of the genuine phenomena that may constitute the bases of the lived world, are substituted in Hartmann's thought by a basic ontological structure which underpins and justifies historical processes.

3. If the unity of the world constitutes a metaphysics assumed to be indispensable to reason, then plurality and heterogenity of beings presents itself as a piece of information that, beyond appearances, settles down in a diversified way of being. The real world is divided into strata (inorganic, organic, psychic, and spiritual), as the distinct sciences

A-T. Tymieniecka (ed.), Analecta Husserliana, Vol. XXXVII, 73—80.

have shown throughout history, each of them having their own categories. These categories, being the most general forms of predication, make up the boundaries which describe the perimeter of each of the strata, although such boundaries cannot be understood as clear ones, for when studying these categories one must keep in mind their autonomy as well as their interdependence. The dimensional categories (space and time), that go across real being, are shown below them; time being the category that necessarily presents itself in all reality. Outside of time there are no real beings.

Spiritual being constitutes the higher stratum. Its limits coincide with those of *culture*. In this sense, culture is not "like a second nature," but properly determines man's being: we cannot cease to be cultural. However, the spirit does not exist as an absolute being (contrary to Hegel), but is incarnated in individuals, who take part in lower strata (Hartmann, 1942, Chap. V).

4. Numerous sciences deal with the lower strata of being; each of them has its own object, and in general they are known as the natural sciences. In a clear opposition much emphasized at the end of the last century, there are the sciences of the spirit (the social sciences and the humanities, such as art, literature, language, ethics, and law) which stand in contrast to the natural sciences; each of them determines with accuracy its object, and all together have great relevance in social life. This separation still used nowadays (with the subgroups being methodologically differentiated), does not have any repercussions on their ontological way of being. All of them, the natural sciences and the sciences of the spirit, are equally part of the spirit, and they are the most representative content of it. They are different as far as the object each deals with is concerned, and the knowledge acquired by each of them, or their representive categories, cannot be transgressed without modification. In other words, they cannot be reduced in vitalist, materialist, or psychologistic interpretations. They are revealed, on the contrary, to have different and exclusive, peculiar features. For example, science is considered to be a product of the spirit, and in this sense it does not differ from the rest of the spirit's creations. Notice the importance that this conception has when judging the historicity of science, so much discussed from other points of view.

But, what way of being undergirds the phenomena proper to these sciences so that we can say that at root they stand in contrast to natural phenomena? The answer to this question is the responsibility of the philosophy of spiritual (cultural) being, taken as a problem, and insofar as this being is understood as the highest stratum of real being, this philosophy partakes of ontology.

Das problem des geistigen Seins (1933), besides giving an account of the previous problematic, lays out the key postulates of N. Hartmann's ontology. Neither spiritual (cultural) being nor the forms in which it takes shape (laws, customs, sciences, arts, etc.) are described from the starting point of subjectivity; on the contrary they are understood as things in themselves, modes of being that, with peculiar features, face a subject in the order of knowledge. And this subject cannot be understood "quam tabula rasa" in the Aristotelian way, but as being at the same time formed by the spiritual community in which it lives, and taking part in the analysis and shaping of culture, taking it as an object present in the intentionality of knowledge, having the hardness of acts felt emotionally to be transcendent; thus is avoided any perspectivism and, on the contrary, the problem of being is emphasized. To Hartmann modern philosophy, based on a theory of knowledge, is obsolete and must give way to a new ontology.

The conceptual systems formed by the sciences of the spirit build on a structure of being as solid as that of the natural sciences, and linked only tó this stratum of being we can build a conception of history and culture which avoids partial interpretive starting points. This does not mean the establishment of a dogmatic ontology, since a deeper understanding of the phenomenon can modify both knowledge of natural processes and that of the explanatory categories. And under these, as a human activity and as an apprehension of reality, science has a non-privileged place together with language, and the arts or the customs of the peoples.

5. Spiritual (cultural) being is formed by three categories: the personal spirit, the objective spirit, and the objectified spirit. What, thus, be the case derives from a factual situation: there are groups of phenomena whose ultimate nature is not identified with that of any other stratum, but which differ among themselves and appear with some independence, though they make up a unitary form of being (*Seinsformen*). "Having a history" belongs to the objective spirit; but, as in historical methodology, we basically take off from the facts and the data; we analyze here this seldom studied aspect of Hartmann's thought, the objectified spirit,

that is, the ontological reflection on the data with which the historian works.

THE OBJECTIFIED SPIRIT IN THE HISTORY

Historical science, like any other sort of human knowledge or activity, is realized in the present – no matter how difficult to determine and ambiguous in its use this concept may be, given that it may in practice mean periods of time of different length. Thus, since history constitutes knowledge of the past, in a way that past *must be* in the present. If this were not so, we could not reach any knowledge of the past. History could not be possible. In this sense history is precisely formed by a sort of preservation of the past in the present.

From a different point of view, one could ask whether it is possible to establish any propositions about the past, and whether these propositions can provide objective knowledge. We would ask such questions about the nature of historical knowledge, and in regard to them historians and philosophers have given various answers which stand in more or less close relationship to the general problem of the knowledge. In their major lines, these positions compare scientific to historical knowledge, and the result depends on the conception one has of the former (Danto, 1965). Hartmann does not place himself on this level, but refers back to the ontological level, where the problem of historical *being*, with its own categories, emerges, for only in them can history be explained.

How does the unreal historical past extend to the present? Hartmann rejects determining relationship as a procedure (1933, Intr. 16), because in that kind of nexus, the cause is consumed, it does not continue in the effect, inasmuch as he expects that in the historical datum the presence of the events subsists; and he points out several forms of the past's permanence: in a "tacit" way (*stillschweigende*), when it is alive in us without our feeling it as the past; or in a "perceptible" (*vernehmliche*) way, when the past as such is present in the consciousness we have of it. These forms of permanence define the dynamic process of the spirit, either through continuance, the still "being valid" of some cultural relationships (for example, a state organization), or through the adaptation made of old relationships to the new way of configuring the spirit (1933, Intr. 17). In historical methodology these kinds of permanence would form knowledge in general, knowledge not based on sources, that from which the historian necessarily begins his research (Topolsky, p. 175); and that ontologically Hartmann locates in the objective spirit, which it nourishes, but from which at the same time the personal spirit feeds.

Among the "perceptible" types of the past's permanence, we are interested in emphasizing continuance through a *medium* of much vaster and precise duration than the living carriers of the spirit (persons), or than the living objective spirit itself in whose bosom different fields, without precise limits (language, customs, technology . . .), emerge, endure and perish. This continuance of the past, through a persistent *medium*, is in harmony with the documentation used by historians, with the "data" in the perspective of historiography, and with the category of the *objectified spirit*, which is among the categories shaping spiritual being, according to Hartmann. And though this category holds the last place among the constituting elements culture, nevertheless for historiography it is the basic concept around which it is structured; from this we see the importance to the methodology of history of its ontological support.

Objective spirit has the peculiarity of producing forms that get free of the creator (the personal spirit and the objective spirit) when they are fixed to outer material elements, and they remain integrated in the development of history. Written documents and masterpieces are obvious instances of this objectivation (1933, 44, a). But, as the creation of a spirit whose process has already finished, they do not constitute any living reality, they do not take part by themselves in the spiritual form of being, nor, accordingly, in history.

What remains of the law of a period or of its science, once the objective spirit, which nourished them, has died? This question seeks what remains in the form of the laws, customs, or written documents that are the data or sources of history, and which Hartmann calls objectified spirit. This form of culture fixed to a material element (for instance, a written scientific theory or a masterpiece) is not a living spirit, nor is it identified with the material elements (canvas, marble, parchment, etc.) out of which it has been created; consequently, it does not take part in any of the strata of reality, therefore it is not real: it does not impose itself with the hardness and imperiousness of the real, with its demonstrable presence. Its nature consists in being an objective "imprint" that the spirit leaves behind (1933, 44, 5), and that has been printed on a more lasting material than the life of spirit.

But this "imprint," while it is the spirit's work, is itself not spirit, it is only its expression, and, precisely, the expression of a spirit already dead; neither is it just material (paper, ink, bronze, etc.) with an imprint of the spirit, for then we would not reach the phenomenon of the objectivation. It is a peculiar phenomenon, unexplainable from other categories of its own stratum or from other ways of being; therefore, it is the original category of the spiritual entity.

Now, if "objectivation" is not thoroughly explained setting out from those two extreme points, of which it shares some thing (being something more than just participation and something less than lasting in the life of a spirit which has already passed), it must find its justification in the living spirit. In fact, the key moment for the objectified spirit is when a living spirit recognizes it. For example, the idea of the State that is preserved in Plato's *Republic* is not the same as the one Plato and his pupils had, nor does it consist of some simple written texts; they shape a new idea operant in the living spirit (personal or common) of the moment.

A new idea is one that the living spirit recreates after recognizing it through a medium. Hence, as important as the endurance of the material support, is the new living spirit which "reads" and recreates the imprint of the past spirit (1933, 46, b), so that it reproduces it again. "We can only talk of a being with spiritual features, in the sense of its historical permanence, in those periods when there is a spirit capable of recognizing it, agreeing with it, and to whom it becomes a spiritual feature (contents), and who makes it its own" (1933, 49, f).

The data called historic keeps, in a way, an unchangeable neutrality for the subject who contemplates them. We cannot give an arbitrary interpretation to them; they can, of course, hide their deepest meaning or even become opaque in front of the historian. But, at any rate, to be integrated in the subject of history, they need a creative mind that introduces them into the spirit in which the historian participates. In history, as well as in science, the object is not given in its totality, a creative mind assimilates it as a living element. Neither is it created in an absolute way because it depends, if it is authentic, upon the imprint that the past spirit has communicated. Therefore, there is not a written history, but one drawn up every moment, not in an arbitrary way, but by reading with increasing attention and capacity the imprints left by previous cultures.

What endures "objectified" in the material and what makes the living spirit recreate it? Hartmann consciously speaks in metaphoric terms (instance, exigency, stimulus, provocation), being aware of their inadequacy, and in any case "they are not expressive of the effect of the real upon the living spirit." (1933, 50, d) The imprint of man (of the person) already provides the recognized isntance renewed by the living spirit (person, objective spirit) through some agreement which must be always given between them. Hartmann's imprecision here is remarkable; it is justified in that it corresponds to the irrational (unintelligible) limits among the categories of various strata; but the phenomenon appears with unavoidable (elemental) power: a spiritual mark imprinted on a material which a living spirit has to recreate according to different levels of profundity. It is clear in this context that the spiritual mark is not real (since it is not spirit but a product of it) and it only acquires reality through a living spirit, for historicity pertains to the objectified spirit, as much as to the spirit that impresses its mark as to the one which recreates it.

This duality makes evident some very important aspects in historiography: the independence of the objectified spirit from the historical consciousness in a given period, and the variation that it undergoes in time through the different interpretations that accord with the changes of spirit of a period. To know the spirit of a time, we will have to appeal to all vestiges it has left (all manner of "sources") and also to the interpretation of our own period will excel previous ones, only if the interpreptation of our own period will excel previous ones, only if the agreement between the vestiges and our spirit reaches different levels than those which previous historians have succeeded in reaching and objectified. Or, if the development of historical sensitivity achieves new levels or echoes other documents unnoticed until now (1933, 51, c). In any case, the being of the objectified spirit not only shapes a history, but it shows itself to be historicity (a temporal succession of recreations), while it holds a relatively timeless aspect, the one corresponding to the vestigial imprint left on enduring material. This knowledge of something persistent inclines us to think that "objectivation" escapes from the typical historicity of the spirit, but we have just seen, on the contrary, how its *recreation* places it precisely in another line of history, the one that corresponds to the objective spirit.

This way of being which is peculiar to the historicity in cultural data, besides pointing to a typical ontological attitude, can also be useful in furnishing evidence in a problem so much discussed by historians: the continuity or discontinuity of historical process.

In Hartmann's work this potential debate would be resolved, if we kept in mind that historical continuity reflects the superexistent (*superexistenz*) character of the objective spirit. This category does not live within individuals, but "it passes 'through' them in a mercurial plurality and continually reproducing itself" (1933, 30, b); it forms the common spirit in which individuals develop. When dealing with the "superexistence" of the objective spirit, Hartmann compares it to the "species" in organic life, though only for the sake of a better comprehension, since the objective spirit maintains continuity only insofar as it preserves tradition and, above all, when it maintains it tacitly.

Discontinuity would be the reflection of the objectified spirit. In a vestigial print left on a durable material, tradition can be broken when detached from the living being and from the changes produced in it. The objective spirit consists of a process of new relationships and creative acts in which the imprints of the past are not necessarily integrated. When coming back to them it is from "other times," with consciousness of a cut and of having reached levels at different profundity.

What remains as continuity in history and what is grasped as discontinuity? The assumed and actual tradition in the present objective spirit reveals itself as a continuity above individuals and their works; the imprint of the spirit in matter, detached and objectified, points to different moments in comprehension.

University of Valencia

BIBLIOGRAPHY

Aster, Ernst von. *Introducción a la Filosofía Contemporánea.* Madrid: Guadarrama, 1961.

Bloch, M. *Apologie pour l'histoire ou métier d'historien.* Paris: 1949.

Breton, St. *L'être spirituel. Recherches sur la philosophie de N. Hartmann.* Paris-Lyon: E. Vitte, 1962.

Carr, David. *Phenomenology and the Problem of History.* Evanston: Northwestern Univ., 1974.

Danto, A. *Analytical Philosophy of History.* Cambridge: 1965.

Hartmann, N. *Das Problem des geistigen Seins.* Berlin: Walter de Gruyter & Co., 1933.

Hartmann, N. *Neue Wege der Ontologie.* Stuttgart: Kohlhammer, 1942, 1949, 3rd ed.; Spanish translation, Ed. Sudamericana, 1954.

Hartmann, N. *Philosophie der Natur. Abriss der speziellen Kategorienlehre.* Berlin: Walter de Gruyter & Co., 1950.

Montero Moliner, Fernando. *Retorno a la fenomenologia.* Barcelona: Anthropos, 1987.

Topolsky, Jerzy. *Metodologia de la Historia.* Madrid: Cátedra, 1982 (1st ed. 1973).

OCTAVI FULLAT I GENIS

HOMBRE Y CIVILIZACIÓN: 1492, LA EDUCACIÓN IMPOSIBLE

INTROITO

De alguna forma hay que entrar — *Introitus* fue, para los romanos, tanto la entrada como la acción de entrar — en el tema de la presente ponencia. Mi ingreso en ella se hace a modo de advertencia y de aviso. ¿En torno a qué levanto la caza?, sobre el valor epistemológico de mi discurso. Este no constituye ni una demostración — *epistéme* — ni tampoco una simple opinión razonable, o proceso mental argumentativo o tal vez retórico y persuasivo — *eulogos* —. Menos, todavía, pretende ser una arbitrariedad — *dislogos* — seductora. ¿Cuál es, pues, el propósito de la presente enumeración u homilía? — el verbo griego *homileo* significó conversar o frecuentarse; ¿acaso no conversamos y nos frecuentamos aquí? —.

Mi recitación organiza una hipótesis, pero en modo alguno demuestra una tesis. En la lengua griega, el verbo *títhemi* indicó instituir, erigir, poner. *Thésis* fue la "acción de poner", la "acción de establecer" algo. Lo mío, hoy, no forma tesis alguna; no dispongo de pruebas suficientes como para dejar sentado o fundamentado lo contenido en la proposición que inaugura esta ponencia. Mis pretensiones son más humildes — encogidas, empero, a la fuerza-. No establezco una tesis, sino simplemente una hipótesis; es decir, "algo puesto" — *Thésis* — "debajo" — *hypo* — . El mío no es un enunciado visible, palmario y saliente, sino recatado, cauto y circunspecto. Pero, eso sí, tiene la pretensión de disparar la curiosidad y la comezón intelectuales para que otro valide o invalide mi barrunto. Esta hipótesis constituye una ficción cómoda que aspira a que el fenómeno de la educación iberoamericana, llevada a cabo por los castellanos — empresa iniciada en 1492 y en trance, ahora, de liquidarse — , resulte más comprensible. Mi labor es levantar la pieza, pero no cazarla. Soy podenco; que no montero o venador. Que otros se pronuncien téticamente; después, esto sí, de las comprobaciones pertinentes. *Faxint dii* — para repetir a Cicerón — .

A-T. Tymieniecka (ed.), Analecta Husserliana, Vol. XXXVII, 81–97.

1. EL PROCESO EDUCATIVO: SU "EIDOS"

No es factible iniciar la aventura que nos hemos propuesto sin antes establecer el significado del término *educación*. Nos interrogamos acerca del *eidos* educacional y no indagamos sus múltiples presencias a los sentidos — *aísthesis* —. El *eidos* de algo — ya desde Sócrates — fue *lo que*, o *el qué*, algo es. El acto educante que vi ayer en aquel local es distinto del que contemplo hoy en este patio; ni el uno ni el otro constituyen el *eidos* de la educación, sino sensibilizaciones de la misma. El *eidos* perseguido intelectualmente nos precisará el aspecto que la educación ofrece de modo inexorable al intelecto; el *eidos* de educación es la presencia misma del educar al acto de inteligir. Iniciemos la pesquisa.

El hombre es un mamífero, un portador de mamas. Se halla inserto en la escala zoológica. Pero, con esto no basta para hacernos cargo de él. El ser humano es *animal educando*, y no *animal educable*. El delfín, en cambio, es educable; la hormiga igualmente. Uno y otra pueden aprender conductas de nuestra especie, pero tales conductas no los mejoran ni en cuanto delfines ni en cuanto hormigas. No van más allá de ser extravagancias para sus respectivas especies. No, así, con la bestia humana. Esta o se educa, entrando en civilización, o bien se queda en bestia sin alcanzar, empíricamente, el estatuto antropológico. El hombre no es educable, sino educando. Pero, ¿en qué consisten las secuencias que lo educan?: en el conjunto de transacciones que se dan entre el código genético de la especie *homo sapiens* y una civilización, la que sea. Denomino *procesos educativos* al conjunto de transacciones apuntado. En cambio, calificco de *procesos madurativos* a las transacciones que se dan entre un código genético del *homo sapiens* y el medio ambiente no humano. El término *crianza* de otras épocas englobaba ambos procesos. Actualmente los discernimos.

Todo proceso educativo modifica la conducta tanto del educando como del educador. Más la primera que la segunda, sin embargo, dadas las fuerzas y tecnologías de que dispone el educador. Al educar, modificamos las informaciones, las actitudes y las habilidades de alguien, capacidades que brotan del código genético, objetivado en bioquímica de cerebro y en endocrinología, entre otros extremos.

Puede suceder que cuando se emprende un proceso educante, aquel que lo recibe haya estado ya educado según las pautas de otra sociedad. No cambia, a pesar de esto, la anterior definición de proceso educativo aunque sería más pertinente, en tal supuesto, referirse a *re-educación*. "Campos de reeducación social" llaman en Vietnam a aquellos centros donde se encierra a gentes educadas según la mentalidad burguesa y a quienes quiere convertirse en comunistas. Los "Ejercicios espirituales" de San Ignacio fueron igualmente procesos de reeducación Esto, sin embargo, no modifica en nada la anterior definición de procesos educativos.

Al educar, pues, uno educa a otro para algo. La educación resulta ser, en consecuencia, una relación — ni una cosa ni una propiedad — entre: (1) elementos modificadores de conducta humana; (2) elementos modificandos en su conducta; (3) elementos en función de los cuales se intenta la modificación de conducta.

Tal como ha quedado ceñida intelectualmente la educación, coincide, ésta, con el concepto de antropogénesis — *ánthropos*, hombre, y *génesis*, generación o engendramiento —, concepto que tanto puede aplicarse al individuo — biografía — como al colectivo, sea éste limitado — sociografía — o bien integrador — historia —. Esto me obliga a establecer una distinción en el seno de los procesos educativos. Los hay que son intencionados, como es el caso de aquellos que tienen lugar en las instituciones sociales de la familia y de la escuela; otros, por el contrario, no son intencionados intentando otros propósitos cuando, sin pretenderlo, educan, o sea modifican conductas. Tal es el caso de los procesos disparados por institucionese religiosas, políticas, sindicales, productoras, periodísticas . . . .

Resulta indispensable apuntar finalmente a una cuestión peculiar: ¿no es posible otra educación que la señalada en la definición general?. En el supuesto negativo, se concibe al *ánthropos* a modo de robot programado por el código genético y reprogramado socialmente. En el supuesto de que uno se la juegue a favor de la respuesta positiva, resulta imaginable otro modelo educacional, aquel en que la modificación de la conducta se debe también al proyecto y a la decisión del propio educando. Ahora bien; esto segundo supone ya apostar metafísicamente por cierta autonomía del educando con respecto al mundo material. Resulta harto difícil representarse una materia, incluida la biosociológica, allende la necesidad.

Este apunte en torno al *eidos* educacional nos permite ya proseguir el discurso con coherencia.

2. LA CIVILIZACIÓN: CONTENIDO EDUCACIONAL

Cuando se educa, indicaba, se reprograma un código genético, desde un medio humano, con ánimo de que el resultado sea un individuo conformado a los modelos de dicho medio. Educar es, para comenzar, *tradere* y, en consecuencia, tradición y entrega. Lo traspasado hace que uno sea griego o romano o feudal o renacentista o ilustrado . . . , o bien que sea maya, azteca o castellano.

Cuando se considera un medio humano que constituye ya un amplio espacio antropológico estructurado y coherente, podemos referirnos a *civilización.* Siempre nos educamos en el vientre de una civilización. El hombre es tal porque es *zoon politikón,* animal cívico o social. Jamás el ser humano es natural. El diálogo *Protágoras,* de Platón, ya describe esta elementalidad antes de Aristóteles. Dice así el diálogo:

> Como no era del todo sabio Epimeteo, no se dio cuenta de que había gastado las capacidades en los animales; entonces todavía le quedaba sin dotar la especie humana, y no sabía qué hacer . . .
>
> Prometeo, apurado por la carencia de recursos, tratando de encontrar una protección para el hombre, roba a Hefesto y a Atenea su sabiduría profesional junto con el fuego . . . y los ofrece como regalo al hombre . . .
>
> Los hombres ya intentaban reunirse . . . , pero cuando lo hacían se atacaban unos a otros al no poseer el saber político . . . Zeus, entonces, temió que sucumbiera toda nuestra raza, y envió a Hermes que trajera a los hombres el sentido moral — *aidós* — y la justicia — *dikayosyne* —, para que hubiera orden en las ciudades y ligaduras acordes de amistad. — 321, c hasta 322, c — .

Platón advierte, aunque sea literariamente, que el *ánthropos* despunta y se singulariza, de forma esencial, con respecto al resto zoológico, merced a la *tékhne* — artes para tirar adelante con la vida, siempre tan amenazada — y a la *sophía politiké* — sabiduría social — . El hombre deja de ser simple bestia, en consecuencia, gracias a la producción y a la convivencia. El autor del *Protágoras* nos ha introducido bellamente en el concepto de civilización, tan importante, ésta, en todo proceso educante. El botín del robo de Prometeo — a los dioses — y el obsequio de Hermes, en nombre de Zeus, son cosas divinas y no zoológicas. La herencia biológica no las encierra; solamente la educación hace entrega de las mismas.

Resulta cómodo entender una civilización como un conjunto formado por tres ámbitos: la cultura, la técnica y las instituciones. — No faltan autores que denominan *cultura* lo que aquí he llamado *civilización* — . Lévi-Strauss, en *Structures élémentaires de la parenté* — P. U.

F., pags. 3—10 — sostiene que: "La distinction entre *état de nature* et *état de société* . . . présente une valeur logique qui justifie pleinement son utilisation".

¿Por dónde transcurre la frontera entre naturaleza y sociedad?; responde en las mismas páginas con las palabras siguientes: "Ce qui est constant chez tous les hommes échappe nécessairement au domaine des coutumes, des techniques et des institutions par lequelles leurs groupes se différentient et s'opposent."

Aunque sólo sea por utilidad lógica del discurso, conviene distinguir entre lo natural humano — "lo que es constante en todos los hombres" — y lo histórico humano — "aquello que diferencia y opone a los distintos grupos" — . Naturaleza y civilización. Esta segunda especifica al hombre y, al propio tiempo, constituye el contenido de los procesos educadores.

Toda civilización — tolteca, inca o romana — abarca una *cultura* o manera de interpretar la circunstancia. El mundo del hombre no es el mundo *tout court*, sino la "manera-de-ver-el-mundo" que posee un grupo humano. Al ser *animal loquens* y *animal symbolicum*, el hombre no vive solamente de la presentación del mundo, sino de la "re-presentación" del mismo. Las civilizaciones precolombinas tenían sus culturas, tan distantes, por cierto, de las culturas europeas. Los lenguajes evidencian y constituyen el síntoma del hombre como animal culto.

Toda civilización — chichimeca, olmeca o egipcia —, además, abraza unas *técnicas* o maneras de modificar el mundo. Medicina, agricultura, arquitectura, ganadería. El hombre es *animal faber* que no se contenta con la naturaleza y trabaja para cambiarla. Técnicas hay que son útiles — tecnología — ; otras, en cambio, son gratuitas — arte —. No hay civilización sin instrumentos.

Toda civilización — mixteca, de Teotihuacán o bizantina —, finalmente, engloba *instituciones sociales*, sean de crianza, jurídicas, militares, religiosas, docentes a bien políticas, instituciones que constituyen las maneras como un grupo se instala colectivamente en el mundo. El hombre es *animal sociale, animal societatis.*

Cuando alguien se educa, se incorpora a un civilización — cultura, técnicas, instituciones —, sea, ésta, maya, inca o feudal europea. El aprendiz de hombre es, inexorablemente, aprendiz de un modelo geohistórico de hombre, y no de todos a la vez o bien de un modelo universal de *anthropos*, que no existe. La sociedad inculca su civilización a los recién llegados a ella a base de proporcionarles informaciones, de despertarles actitudes y de facilitarles el aprendizaje de habilidades. Cada civilización marca los contenidos de su actividad educadora haciéndose suyos los códigos genéticos de la especie humana que se han puesto a su alcance.

3. ESTRUCTURA SISTÉMICA DE TODA CIVILIZACIÓN

Cuando los súbditos de la Corona de Castilla llegan a las Américas encuentran gentes educadas según los moldes de civilizaciones precolombinas. Los conquistadores alcanzan aquellas tierras, educados, a su vez, en conformidad con los modelos de las civilizaciones europeas. El ocupante del continente descubierto pretende *reeducar* a los llamados *indios*; es decir, ambiciona modificar las informaciones de éstos, sus actitudes y sus habilidades. ¿Qué sucede en circunstancias parejas?; ¿es posible el mestizaje civilizatorio?; ¿siempre?, ¿en qué condiciones?.

Mi presupuesto es que una civilización forma una estructura o sistema o cosmos — *Kosmos*, del verbo griego *Kosmeo*, ordenar, significó orden — . Una civilización no es un caos — *Kháos*, en la lengua griega, significaba grieta, abismo, desorden, mezcla confusa; se relacionaba con el verbo *Khaíno*, agrietarse la tierra, abrir las fauces, quedarse estupefacto —. Al pronto una civilización se planta ante el observador novicio como caótica y el recién llegado a ella queda atónito y sobrecogido, pero a poco que uno se esfuerce en descubrir la lógica de aquella, el *Kháos* inicial va mudándose en *Kosmos*. Acaba comprendiéndose que cade elemento de la civilización se intelige desde los restantes, no ocupando gratuitamente un lugar en la estructura civilizada.

Ya Montaigne cayó en la cuenta de que cualquier sociedad de hombres configura un sistema — *syn*, en griego, juntamente; *stema*, trama; *systema*, algo bien tramado —. Escribió: "La société des hommes se tient et se coud à quelque prix que ce soit . . . (Ceux-ci) trouvent d'eux-memes la façon de se joindre et de se placer les uns parmi les autres, mieux que l'art n'aurait pu le faire." — *Essais*; libro 3, cap. 9 —.

Constata, el agudo Montaigne, que los diversos elementos de una sociedad — creencias, costumbres, instituciones . . . — se organizan entre ellos al margen de los individuos que la forman. Los hombres quedan organizados a pesar suyo.

Me inclino a pensar que las características de cade sociedad configuran un sistema lógico de elementos indispensables. Si no fuera así, no contaríamos por ejemplo con una etnología, habiéndonos de conformar a lo sumo con una etnografía. No hay ciencia posible de lo azoroso, del puro acervo, de la hacina; sobre esto sólo se produce fantasía y capricho. Lo inteligible está presidido por el orden de la lógica. Sin *logos* no hay saber universal y necesario.

Una civilización constituye una estructura. Y por esto Lévi-Strauss sostiene que es: "la logique des oppositions et des corrélations, des exclusions et des inclusions, des compatibilités et des incompatibilités, qui explique les lois de l'association, non le contraire." — *Le totémisme aujourd'hui*; P.U.F., 1962 pag. 130. — .

En cualquier civilización descubrimos sistemas de relaciones y dependencia de las partes con respecto al todo. La civilización es una estructura porque sus elementos constituyen una totalidad. Las propiedades de los elementos dependen de los caracteres de la totalidad. Esto es así en la civilización Maya o bien en la germánica. La fonología estructural nos sirve, como modelo, para entender la civilización. La hipótesis fundamental de la fonología estriba en que cada fonema se define desde el contexto en el que aparece; puede identificarse un fonema en la medida en que lo relacionamos con las otras entidades pertinentes de aquella lengua.

El método para inteligir una civilización comprenderá tres momentos: (1) observación de la realidad; (2) construcción de modelos; (3) análisis de su estructura. Esta es lo inteligible de la realidad social. Al descubrir la estructura de una civilización, se introduce un orden explicativo en la incoherencia de los fenómenos. La inteligibilidad de una civilización es autónoma; es decir, le viene de su estructura, la cual está dotada de una ley de composición interna. Dicha estructura de la civilización, sin embargo, no es un dato que pueda constatarse, sino algo que es necesario reconstruir deductivamente, como sucede de modo paralelo con la causalidad, en física.

Piaget determina la estructura a partir de tres caracteres. Escribe: "Une structure comprend les trois caractères de totalité, de transformations et d'autoréglage." — *Le Structuralisme*; P.U.F., 1970; pags. 7—16 —.

Cualquier civilización precolombina, como igualmente la castellana, está, ciertamente, constituida por elementos; ahora bien, éstos se hallan subordinados a leyes que caracterizan el sistema en cuanto tal. Sus leyes de composición son estructurantes y, por tanto, transformadoras, en el seno, siempre, de la estructura. La autoregulación de un espacio civilizado convierte a éste en conservador de sí y en cerrado.

Lévi-Strauss, en *Anthropologie structurale*, concibe: "les règles du mariage et les systèmes de parenté comme une sorte de langage, c'est-à-dire un ensemble d'opérations destinées à assurer, entre les individus et les groupes, un certain type de communication" — Plon, 1958; pag. 69 —.

Mi hipótesis está en que la civilización, toda, queda organizada al modo de los sistemas de parentesco y de los sistemas lingüísticos — parcelas de aquella — . El concepto de *episteme* elaborado por Foucault en *Les mots et les choses* — Gallimard, 1966 — ilustraría mi conjetura. La *episteme* es un *a priori* histórico que permite hacerse cargo de un momento de la historia. Cada diapositiva histórica constituye un conjunto lógico.

La civilización *náhuatl,* de Mesoamérica, que abarcaba aztecas, tezcocanos, cholultecas, tlaxcaltecas . . . , configuró un sistema antropológico con sus subsistemas educativo, ético, filosófico, jurídico, social, científico . . . . Los *nahuas,* a través de los siglos — empalmaban con las culturas de Teotihuacán y de Tula —, perfilaron una estructura cohesionada de civilización. Esto nada tiene que ver con que entre ellos hubiera grupos enemigos, como los aztecas y los tlaxcaltecas, ya que nos referimos a una cohesión inteligible cultural y no a la práctica política, la cual, sin embargo, se explica igualmente desde el todo *Náhuatl.* La cultura de los *nahuas* estableció una estructura lógica que ofrecía resistencia al intercambio, exactamente como no importa qué otra cultura.

4. CLASES DE CIVILIZACIONES Y SU GRADO DE ESTRUCTURA

La palabra *civilización* cubre un abanico semántico. A partir de lo que poseen en común todas las civilizaciones interesa ahora establecer distinciones a fin de avanzar en el discurso emprendido. Atendiendo a su grado de independencia, resulta práctico clasificar las civilizaciones — de mayor a menor autonomía — en: (1) *macrocivilizaciones* — e.g. la del Trigo, u occidental; la del Arroz, o asiática; y la del Maíz, o precolombina — ; (2) *mesocivilizaciones* o subconjuntos de las anteriores — e.g. culturas latina, germánica, anglosajona . . . , dentro del ámbito de la Macrocivilización del Trigo — ; (3) *microcivilizaciones* o subconjuntos de los subconjuntos — e.g. culturas catalana, castellana, veneciana, napolitana . . . , en la esfera de la mesocivilización latina —.

Estimo que no contamos con una *civilización universal,* considerando que ésta no va más allá de ser un sueño provocado por el deseo de *beatitudo* que habita en los hombres a causa de su conciencia de fracaso. El *to Agathón*, la Civilización de todos, no preside la historia de las civilizaciones en contra del *análogon* platónico. En el plano de los hechos, el concepto de *Civilización* con mayor extensión lógica — la realidad que cae bajo dicho concepto — es el de *macrocivilización*. Esta constituye la màxima estructura posible de lo civilizado. Este hecho le proporciona una coherencia integradora total que la convierte en cerrada sobre sí misma hasta tal punto que resiste al intercambio con otras macrocivilizaciones. Alternar substanciosamente con otra macrocivilización implica estrenar un duelo con ella a vida o muerte.

Imagino que una macrocivilización encierra una estructura autosuficiente y excluyente gracias a que en el terreno de los hechos se ha construído en un casi aislamiento total. El Maíz y el Trigo se desconocieron durante milenios, tiempo suficiente para que se organizaran como sistemas con lógica interna autónoma. Diría que el triángulo formado por Jerusalén, Atenas y Roma da origen al Trigo con todos sus subsistemas. Profetismo, precisión y eficacia. Tal vez la recta que va de Teotihuacán al Cuzco engendró el Maíz y a las subestructuras respectivas. El mito lo informa todo.

Así como en la parcela del arte, el México antiguo se organiza en estilos interrelacionados — el estilo olmeca, el totonaca, el teotihuacano, el zapoteca, el tarasco, el maya . . . , así imagino también al conjunto de la macrocivilización del Maíz. Sospechas, desde luego; pero que se me antojan pertinentes. Un trabajo empírico debería relacionar, por ejemplo, las piedras-calendario con inscripciones en caracteres jeroglíficos procedentes de los santuarios mayas de Palenque — que pueden fecharse en el 23 de diciembre del año 691 — con otras piezas procedentes, por ejemplo, del Machu-Pichu. Sólo después de mucha observación empírica nos encontraríamos en condiciones de reconstruir la macroestoructura que haría inteligible al Maíz todo.

El libro de Miguel León-Portilla, titulado *La filosofía Náhuatl* — Universidad Nacional Autónoma de México; 1983 —, constituye un buen ensayo, aunque regionalizado, de lo que ando sugeriendo para que un día pudiera validarse o invalidarse mi hipótesis. A falta de datos suficientes para configurar la estructura de la Macrocivilización del Maíz, el libro indicado nos permite apuntar modelos, cuando menos, de la civilización Náhuatl.

El espacio cultural de los *nahuas* formó un sistema en el que se organizaban, estructuralmente, arquitectura, escultura, pintura de códices, la ciencia del tiempo expresada en sus dos calendarios, una religión compleja, un derecho severo, un comercio organizado, una milicia poderosa, una institución educativa y unos saberes botánicos con fines terapéuticos. Podría probablemente establecerse la estructura de la mesocivilización Náhuatl teniendo presentes su cultura, sus técnicas y sus instituciones. El libro de Fray Bernardino de Sahagún, *Historia General de las Cosas de Nueva España* — Edición Porrúa; México, 1956 —, redactado durante los primeros tiempos de la conquista, constituye un buen instrumento de trabajo en lo tocante a la cultura Náhuatl; pero a todas luces resulta insuficiente.

En lo que se refiere a la concepción antropológica, la obra de León-Portilla proporciona material interesantísimo que permitiría ya elaborar un modelo; no, claro está, la estructura de esta mesocivilización. He aquí algunos textos, sacados de la Colección de cantares Mexicanos — donde se recogen breves poemas del mundo Náhuathl —, altamente significativos:

> "Sobre la tierra, ¿acaso puedes ir en pos de algo?".
> "¿Aquí he venido sólo a obrar en vano?'
> No es ésta la región donde se hacen las cosas.
> Ciertamente nada verdece aquí:
> abre sus flores la desdicha."
> "¿Acaso son verdad los hombres?.
> "¿A dónde iré?,
> a dónde iré?.
> El camino del dios de la dualidad.
> ¿Por ventura está tu casa en el lugar de los descarnados?"

Los *tlamatinime* — plural de *tlamatini* —, sabios o pensadores *nahuas*, a quienes debemos las reflexiones cosmológicas y antropológicas, formaron la *paideia* de la cultura Náhuatl.

Al llegar los soldados, los escribanos y los eclesiásticos de la Corona de Castilla violentaron aquella estructura del Maíz iniciando su derrumbamiento. No hay por qué acusarlos, sin embargo; eran simples instrumentos del choque entre dos Macrocivilizaciones, tan cohesionada cada una consigo misma que no permitía la cohabitación con la otra. Resultaba imposible una hiperestructura que abrazara a ambas.

Cada Macrocivilización intentó educar a la otra — es decir; procuró modificar sus informaciones, sus actitudes y sus psicomotricidades —, sometiéndola a los propios parámetros de interpretación, de reforma y de instalación. La del Trigo comenzó con gran ventaja: era agresiva, precisa y eficaz, amén de soñar en profetismos salvadores. La macrocivilización del Maíz jugó, desde el inicio, con inferioridad y mengua: fue la atacada y, además, estaba en demasía pegada al mito y al fatalismo.

Tan seguros andaban los del Trigo de su superioridad cultural, tecnológica e institucional, que José de Acosta en su *Historia natural y moral de las Indias* — F.C.E., 1962; pag. 320 —, redactada en pleno Renacimiento, no dudó en escribir: "A los indígenas es necesario enseñarlos primero a ser hombres y después a ser cristianos".

A los *indios* se les enseñaría a vivir con "policía y buenas costumbres"; pues esto era para los castellanos, avanzados del Trigo, ni más ni menos que humanizarlos. Los naturales debían someterse al sistema político, económico, social y culto de los europeos.

5. LA EDUCACCIÓN DEL MAÍZ POR PARTE DEL TRIGO

Los *indios* se hallaban educados en conformidad con la Macrocivilización del Maíz en sus diversas mesocivilizaciones y microcivilizaciones. Y en esto que llegan las tropas de la Corona de Castilla, juzgan como bárbaros — ineducados — a aquellos sujetos y, acto seguido, estrenan un fiero y total proceso educador siguiendo las pautas de la Macrocivilización del Trigo. Estructura contra estructura. Si mi hipótesis — formulada en los puntos tercero y cuarto — resulta aceptable, no es posible síntesis alguna entre ambas Macrocivilizaciones. El verbo griego *syntíthemi* significó reunir, ajustar, construir; *sýnthesis* indicó arreglo, combinación, composición. No había compostura posible, ni acuerdo, ni conciliación, ni acomodo, entre Trigo y Maíz, tan autónomas y autosuficientes eran una y otra Macrociviliszación. El Trigo no educó al Maíz; no hizo otra cosa que desnucarlo y acabar con él.

La contienda y el acometimiento fueron a muerte. Unos atacaban; los segundos resistían. El mestizaje macrocivilizatorio es inviable, no encajando ninguna pieza de una estructura en la de enfrente. Cada penetración del Trigo constituyó un derrumbe y un ocaso y un desmoronamiento para el Maíz. Ensayos como el del Obispo "Tata Vasco" — Vasco de Quiroga —, en Michoacán fracasaron absolutamente. Pátzcuaro no va más allá de ser el cadáver de una ilusión imposible. El *Animal symbolicum*, el *Animal faber* y el *Animal civitatis* del espacio macrocivilizatorio precolombino se defendió — en la medida en que formaba una estructura lógica —, de la cultura, de la técnica y de las instituciones sociales procedentes del Mundo avasallador del Trigo. En la actualidad, El Maíz está al morir. Tan exhausto y desmarrido anda que en el día de su defunción total, nadie va a enterarse. Cada vez más, allá, se interpreta el mundo — ciencia y moral, por ejemplo —, se transofrma el mundo — medicina y microelectrónica, pongamos por caso —, y se instalan grupalmente en el mundo — partidos políticos y sistema docente, supongamosegún los modelos que los súbditos del Reino de Castilla traían de Europa, la cuna del Trigo.

A falta de estudios estructurales de la vasta Macrocivilización del Maíz, una vez más me sirvo de la mesocivilización Náhuatl para apuntalar el modelo de lo que sería mi discurso. Los nahuas contaban con la *Tlacahuapahualiztli* o arte de criar y educar hombres — *huapahualiztli* significó *crianza*, o educación, mientras *tlaca* quería decir *hombres* —. Educaban "transmitiendo sabiduría a los rostros ajenos"; esto significaba *Ix-tlamachiliztli* — León-Portilla; o.c., pag. 221 — . El proceso educador se iniciaba en la familia; en ésta aprendían la fortaleza y el autocontrol a base de privaciones. Más tarde, hacia los quince años, ingresaban o en internados — *Telpochcalli* — donde se les formaba para ser guerreros, o bien entraban en monasterios — *Calmécac* — en los que se les preparaba para ser sacerdotes. En estos segundos centros, los estudios eran superiores y en ellos los *tlamatinime* — los "sabios" — comunicaban lo más elevado de la cultura náhuatl.

Dicha cultura, o hermenéutica del mundo, se organizaba en sí misma configurando un conjunto lógico y al propio tiempo se unía a través de oposiciones y de conexiones con los ámbitos tecnológico e institucional. La defensa que los indígenas hacen de su poligamia muestra una institución que se estructura en la totalidad de aquella mesocivilisación; la practicaban según Fray Juan de Zumárraga: "porque no tienen otra renta sino lo que las mujeres les ganan con su labor para se mantener y en satisfacción de sus trabajos les pagaban con sus mismos cuerpos, y que no pueden dejar esta ley en que fueron criados." — Revista *Historia Mexicana*; Nº 88; México, 1973; pag. 463 —.

En lo que es, actualmente, Iberoamérica, un conjunto de subconjuntos organizaba un sistema de civilización que sólo se modificaba por autoregulación de aquella misma estructura de estructuras. Llegados los representantes del macrosistema occidental, introdujeron, éstos, una cuña mortal en la Civilización del Maíz. Esta no pudo ya continuar autoregulándose como estructura lógica interna y comenzó su período agónico y su progresiva consunción.

Los castellanos, ejecutores fatales de la Macroestructura del Trigo, acorralaron los reductos de los mesosistemas civilizados americanos — algunas de estas bolsas aún malviven — y emprendieron al propio tiempo la educación de los cerebros tiernos, más plásticos en vistas a la nueva comprensión del mundo, a las técnicas recién llegadas y a las asociaciones venidas del Viejo Mundo. Documentos múltiples y dispersos proporcionan información empírica de esta educación que sólo triunfó a base de asesinar, antes, al educando — entendido, éste, como alguien codificado ya por una sociedades organizadas según otra civilización —.

José de Acosta (1539—1600), consigero de Felipe II, gran renacentista, escribio en su libro *Historia natural y moral de las Indias* — 1590 — lo que sigue, refiriéndise a los indígenas: "Es necesario enseñarlos primero a ser hombres y después a ser cristianos". — F.C.E., México 1962; pag. 320 —.

En la misma línea de pensamiento se expresa el virrey Francisco Toledo en carta al rey: "Mande V. M. proveer que en ninguna manera se bauticen los indios ... sin que primero se les infunda y enseñe la natural política y civil ... ; sin enseñarles primero a ser hombres". — Revista *Historia Mexicana*; N° 88; México, 1973; pag. 439 —.

En el libro de Christian Duverger, *La conversion des Indiens de Nouvelle Espagne*, — Seuil; París, 1987 — se hallan múltiples referencias en torno a lo que venimos sosteniendo, todas ellas circunscritas al siglo XVI, época en la que centra su historia de la cristianización de los indios mexicanos.

En le obra de Frank Moya Pons, titulada *Después de Colón* — Alianza Editorial, 1987 —. se nos cuenta la formación de la primera sociedad europea en tierras americanas. Desde comienzos del siglo XVI, las diferenciaciones institucionales y políticas que el autor detecta en la Española no son más que acomodaciones, o autoregulaciones, que la gran Civilización del Trigo hace en contacto con otra geografía. De hecho, los colonizadores recrearon en Santo Domingo una sociedad tan parecida como fue posible a la vieja sociedad castellana.

La metodología usada para que triunfara El Trigo consistió en educar, separadamente, a los niños y adolescentes de las principales familias indígenas. Los primero franciscanos que alcanzaron Tenochtitlan, en Nueva España, pidieron a los principales y a los sacerdotes de aquellas tierras que les entregasen sus hijos a fin de instruirlos. Para lo cual, según Gerónimo de Mendieta: "Es necesario que vosotros nos deis y pongáis en nuestras manos a vuestros hijos prequeños, que conviene sean primero enseñados. ... Vuestros hijos, como niños y tiernos en la edad, comprenderán con más facilidad la doctrina que les enseñaremos". — *Historia eclesiástica indiana*; México, 1971; pp. 214–215 —.

No era cuestión únicamente de aprender la cultura de los conquistadores, mas también sus tecnologías. Este último aprendizaje se llevaba a término aprendiendo oficios mecánicos y otras pericias al lado de expertos. Esto se inició institucionalmente, en México, a partir ya de 1530. Según el mismo Mendieta, los *mozos grandecillos* aprendían los oficios de sastre, zapatero, carpintero, de lapidario, de orfebre, cantero, alfarero, de teñidor, de curtidor, de fundidor de campanas, de herrero, de bordador, de pintor, de escultor. ... Los más aprovechados encontraban fácilmente trabajo, tan grande era la demanda de aquellos castellanos a los que no alcanzaban los bienes que llegaban de España. Los indígenas jóvenes se entregaban, cada vez más, al servicio de todo lo nuevo, renunciando a su mundo.

En esta misma línea metodológica se sitúa la carta que el Virrey de Nueva España envía a la Emperatriz con fecha de 30 de marzo de 1531: "Y uno de los principales intentos que tenemos *la perpetuidad de todo* es enseñar a los naturales a vivir políticamente. Y aun nos hemos puesto en pedir a los señores indios de esta ciudad que nos den mancebos hábiles para los poner con oficiales castellanos en todos oficios para aprendices, como se hace en esos reinos." — Revista *Historia Mexicana*; Nº 88; México, 1973; pag. 451 —.

A muchos niños, en México por lo menos, se les internaba y sometía a régimen monacal. Según Bernardino de Sahagún e igualmente según Mendieta — obras ya citadas —, el hecho de colocar a los pequeños en internados con un régimen de vida propio de frailes tenía como finalidad desvincularlos y hacerles olvidar las costumbres y las ideas de sus antepasados.

Tan *fervorosos* salían, aquellos jóvenes, de los internados que según narra Pedro de Gante: "Nosotros con ellos — aquellos jóvenes — vamos a la redonda destruyendo ídolos y templos por una parte, mientras ellos hacen lo mismo en otra." — Revista citada, pag. 444 —.

El propio Mendieta describe como el 1 de enero de 1529, en Tetzcoco, tres frailes iniciaron una campaña destructora de templos e ídolos, campaña seguida por los muchachos indígenas formados por los conquistadores. Los indígenas adultos "quedaban espantados y abobados, y quebradas las alas, como dicen, viendo sus templos y dioses por el suelo" — Obra citada de Mendieta; pag. 228 —.

Desde 1529 se comienza la educación institucional de las muchachas indígenas, en Nueva España. También se las encierra a manera de monasterio de monjas, tal como consigna Zumárraga: "Mucha cantidad de mujeres doncellas . . . , que aunque no son monjas profesas . . . , hay clausura . . . , y aquel monasterio y mujeres tienen a cargo una matrona, mujer honrada, de nuestra nación." — Revista citada; pag. 453 —.

La misma Emperatriz manda seis mujeres a México para la formación de las niñas indígenas. Llegan a Nueva España a principios de 1531. Un año más tarde habíanse fundado ya internados femeninos en ocho poblaciones.

En aquel contexto alcanza todo su dramatismo el descubrimiento antropológico-existencial que habían hecho los *tlamatinime* unos años antes de la llegada de los europeos: "¿Son acaso verdad los hombres?". Colón apostilló un ejemplar de *El Libro de Marco Polo*, impreso em Amberes en 1485 — Alianza Editorial; 1987 —; la diferencia entre el viaje del veneciano en la macrocivilización del Arroz, realizado en 1275, y la incursión de las naves colombinas radica en que el primero no pudo introducir perturbaciones en la estructura del Arroz mientras las segundas inician la desestructuración irremediable del Maíz. Aunque fuera verdad lo que comenta el Inca Garcilaso de la Vega, en 1606, cuando escribe: "Las riquezas que el Perú ha enviado al mundo viejo y derramado por rodo él, antes le han dañado que aprovechado . . . Han crecido las rentas de los ricos . . . y también han crecido las miserias de los pobres." — Revista *Humboldt*; Nº 91; 1987; pag. 31 —.

Aunque fuera cierto esto, decía, nada tiene que ver tal hecho con la imposible educación del Maíz por parte del Trigo, el cual prosigue autoregulándose como estructura, asimilando, esto sí, las influencias recibidas.

Resulta del todo improcedente pasar por alto, en este trabajo, los escritos del obispo dominico Fray Bartolomé de las Casas (1474—1566). Fue obispo de Chiapas, en México. Su *Brevísima Relación de la Destrucción de las Indias*, dedicada a la Católica Majestad del rey Carlos I de España y Carlos V de Alemania, emperador del Sacro Romano Imperio de nación Alemana, constituyõ un escrito que desde el primer momento despertó vivas polémicas. El texto original se publicó en Sevilla en 1552. Siguieron traducciones múltiples. El Santo Oficio prohibió el libro en 1660. Las Casas vivió más de cuarenta años en las colonias americanas en contacto directo con los indígenas. Su documento es vivo. En los treinta años que separan la llegada de Cortés de la composición de la *Brevísima Relación*, la población central de México disminuyó de 25 millones a 6 millones. Las epidemias, las enfermedades virales, la malaria, el hambre, los trabajos forzados, los asesinatos . . . explican tal destrucción demográfica. Los portadores de la Macroestructura del Maíz desaparecen físicamente. Las Casas increpa al colonizador, no tanto por pretender el poder, cuanto a causa del orgullo de creerse en su derecho. Las Casas concluye su *Relación* afirmando que los malvados españoles, obcecados, dejados de Dios y entregados a falsos razonamientos, no quisieron comprender que los indios habrían tenido la más justa y válida razón — de haber estado armados de otra manera — para destrozarlos y echarlos de sus tierras según las leyes naturales, divinas y humanas, y del modo más legítimo. No entro en el valor de este documento por ser lego en la cuestión. Pero, me ha servido para subrayar una vez más la imposibilidad de juntar Maíz y Trigo con ánimo de ofrecer unos contenidos educacionales nuevos e inéditos.

Sahagún recopiló en *Los Coloquios* — Obra citada de León-Portilla; pags. 130—135 — la documentación que halló en Tlatelolco sobre las discusiones tenidas por los doce primeros frailes, llegados en 1524, con los indios principales y sus sabios. He aquí algunos textos en que hablan, significativamente para mi hipótesis, los indígenas:

> Tal vez a nuestra perdición, tal vez a nuestra destrucción, es sólo a donde seremos llevados . . . .
>
> Déjennos ya morir, déjennos ya perecer, puesto que ya nuestros dioses han muerto . . . .
>
> Y, ahora, nosotros ¿destruiremos la antigua regla de vida? . . .
>
> Es ya bastante que hayamos perdido, que se nos haya quitado nuestro gobierno . . . .
>
> Sólo seremos prisioneros. Haced con nosotros lo que queráis.

La Macrocivilización del Maíz estaba sentenciada a muerte. Año del Señor de 1524. Aquella estructura no podía continuar autoregulándose. Los *Tlamatinime* — sabios — fueron abatidos, los códices quemados, los templos convertidos en montones de piedras. No era posible otra educación que la procedente de Europa.

ITE. LECTIO EST.

Ha concluido la exposición de mi barrunto, de mi presuposición. *Lectio est. Lectio* fue, para los latinos, lección y lectura. He terminado una y otra. Simple hipótesis o conjetura o supuesto o sospecha. Poca cosa. Por tal razón he escrito igualmente *Ite; eo, is, ire* indicó andar, marchar, caminar. El imperativo. Se trata de algo perentorio, exigente, indispensable.

Caminar, peregrinar, avanzar. ¿Hacia dónde?; hacia la resolución del dubio de los *tlamatinime* nahuas, incertidumbre que inunda de tristura antropológica: "Sólo venimos a soñar, sólo venimos a dormir: no es verdad, no es verdad que venimos a vivir en la tierra."

La vida del hombre en *tlaltípac* — encima de la tierra —, ¿es acaso estúpida y abiertamente insensata?.

La validación de mi hipótesis daría pie, en principio, a ser pesimista antropológicamente. La Macroestructura del Trigo puede tan sólo destrozar la Macrocivilización del Maíz, pero jamás educarla. Sin embargo, quedan, trémulas, las carnes y las conciencias de cade quien. ¿O, acaso, los individuos son inexorable y totalmente juguetes de las estructuras y de sus autoregulaciones históricas?. Con todo, mucho me temo que, de ser verdad mi sospecha, la buena voluntad y el esfuerzo de algunos recién llegados del Viejo Mundo no pueden torcer el sistema, o conjunto lógico, que organiza e independiza a una vasta civilización. Podría suceder, no obstante, que mi supuesto quedara invalidado y entonces. . . . Lo que sí me parece más pertinente es apostar por la libertad singular, eficaz en las relaciones interindividuales. No faltaron castellanos en aquellos comienzos del siglo XVI que trataron individualmente de llevar a cabo un mestizaje cultural y, en consecuencia, una mutua educación. ¿Que esto carece de peso sobre el curso de la Historia?; éste no es problema de cada cual. No nos incumbe. Lo de uno está en ser fiel o su propia conciencia. No tenemos espaldas suficientes como para cargarnos con el mal del mundo entero.

La Universidad Autónoma de Barcelona

ANNIBALE SALSA

PHENOMENOLOGY AS A THEORY OF CULTURE

A Remedy for the Entropy of Knowledge

The main debt which we European men of thought owe to Edmund Husserl is principally to be found in the theoretical/moral testimony with which the Moravian philosopher has restored modern man's, and — I should like to add — post-modern man's, faith in the unexplored possibilities of philosophy as "universal knowledge," and this, at a most crucial moment in the decline of Western identity. The end of the great philosophical systems founded on the intellectual speculation and arrogant intellectualism as well as on the triumphal success of the natural sciences prefigured, within the ambit of the philosophical *episteme* of the last 150 years, the beginning of a *regressus in infinitum* the results of which are not easy to foresee.

The monopoly of rigor and the scientific approach is transferred to a projecting form of abstract reasoning which, in the multiplicity of the deductive systems and of its own inferential rules, incontrovertibly decrees the paradigm of the certainty of facts. Formal legality is thus guaranteed and the consequent scission between science and life paints a disturbing landscape for the *Heimwelt* of contemporary man. It is known how Husserl's intuitions and successive development of phenomenology have pointed to some alluring interpretative paths for the psychological and psychiatric sciences, accelerating the process of the recomposition of the subject as an *in-dividuum.* Having said this, it is my intention on this occasion to utilize these interpretative categories from an antidualistic/antidivisionistic phenomenological standpoint. These categories have already seeded the ground of the sciences of the psyche, and there may be a greater role for phenomenology as the "philosophy of culture": i.e., through its investigation of the horizon of the interaction between man and the world, the individual and the environment, life and thought, the "artificial" dimension (science and technology) and the "primordial" dimension. The work which phenomenology has undertaken of stitching up the "tear" in modern scientific-philosophical culture has required a reworking of the themes of the relationship between the natural sciences (*Naturwissenschaften*) and the sciences of the spirit (*Geisteswissenschaften*), and this in turn has

A-T. Tymieniecka (ed.), Analecta Husserliana, Vol. XXXVII, 99–106.

required the definitive abandonment of Cartesian and neo-Cartesian paradigms. It is urgent, therefore, that we radically rethink the way in which knowledge is formed starting from the direct experience of things (*zu den sachen selbst*), of real and ideal meanings, and of all those sense experiences which find in the world-environment (*Umwelt*) an inter-active container of referential links. Human life crosses realms and areas of the consciousness of things which in their separate existence as objects adopt the *ontisch* status of disjointed parts, drifting "still lifes." The material nature of the world, permeated with the blind dynamism of the forces at play in it, tends towards the destructuring of any immanent order whatsoever. We know from the laws of physics (the second Law of Thermodynamics) that "entropy, disorder, increases in every isolated system"[1] and that, therefore, if life and the world-container (*oikos*, the house) were not assumed to be related systems, a scientific order could not be created. Husserl's phenomenology is the "rigorous science" (*strenge Wissenschaft*) of a system of relationships. Relationships between essences but also between concrete, dialogue creating "in-dividuals" committed to weaving schemes having pluri-dimensional meanings and perspectives.

The refounding of Western thought which Husserl set in motion traces a completely original way of restating the problem of the relationship between perceived experience and organized knowledge, structured in the form of categories. The "principle of all principles" of phenomenology (*das Prinzip aller Prinzipien*): the "original vision," in guaranteeing the pristine evidence of "things themselves," re-establishes the connection between consciousness and the world, temporality and spatiality, the historicity of lived experiences (culture) and the hyletic nature of things (material nature) in the dialectics of the "internal-external," "egological pole-world," "immanence-transcendence" relationships that we can identify as a consequence of the theoretical fulcrum of Husserl's skepsis. The goal is still that of safeguarding knowledge from the dangers of naturalistic objectivism, which aims at domination (*ge-stell* in Heidegger's sense) over a nature permeated with the forces of entropy.

Modern natural science, by making Thales' rule for the investigation of nature its own, *iuxta propria principia*, increases the will to dominate the universe, assured by the reinforcement of the objectivistic paradigm and by the oblivion into which the subject 's vitality has sunk. The "man-nature" relationship, a condition of the experimentability of the world, is dangerously weakened giving rise to that vertical split between the two moments which contemporary Western civilization can no longer manage to put back together.

The effect of the naturalistic option of modernity has, therefore, been the loss of the intimate nature of things to the exclusive advantage of a process which renders the objective/factual data of natural, the universe (the worldness of the world), the absolute processes of which are now met through the mediation of the artificial, of *techne*. The vital "spiritual" force of nature, which includes within itself innumerable immaterial virtualities, is mechanically subjected to the rule of a technocratic rationality in determined pursuit of the creation of infinite technico-structural units produced by a naturalized consciousness, extraneous to its own specific nature. Husserl's proposal for the re-founding of the cultural project so that it is able to re-establish a balance between nature and culture and protect from the temptations of the naturalistic *Weltanschauungen*, represents the most rigorous effort made in contemporary philosophical science to repair the gaping rents which have appeared in the universe of Western values.

In *Philosophie als strenge Wissenschaft*, the programmatic manifesto of Husserl's phenomenology, we read:

> The characteristic of extreme and consistent naturalism, from popular materialism to the most modern sensorial monism and to energetics, lies in the *naturalization of consciousness*, on the one hand (the naturalization of consciousness, immanent-intentional units included), and in the *naturalization of ideas* on the other hand with the consequent naturalization of absolute ideas and of norms.[2]

This statement throws into relief the awkwardness with which naturalism has subjugated the sphere of the conscious as to the source of intentional acts and with that the one-dimensional character of the theoretical proposal responsible for the self-destructive results of the "European sciences." Husserl repeatedly restates the central problem of classical Western philosophy, that of the universal foundation for knowledge, for science, for truth. But the thinker of Freiberg evaluates this problem using renewed methods and language, thus evoking a kind of radicality which brings all forms of knowledge into play. This type of radicality found its reply, from the end of the Renaissance, in the scientific-naturalistic standard, the dispenser of the successes of the new discoveries and of that insane certainty with which the new science was gradually to push Western man towards "liberation" from metaphysics and philosophy *tout-court*.

Husserl specifically says:

> it is precisely in the energy with which naturalism seeks to create the principle of a rigorous scientific method in all spheres of nature and the spirit, in theory and in practice . . . [that] its merit lies . . . as does its greatest power in our age. Is there perhaps in all of modern life a more potent idea, which progresses more irresistibly, than the idea of science? . . . The idea of science, thought of in its ideal perfection, is no more then than that reason which cannot recognize any authority above or beyond itself.[3]

The discursive reason which natural science assumes as its own is repudiated by Husserl in that it is abstract and detached, unable to take account of the origin of the sense formations which are the basis of all of the procedures of meaning-bestowal. Concrete culture, or that collection of rules and ways of being accepted in the immediacy of the life-world, finds in lived experiences the locus of its own foundation of sense. Scientific laws, moral ideals, and the social rules governing behavior must seek legitimization in the pre-scientific and ante-predicative dimension of experience.

The "crisis of reason" with which Husserl found himself grappling in the course of his experience as a thinker basically constitutes a confirmation of the petering out of the Western thrust towards *Kultur*, which has been reduced in importance by a naive faith in *Zivilisation*, the bearer of empty norms and rigid forms. It is precisely that freezing of "lived experiences" (*Erlebnisse*) in the categories of a rootless rationality which is at the base of the divergency between life and thought and the pathological disassociation which has accompanied European civilization in its march towards self-destruction. A profound dichotomy between "legality" and "legitimacy" characterizes our epoch, permeated as it is with outbreaks of excess that have uncertain or nihilistic outcomes. The up-to-date nature of a "dated Nietzsche," with all that accompanies his delirious and extravagant proposals, or the increased theoretical interest in Heidegger's "ontological difference," give an exact measure of the course followed compulsively by the philosophy of crisis.

Heidegger's *Überwindung* ("going beyond") or Derrida's "metaphysical deconstruction" leave only few openings towards a rebirth of faith in philosophy as a rigorous first science.

Looking on this desolate and resigned panorama of *décadence* makes the constructive message of phenomenology stands out with even greater authority, and it confirms the richness of Husserl's teaching. The Czechoslovakian phenomenologist J. Patocka states:

> The Nietzschian attack on contemporary European civilization which its author deems nihilistic, is, nevertheless, nihilistic itself, and it considers the passage towards nihilism to be a proof of its own sincerity and merit. Such a radical stance remains radical to this day — while the titanic action of individuality seems merely comic; and his criticism of progress and the Enlightenment as forms of crypto-nihilism is still valid. Consequently the diagnosis of nineteenth century European society as nihilistic sums up all the crises of the age: the political and social crises have their roots in the moral crisis.[4]

It seems to us that phenomenology is all the more effective as a remedy against the definitive loss of the inheritance of our culture, it being rooted in classical Hellenism and the great problems to which that movement brought our attention. The thematic treatment of "original being," of the concreteness of the *logos*, of the epistemic foundations of knowledge eludes the fragile grip of the vacillating strategies of "weak thinking." These things require that *vis theoretica* of which phenomenology has become the upholder thanks to a methodology which commences at the roots in an effort to remove presuppositions, prejudices, or preconceptions which may obfuscate clarity of meaning in any way whatsoever. On adopting phenomenology as a general theory of the life-world, the distinction between the abstract legality of norms and the concrete legitimacy of behavior becomes easily comprehensible.

The divergency — postulated above — which Mann saw between *Kultur* and *Zivilisation*[5] thus assumes a clear philosophical relevance which goes beyond any literary tonality. It opens up interesting perspectives on the field of the sociology of consciousness, on political science, and on the general theory of law. As V. Belohradsky tells us:

> Phenomenology is a radical criticism of European reason in the sense that it refuses to recognize any pretensions to legitimacy until such pretensions are founded on what Patocky calls "that reflection upon life from which all knowledge springs," and that is upon the reflection on the necessary content of every personal consciousness.[6]

The civilization of crisis with which Husserl came to grips, is the result of a profound "de-legitimization" of discursive reason and a fundamental mistrust of philosophy. It seems, that is, that Western knowledge is no longer able to bear the "challenge of complexity" thrown down by the mathematical sciences and the applied technologies. Faced with a marked destabilization of the *logos* and of civilization, the emergence of discontinuity and of the hyper-segmentation of knowledge determines the failure of the unifying perspective of theory to come up to the mark. A new sort of entropy pervades the spaces within our culture producing the pulverization of disjointed branches of knowledge, multi-perspective polycentrisms, the dismembering of meanings. Western man's falling back on theory and practice is in fact measurable in certain forms of individual and social pathogenesis, in the neurosis of our epoch which has forgotten where its own origins lie. E. Paci, thus, writes:

> The sciences are in crisis precisely because they exist in an irrational human society which makes use of rational means, that is to say, the scientific method reduced to the level of pure instrumentality.[7]

The gap between the rationality of working methods and the irrationality of the regulatory principles of our civilization represents the real element behind the ethical-existential malaise afflicting the late-modern age. This malaise Husserl diagnosed with the perspicacity characteristic of a thoroughbred thinker who is not prepared to offer any new hyper-rationalistic or anti-rationalistic therapeutic panaceas.

It would have been too simplistic to join the chorus of "complaint" sent up by the masters of suspicion, who surrendered to the victory of "anti-reason," and negative thinking which they saw rising phoenix-like from the ashes of metaphysical reason. Instead phenomenology is one of the few voices which, amid the disturbing scene of philosophical raving, may still be heard to endorse the possibilities of classical reasoning. In the presence of an increasingly steadfast apotheosis of the negative, Husserl testifies to the radicality of rigorous philosophizing in no ingenuous manner, aware as he is of the problem posed by the primary foundations of research. There are two theories of knowledge, therefore, which now dominate the world stage unchallenged and which battle for the hold on legitimacy. The scientistic ideology engaged in renewing the sacrificial rites of the "progressive progress" of scientific rationality, on the one side, and the catatonic impotence of the irrationalistic solutions, on the other, both paint a picture of grave uncertainty on the theoretical plane which is not undone by insecurity in the sphere of behavior. The *naif* optimism of the first and the disconsolate pessimism of the second give value *a fortiori* to Husserl's proposal for a program marked by great theoretical commitment and high moral significance. In my opinion, since the value of these proposals remains clear to this day, phenomenology as a general theory of culture represents an ever so topical chance to revisit and rethink theory.

Given that Husserl was, above all, concerned with subjecting Western science to a thorough review capable of investigating all the formations of sense, the inheritance which we must gather from the lessons of the phenomenologist must take the form of a determination to deepen the new possibilities of phenomenological theory in relation to new fields of knowledge.[8] The sciences of the psyche, as we have already observed, have made use of phenomenological research in relation to the themes of "comprehension" and "corporeity." The themes related to the life-world (*Lebenswelt*) certainly await new explanations through a comparison between facts and values, things and norms, cognitions and behavior. But, also, the sciences of man engaged in investigating the bonds linking a culture and the terrain in which it is rooted (as a bulwark against the "atopia" or "de-territorialization" typical of the "global village" evoked by Marshall McLuhan), geographical space and its "material lived experiences," the metaphor of "another place rediscovered," can all reveal useful methodological paths. The notions of "geographical landscape" and/or "historical landscape" themselves are enriched with an unusual vein of significance when the referents of the signs resulting from the interaction of and the contributions made by material cultures are deciphered.

The paths taken by Merleau-Ponty and Heidegger reveal interesting approaches to this kind of problem. It is sufficient to think of the Introduction to *Phénoménologie de la perception*, where the French philosopher contrasts an abstract, scientistic notion of geography with the environmental lived experiences which ought to constitute the legitimization of its sense.[9]

Through Heidegger's intuitions it is possible to follow a guiding thread to a view which tends to see cultural lived experiences bound to the spatiality of the world as an onto-phenomenological problem. For example, we may consider the following notes from the Darmstadt conference on "Constructing – Living – Thinking":

> In no case . . . are number-measurements and their dimensions also to be considered the *foundations* of the spaces and the places which can be measured with the use of mathematics by virtue of the sole fact of their being applicable universally to every extended thing. . . . The spaces through which we move every day are disposed and opened by places. . . . If we pay attention to these relationships between places and

spaces, and between spaces and space, we also find a basis for reflection upon the relationship between man and space.[10]

As can be seen in this perspective too, the interaction between man and space goes beyond considerations of simple measurability deduced from calculations. It excludes every possible recourse to "the essential nature of things" of naturalistic objectivism. The spatiality of experience which, in Husserl's sense, is constituted by the original proofs, represents the ground (*Boden*) of a rigorous resetting of the "geo-metry" and "geo-graphy" of the surrounding experimentable world (*Umgebung*). On this *Boden*, it is possible to construct a new grammar for the interpretation of the world which is still able, because it can dispel the mists of forgetfulness that obscure origins, to grasp the difference between the subjective element of the "life-world" (the concrete *a priori*) and the empty notionism of idealizing "sub-structions." Phenomenology opens up inexhaustible possibilities for "the world-surrounding-the-things-of-the-culture" (*Kultur-Sachen-Umwelt*), which, although it distinguishes between the sphere of necessity (material culture) and that of immaterial good, restates the primacy of that "unity of spiritual life" which is the antidote to the "crisis of the European sciences."

University of Genoa

NOTES

[1] Cf. N. Wiener, *The Human Use of Human Beings. Cybernetics and Society* (1954). Italian translation by Boringhieri (Turin: 1966).

[2] E. Husserl, *Philosophie als strenge Wissenschaft*, Italian translation by F. Costa (Torino: Paravia, 1958), p. 12.

[3] *Ibidem*, pp. 14—15.

[4] J. Patocka, *Saggi eretici sulla filosofia della storia* (Bologna: CSEO, 1981), p. 120.

[5] Cf. T. Mann, *Considerazioni di un impolitico*, in "Opere complete" (Milan: Mondadori, 1968).

[6] V. Belohradsky, *Il-mondo-della-vita: un problema politico* (Milan: Jaca Book, 1981), p. 69.

[7] E. Paci, *Funzione delle scienze e significato dell'uomo* (Milan: Il Saggiatore, 1963), p. 170.

[8] In this case we are referring to bioethics, human ecology, anthropogeography, the semiology of culture, etc.

[9] M. Merleau-Ponty, *Phénoménologie de la perception*, Italian translation by A. Bonomi (Milan: Il Saggiatore, 1980), p. 17.

[10] M. Heidegger, *Vortrage und Aufsatze*, Italian translation by G. Vattimo (Milan: Mursia, 1976), p. 104.

PART TWO

HUSSERLIAN AND POST-HUSSERLIAN APPROACHES TO AESTHETICS

SAID TAWFIK

THE METHODOLOGICAL FOUNDATIONS OF PHENOMENOLOGICAL AESTHETICS

It is not the purpose of this study to rethink the phenomenological method itself, but to consider the way it is applied to the field of aesthetics, and to bring out the main problems and results of that application. In other words, we shall attempt to review the methodological foundations on which phenomenological aesthetics relies when confronting art phenomena.

It is known that Husserl's main contribution to phenomenological literature lies in his originating and outlining a method for thought and philosophizing. His main purpose was not to give us philosophical instructions, but to teach us how to philosophize. On the other side, one can say that the main purpose of Husserl's followers lies in their elaborating and employing this method in various fields of philosophy. The field of aesthetics in particular has become one of the most fruitful areas in which the seeds of the phenomenological method have been sown.

There is a good reason why the field of the phenomena of art was convenient and fertile ground for the employment of the phenomenological method. We should remember in the first place that the phenomena which phenomenology studies are not something other than conscious experiences, for phenomena are what appears to consciousness. That is why intentional experiences, of which the main feature is "a consciousness of" or "an appearing to" are called "phenomena."[1] Similarly, art phenomena are experiences of a certain type of objects which we call aesthetic objects. We should also remember that phenomenology presents itself only as a method for reformulating our perceptive and imaginative experiences and, in general, all of our conscious experiences. Since artistic or aesthetic experiences (which may appear in a creative aesthetic process or in a perceptive aesthetic experience) are among these experiences, it was natural that an intimate encounter between art as a domain of experience and phenomenology as a method for understanding experience would occur — even though it was an encounter which came later than expected.

It is worth noting here that there is a certain affinity between

A-T. Tymieniecka (ed.), Analecta Husserliana, Vol. XXXVII, 109–121.

aesthetic vision or experience and the phenomenological method itself. Husserl himself noticed that the way in which the phenomenological reduction proceeds is an analogue of the core of the aesthetic experience.[2] However, this issue lies beyond our purpose, and it is more important for our purpose to turn now to consideration of the way in which phenomenology deals with artistic phenomena, of what distinguishes phenomenological aesthetic research from traditional aesthetics.

Undoubtedly, the methodological character of the research is the central and main feature which distinguishes the phenomenological approach to aesthetic phenomena from the traditional aesthetic attitudes, for the principles on which phenomenological aesthetic research relies are derived from the phenomenological method itself, while traditional attitudes rely on non-methodological principles derived from the metaphysical attitude which the philosopher adopts. In considering aesthetic phenomena or any other phenomena, the phenomenologist does not proceed from a theoretical framework or with any philosophical presuppositions, but he rather proceeds in his research from what we may call "vacant methodological frames," i.e., frames which have no pre-determined content.

Accordingly, phenomenological aesthetic research dismisses all the metaphysical interpretations of art and rejects any interpretation that determines a priori criteria for aesthetic experience, for such interpretations are not methodologically abstracted, verified, and refined through descriptive analysis, but, rather, serve metaphysical purposes. This holds good for the Kantian critique of aesthetic judgement, the Hegelian dialectical interpretation of art, and the contemplative intuitive aesthetic attitudes of Schopenhauer, Bergson, and Croce. Such systems were subordinated to the metaphysical foundations on which they rely. In the faculty of feeling Kant has found a mediate link between the domain of necessity and that of freedom. Hegel's view of art as a sensuous manifestation of the Idea was demanded by his metaphysical dialecticism, since he considered art as a moment in the temporal process of the spirit. Schopenhauer found in aesthetic experience pure intuitive metaphysical contemplation free from the servitude of the restless will and the limitations of time, space, and causality. In agreement with Schopenhauer, Bergson and Croce considered art as a kind of intuitive apprehension in opposition to the logical and scientific knowledge which depends upon abstract concepts. In spite of the fact that these attitudes are completely different in their details, there is a common feature which connects them, that is, each of them was the product of the systematic view of the world which the philosopher's imagination and contemplation drew. This also may hold good for the symbolic attitude which interprets art as part of a network of symbols which mediates between man and the world with its objects, and for the pragmatic attitude which derived its conception of aesthetic experience from a predetermined meaning of experience in general and, thus, it could not understand aesthetic experience apart from the context of everyday life or practical experience.

Generally speaking, phenomenology rejects such attitudes because they adopt interpretations that depend upon presuppositions, and instead of seeing the meaning in experience, they impose meaning on experience. This does not mean, however, that these interpretations are necessarily false and of no value, for they, undoubtedly, contain many correct and careful insights concerning the metaphysical significance of art and aesthetic experience. It is worth noting here that phenomenology also tries to reveal the metaphysical significance of art and aesthetic experience. In proceeding in this task, however, phenomenology does not rely on any metaphysical foundations, but only on what can be seen in, and only in, experience. Thus, one can say that phenomenology does not reject these interpretations of traditional attitudes for their content, but for their lack of methodological procedure. Consequently, this phenomenological rejection does not mean that phenomenology tries to refute these attitudes or to combat them drawing upon proofs and demonstrations, for what matters to phenomenology is not an interpretation or a point of view which has to be demonstrated, but rather different point of departure. Phenomenologically, this means that phenomenology "brackets" these attitudes.

It is also important for our purpose to note that the phenomenological rejection of the attitudes in question due to their lack of methodology does not imply that any methodological attitude or theory is accepted by phenomenology. Methods of research into psychological attitudes, for example, are not the methods which phenomenologists intend when they look forward to making aesthetics a methodological science. Psychoanalysis and experimental psychology are in their approaches two main examples of these attitudes rejected by phenomenological aesthetics:

Within the psychoanalytic attitude, the process of artistic creation is interpreted by probing to the unconscious life of the creator, and the work of art is considered as a kind of dream or an unconscious explosion that reaches into conscious life, that is, as a kind of projection or sublimation of the content of the unconscious. According to the phenomenological standpoint, this attitude confines itself to the borderlines of the creative process and neglects the aesthetic experience as appreciation, which is a main concern of phenomenologists working in the field of aesthetics. Even with regard to the interpretation of the creative process itself, the psychoanalytic attitude commits a great error when it interprets the work of art with respect to the artist and his private life, that is, to an empirical ego or subject. Thus, it transcends the given, that is, the work of art, and reaches to something which lies beyond our experience; it transcends the aesthetic and reaches to the non-aesthetic. If we are to stay in the domain of what is given to our experience, we should consider the artist solely through his work; for the artist, according to phenomenological aesthetics, should be nothing more than his artistic product. We are to then recognize his intention or his phenomenological ego through his work, through the given which is the artist's intention as it is manifested in a physical expression. In reading Stendhal, Merleau-Ponty says, "I get closer and closer to him until in the end I read his works with the very same intention that he gave to them."[3] This means that I recognize the writer's (or the artist's) intentions, his vision of the world and his style of expression or creation, only through the context of his creative product. Consequently, there is no justification for assuming two kinds of intentions, one of which translates the other, for in that case we would assume two subjects: one of them would live outside the work of art, that is, the artist with his psychological intentions contained in his fancies, motives, and motions, and the other would be the artist as he expresses these psychological intentions. Generally speaking, the interpretation of the work of art or the aesthetic process through examination of the life of the artist or of his private experience is a matter which phenomenological aesthetics dismisses from the field of aesthetic research and "brackets."

On the other hand, the attitude of experimental psychology has attempted to constitute an aesthetics, in opposition to the contemplative aesthetics which relies on general theories and metaphysical presuppositions, on an empirical method which begins with particular facts from which it abstracts through generalization laws which interpret aesthetic phenomena. This attitude prides itself in having presented a scientific descriptive study derived from experiment. But this is exactly what phenomenological aesthetics rejects, for it does not seek to interpret aesthetic experience, or aesthetic phenomena in general, through generalizations experimentally abstracted, but rather through insight into the essences of aesthetic phenomena. According to phenomenology, the scientific character proper to human sciences is not be attained by following the methods of the exact sciences, for the empirical science of facts should be based on the science of concrete essences. Thus, phenomenological psychology, which is an a priori pure science of essences, is, as Husserl taught us,[4] the basic methodological foundation on which a rigorous experimental psychology can be based.

Above all, it is the characteristic of the phenomenological aesthetic that it does not take its point of departure from any relative subjectivism, whatever its form may be. It, rather, takes its departure from the modest simple fact that there are, on the one hand, works of art, and, on the other hand, experiences of these works. Now, the main question which the approach of phenomenological aesthetics tries to answer is: How do works of arts occur in our experience? However, putting the question as such, is not enough to determine the point of departure of phenomenological aesthetics, for the same question may arouse certain problematic issues. For example, it is taken for granted that aesthetic experience comes to be in a culture world; for works of art live and manifest themselves in that world, and we learn how to recognize and appreciate them through that world in which we live. This means that works of art as cultural products vary according to the variation of the cultural world not only in different ages, but also in one and the same age. This also holds good for one and same work of art, for, as Ingarden remarked, "any work of art (and this operates differently for the different arts) passes through various periods of brilliance, that is periods in which it attracts frequent and correct aesthetic concretions, and other periods when its attractiveness is weakened or even disappears if it is no longer 'legible' to its public."[5] All these facts are true, and phenomenologists acknowledge them. However, they do not see in these facts any justification for adopting historical relativism; for, as Dufrenne puts it, ". . . the historicity of artistic production, the diversity of art forms and of judgments of taste, no more imply a relativism destructive of an eidetic of art than the historicity of the ethos implies a similar relativism in Scheler's eidetic of moral values."[6] The same observation was repeatedly made by Ingarden in several places in his writings. He asserts that the diversity and the deviations in understanding and appreciating the work, and the alternate periods of obscurity it passes through which do not imply that the value of the work is relative or that the work is then deprived of its value, but that this only means that the work has been manifested in false concretions. He declares: "A literary work of art can be expressed for centuries in such masked, falsifying concretizations, until finally someone is found who in one way or another understands it correctly, who sees it adequately, and who shows its true form to others."[7] This means that the work of art, if we are to use Heideggerian terms, still "awaits" those who can understand its truth and "preserve" its essence.

Thus, phenomenological aesthetic research does not take its point of departure from any historical relativism; for "what we discover in history," as Dufrenne puts it, "... is not altogether historical. We are persuaded of this by art itself, which is perhaps a more universal language than rational discourse, and does its best to deny the kind of time in which civilizations perish."[8] This means that phenomenological research tries with the help of an eidetic elucidation of the empirical facts, that is, of the works of art, which manifest themselves throughout history to discover the essences of art phenomena which manifest themselves beyond the limits of history. Thus, the process of ideation begins with the empirical fact only insofar as it is an example through which we can achieve insight into its eidos.

In addition, phenomenological aesthetics rejects relativistic subjectivism in its psychologistic form. The traditional form of psychologism in the field of aesthetics tended to consider aesthetic experience as mere aesthetic enjoyment which involves pleasures, emotions, and pleasant experiences. This enjoyment, according to psychologism, is a condition for the judgment of taste or the achievement of aesthetic experience. In opposition, phenomenologists acknowledge the fact that this enjoyment and experience of pleasure actually occurs in our aesthetic experience, but, nevertheless they do not see in them any condition for our aesthetic judgment or experience. For aesthetic experience is not a psychic subjective process which aims at arousing our enjoyment; rather, this enjoyment is something aroused as a result of our experience of the aesthetic object, and which is to be described only on that ground. Thus, the concept of aesthetic enjoyment in phenomenological aesthetics is radically different from the traditional subjective psychologistic form of the concept. Moritz Geiger's *Beiträge Zur Phänomenologie des aesthetischen Genuses* is considered the pioneer contribution to phenomenological aesthetics in this respect. Through an eidetic descriptive analysis, Geiger was able to refine and distinguish the feelings which occur in aesthetic enjoyment from those of pleasure, joy . . . etc., and to dismiss from aesthetic feeling the non-aesthetic or pseudo-aesthetic emotions which occur in the experiences of the insensitive and unqualified observer. This is why Geiger rejects the experimental aesthetics, closely related to psychologism, which assumes that every aesthetic experience is a kind of enjoyment, whatever this enjoyment may be, and which, thus, neglects the fact that the enjoyment which occurs in our aesthetic experience can only be recognized as enjoyment of an aesthetic object. Despite the fact that Geiger's contribution was partial in comparison to other phenomenological treatments which transcend the limits of aesthetic feeling or enjoyment and extend to other experiences which add to the texture of aesthetic experience,[9] it offered a good model for these other treatments which, then approached this issue in a similar manner. Like Geiger, other phenomenologists reject the psychologistic perspective on aesthetic enjoyment. For example, Ingarden stresses the intentional character of aesthetic feeling, that is, feeling which is directed toward the aesthetic object; and Dufrenne stresses so extremely the non-psychologistic nature of aesthetic feeling that he speaks of this feeling as a feeling of the aesthetic object itself.

However, the fact that phenomenological aesthetics does not consider the aesthetic object with respect to our subjective enjoyment or feelings does not imply that it attempts to understand the aesthetic object through an objective perspective of the beautiful as if the beautiful were a rational concept or an abstract idea which may be classified in categories according to objective criteria. Phenomenological aesthetics does not appeal to the idea of beauty and of the beautiful as a point of departure for aesthetic research; for this procedure is, in fact, the way of traditional aesthetics which was not free of the philosophy of art, that is, from those philosophical concepts concerning art, beauty and the beautiful. That is why Dufrenne says: ". . . it is not we who decide what is the beautiful. The [aesthetic] object itself decides."[10] This means that we can recognize the beautiful only through our recognition of the essence of the object in which the beautiful manifests itself. Phenomenologically, the recognition of the essence of the aesthetic object means the analysis and description of its structure, and this structural analysis will supply us with an "objective" foundation for the description of our "subjective" experience.[11] Only in this sense can we understand the meaning of "objectivity" in phenomenological aesthetics: it is the objectivitity of aesthetic research, and not the objectivity of categories or a priori criteria which determine what is beauty or the beautiful. In fact, considering the beautiful according to pre-determined categories or criteria will involve us in either subjectivity or objectivity, for these criteria will fall into two types: we will judge the beautiful according to either emotional subjective elements or according to elements independent from the aesthetic object and our experience of it. In the first case, we will be involved in an exercise of psychologistic subjectivism, and in the other, in one of dogmatic or naive objectivism.

It is precisely from this point that we can arrive at the main starting point of phenomenological aesthetic research, a point which is radically different from that of traditional aesthetic research in all of its philosophical and psychological forms. This fact may be demonstrated as follows.

It is the character of phenomenological aesthetics that it takes its point of departure from the phenomenological epistomology which attempts to find a third way beyond subjectivism and objectivism in their traditional forms, a way which transcends the dichotomy of subject and object. This means a reformation and a new understanding of experience. Husserl himself criticized modern philosophy which lost understanding of human experience when it "forgot" the experience of the "life-world" (*Lebenswelt*), and thus forgot man as a human-being in the world. In fact, the idea of the experience of the life-world is rooted in the origin of the idea of the intentionality of experience; for both of them spring from the same origin, that is, from the principle of transcending subject/object dualism and the distinction between inner and outer experience, the distinction to which, as Landgrebe puts it, "'the confusions' in philosophy and especially in the theory of knowledge should be traced."[12] But where the idea of intentionality represents the phenomenological principle on the level of consciousness in its relation to the world with its objects where consciousness is vacant and contains nothing but this movement (which makes it ever directed to its objects through an operative intention which tries to understand them), the idea of the life-world represents this principle on the level of the experience of the pre-given world which, if we are to use Merleau-Ponty's words, "is always 'already there' before reflection begins — as an "inalienable presence . . . ."[13]

It was natural that this epistomological difference between phenomenology and traditional philosophy would extend to the treatment of aesthetic experience. In this respect, one can say that phenomenology no longer considers aesthetic experience from either the side of the subject or that of the object but that it understands it, as it can only be understood, through the mutual implication of the subject and the object. With respect to the object, phenomenological research into aesthetic experience breaks entirely with that sort of research which pays no attention to the subject, i.e., it breaks with research into objective characteristics and criteria of the beautiful or of what may affect us as being beautiful. But phenomenological research, at the same time, rejects research that neglects the aesthetic object when considering the subject's role in the aesthetic process whether on the level of aesthetic perception or that of aesthetic creation. In a previous study, "Schopenhauer's Metaphysics of Art," we have shown that the metaphysical attitudes which have considered art to be a mere vision and the work of art to be equivalent to the vision of the artist, not only neglect the aesthetic object as a style of expression which aims at communicating the work to the observer, but also do not do justice to the role of the observer and confine it to the mere reproduction of the artist's vision. In this case, the aesthetic object disappears and its being is lost between two subjects which are, in fact, one subject that has a double face: the one producing a vision, and the other reproducing it; and thus the aesthetic object with its appearance, sensuous surface, and its potential meanings disappears and is degraded to serving as a mere medium or a stimulus arousing the observer to grasp the artist's vision.

In contrast, phenomenological aesthetics transcends both perspectives: the purely objective perspective which neglects the subjects' role, and the subjective perspective which neglects the object — and which we may call "the static subjective perspective," for this later perspective supposes that the work of art is a product of a process of "static creation" on the level of vision or imaginative contemplation, and considers the contemplation of this work to be the contemplation of a merely receptive subject. By transcending the two perspectives, the subject becomes coherent with the object in aesthetic experience, for the aesthetic object becomes that object which is constituted through the experience of the observer, and not the object which is constituted independently and prior to the experience of the perceiver. In this case, the subject has an effective role, and an integration between the role of the artist and that of the observer in aesthetic experience, or between, to use Ingarden's terms, the transmitting intention of the artist and the receptive intention of the observer, becomes apparent. The reception of the work by the observer is no longer a mere reception, but an *intending* reception, i.e., a reception of an experience directed to its object which needs that experience in order to be constituted and manifested. The subject, as well, now retrieves its value and effectiveness when the object retrieves its value and is no longer a mere stimulus, but a crucible in which the subject of the artist and the subject of the perciever are fused, or it becomes, at least, a point at which an intimate encounter between the two intentions is achieved.

Now, we may have observed that the epistomological principle from which the phenomenological treatment of aesthetic experience take its departure is itself the principle which transcends the dualism of subject and object in the understanding the nature of experience in general, and that it is also the idea of the intentionality of consciousness, and that it is finally the idea of the lived experience. Thus, this principle has sometimes appeared in the treatment of aesthetic experience embodied in the intentional relation between consciousness and its object, as we find in Ingarden's treatment of aesthetic experience as an experience of a pure intentional object and in Sartre's treatment of the experience of the work of art as an imaginative intention of an unreal object. At other times, this principle has appeared embodied in the idea of the lived experience, i.e., primary direct experience prior to the act of conscious reflection, the pre-reflective experience of the living and concrete present. This has apparently been the approach taken by Merleau-Ponty and Dufrenne, for they understood aesthetic experience as an experience which starts in the direct presence of its aesthetic object before the body and perception.[14] However, it is worth noting here that the idea of the lived experience is not distinct from the idea of intentionality in its deep sense, but is a development of its significance and a transcending of its limits within the level of consciousness to reach the level of the act and dynamism.

Phenomenological research into the level of the act and dynamism has become a necessary for an adequate understanding of the creative process. In spite of the fact that, aesthetic experience was a common concern of phenomenologists working in aesthetics, there has been a considerable lack in phenomenological literature treating the aesthetic experience as an act of creation. This lack may be ascribed to the fact that most phenomenological aesthetic research has been mainly devoted to the treatment of the work of art as an intentional object or as a correlate of the experience of the receiver. In her criticism of Ingarden's aesthetics, A-T. Tymieniecka has observed that the mere intentional analysis of the work of art and of the corresponding aesthetic experience fails to grasp the creative process that engendered and produced the work.[15] According to Tymieniecka, the treatment of the work of art as a mere receptive experience "can not take into account, nor does it allow an approach to the work of art directly from the point of view of the *act of creation* of the work."[16] However, the problematic question which can be raised at this point is: can we overcome the fact that the work of art is *what is directly given* to our experience and undertake an approach to the work *directly* from the point of view of the act of creation which is not directly given to our experience? To state the question briefly: can we understand the given through the ungiven without any deviation from precise descriptive analysis. It is quite clear that the work of art is a noematic correlate of a first creative act. However, the noematic correlate, as repeatedly asserted by Husserl, has priority within the phenomenological intentional analysis.[17] In fact, we agree with Professor Tymieniecka that the creative process should be interpreted on the level of the phenomenology of the act. And it is my contention that Merleau-Ponty was able to supply us with many insights concerning the creative process only because he deals with this process in the light of his phenomenological treatment of the body as a mobile animate apparatus of intentions and acts[18] (a treatment which as it has been developed by Dufrenne has been devoted mainly to the analysis of the experience of the perceiver). But it is also my contention that the treatment of the creative process should be based on an eidetic structural analysis of the work of art, an analysis which can guarantee a safety factor and a rigorous foundation for any further treatments; for the question with which phenomenological aesthetics should begin is: *What* is the nature, i.e., the essence and the structure, of the *created entity* which we call the work of art? On the basis of the answer of this question, we can answer another one: *How* is this entity *created*?

Without doubt, the development of phenomenological aesthetic research deserves a separate study. We only hope that, our study has laid bare the main foundations on which phenomenological aesthetics builds.

Cairo University

NOTES

[1] E. Husserl, "Phenomenology," in *Husserl's Shorter Works*, Peter McCormick and Frederick Eliston, eds. (Notre Dame, Indiana: University of Notre Dame Press, 1981), p. 22f.

[2] For a description of the affinity between the aesthetic perception and the phenomenological approach, see E. Husserl, *Ideas* . . . , trans. Boyce Gibson (London: Allen and Unwin, 1931), p. 311, and pp. 200—201. See also Fritz Kaufman, "Art and Phenomenology," in *Essays in Phenomology*. Maurice Natanson, ed. (The Hague: Martinus Nijhoff, 1966), pp. 144—156.

[3] Maurice Merleau-Ponty, *The Prose of the World*, ed. Claude Lefort, trans. John O'Neill (London: Heinmann Educational Books, 1973), p. 12.

[4] Husserl, "Phenomenology," *op. cit.*, p. 22.

[5] Roman Ingarden, "Artistic and Aesthetic Values," in *The British Journal of Aesthetics*, Vol. **4**, No. 3, 1964, 201.

[6] Mikel Dufrenne, *The Phenomenology of Aesthetic Experience*, trans. Edward Casey *et al.* (Evanston: Northwestern University Press, 1973), p. liii.

[7] Roman Ingarden, *The Literary Work of Art*, trans. George G. Grabowicz (Evanston: Northwestern University Press, 1973), p. 240.

[8] Dufrenne, *op. cit.*, p. liv.

[9] It is worth noting here that aesthetic experience, as the phenomenological aesthetics of Ingarden and Dufrenne demonstrates, is not a single experience, but a complex of experiences tightly interwoven, and the term "experience" is often used in its single form for the sake of simplicity.

[10] Dufrenne, *op. cit.*, p. lxii.

[11] According to phenomenology, every experience is subjective, not in the sense that the object of this experience is *merely personal or relative* to a subject, nor in the sense that the object is *dependent upon* a subject, but only in the sense that it is *related to* a subject. (For an extended analysis of this point, see Herbert Spiegelberg, "How Subjective is Phenomenology?" in *Essays in Phenomenology*, pp. 137—143.)

[12] Ludwig Landgrebe, "The Phenomenological Concept of Experience," *Philosophy and Phenomenological Research*, Vol. **34**, No. 1, 1973, p. 5.

[13] M. Merleau-Ponty, *Phenomenology of Perception*, trans. Colin, Smith (New York: Humanities Press, 1962), p. vii.

[14] However, this is represented by Dufrenne only on the level of the aesthetic experience of the spectator, in its first moment.

[15] A-T. Tymieniecka, "Beyond Ingarden's Idealism/Realism Controversy with Husserl – The New Contextual Phase of Phenomenology," in *Analecta Husserliana*, Vol. IV, pp. 331—334.

[16] *Ibid.*, p. 332.

[17] See for example: E. Husserl, *Cartesian Meditations*, trans. Dorion Cairns, §§17, 20–21.

[18] However, the main defect in Merleau-Ponty's treatment of aesthetic phenomena lies in the fact that he enclosed himself within the limits of his phenomenology of the body and perception, and thus he confined the structure of the work of art to a mere sensuous gestural expression, and could not understand the compound character of aesthetic experience whether on the level of creation or the level of appreciation. In his last writings, especially in *Eye and Mind*, Merleau-Ponty was trying hard to overcome this position through a bestowal of entological dimensions upon his philosopohical approach.

FILIPPO COSTA

BILD UND KUNST IM HUSSERLS NACHLASS

1. PHÄNOMENOLOGIE UND BILDESPHÄNOMENOLOGIE

Vorliegender Abhandlung liegt die Ueberzeugung zugrunde, daß von einem streng phänomenologischen Standpunkt aus kein vorausgesetzter Schichtunterschied zwischen setzendenden und wertenden Akten bestehen kann. Konsequenterweise sollte der klassisch gewordene Gegensatz von "bestimmenden" und "reflektierenden" Urteilen entschärft werden. Es gibt also keine *Vor*-Behauptungen über das Sein und Sosein der Dinge und keine *Nach*-Behauptungen über ihren Wert bzw. ihre Schönheit, sondern nur einen weiten Bereich von Bewußtseinsmodalitäten und von entsprechenden Inhaltsmodifikationen, unter denen sich Bildungsintention und Bild selbst befinden.

Die Phänomenologie des Bildes ist daher nicht etwa als Anwendung auf dem besonderen Bildbereich einer vermeinten "Ersten Phänomenologie" zu begreifen, sondern sie erweist sich als eine Seite der Phänomenologie *tout court.*

Von dieser einleitenden Stellungsnahme aus ergibt sich unmittelbar, daß, um unser Thema genau zu behandeln, wir zunächst den Ort untersuchen müssen, der dem Bildbegriff gebührt. Aus welchen Gründen darf man einer derartigen Begriff thematisieren?

Diese Frage scheint eine streng systematische Auffassung der Phänomenologie und ihre ausführliche Einrichtung zu verlangen, was sich aber u.E. dem wesentlich offenen phänomenologischen Forschungsstil als unangemessen erweisen würde. Gewiß hat Husserl immer an der Idee einer systematischen Gesamtphilosophie bzw. an einer allumfassenden Vernunftkritik festgehalten. U.a. behauptet er, daß

> an *erster* Stelle die Probleme einer Phänomenologie der Wahrnehmung, der Phantasie, der Zeit, des Dinges stehen.[1]

Die "Ausschaltung des Seins" andererseits, die die primäre phänomenologische Entscheidung ausmacht, setzt zunächst die Gleichsetzung der Werturteile mit den Seinsurteilen und daher die unbestimmte Oeffnung der phänomenologischen Untersuchung bzw. Beschreibung

A-T. Tymieniecka (ed.), Analecta Husserliana, Vol. XXXVII, 123–145.

voraus. Sprachwissenschaftlich ausgedrückt, beduetet somit die streng genommene, auf die Seinssetzung begründete Ontologie die präjudizielle "Schließung" jeder phänomenologischen Untersuchung unter der Idee einer totalitären "Philosophie als strenge Wissenschaft".

Das Ganze des Husserlschen Philosophierens scheint nun von zwei entgegengesetzten Tendenzen beherrscht zu sein: Einerseits von einer nach systematischen Einschließung alles Wissens überhaupt innerhalb der transzendentalen Phänomenologie, und andererseits von einer nach der immer offenen Beschreibung der geschehenden Erlebnisse, in denen sich die unaufhörliche Tätigkeit der universellen Sinngebung entfaltet. Dieser Tätigkeit zugeordnet, erscheint als besonders bedeutsam die *Methode der Unterscheidung*, worunter der uns anliegende Unterschied der präsentierenden Wahrnehmung und der Phantasie auftaucht, die im Bewußtsein als Vergegenwärtigung charakterisiert wird.[2]

Da wir der ersten metaphysisch-systematischen Wendung der Phänomenologie fern bleiben, sind wir dazu geneigt, Phänomenologie und Bildesphänomenologie wesentlich zu identifizieren, und den Ausdruck "Phänomenologie des Bildes" als eine Tautologie anzusehen, indem Phänomen und Bild nicht als verschiedene Sachen, sondern als letzten Endes austauschbare Begriffe zu betrachten sind, die dasselbe bezeichnen. Es gibt daher keine spezielle untergeordnete *Bildes*-phänomenologie: erst durch die im Bildbegriff konzentrierte Phänomenologie kann das Phänomensein des Phänomens überhaupt ins klare kommen. Es sei nebenbei bemerkt daß die Phänomenologie keine Ontologie der Erscheinung sein kann, da der Erscheinung kein *Sein* zukommt, das seinem Sosein oder seinen Beschaffenheiten nach beschrieben werden könnte. Sie löst sich vielmehr ins phänomeno*logische* Reden von irgend etwas auf. Anders gesagt: Das Reden *von* Erscheinung ist nur als Metasprache möglich.

Damit das Bild nicht in einem widerspruchsvollen An-sich entschwindet, müssen alle phänomenologischen bzw. ästhetisch-phänomenologischen "Erheblichkeiten" des erlebten Bildes dem echten phänomenologischen Objekt zugeschrieben werden.

2. GESCHICHTLICHE BEMERKUNGEN

Bevor wir auf die nachgelassenen Schriften Husserls über die Bildesphänologie näher eingehen, halten wir es für angebracht, einige Bemerkungen über ihren geschichtlichen Ursprung vorauszuschicken. Husserl selbst erklärt, seine eigene Bildesproblematik aus Brentanos Vorlesungen entnommen zu haben. Er hätte nämlich um die Mitte der achtzigen Jahren die von seinem Meister gehaltenen Vorlesungen gehört, die die Verhältnisse zwischen Phantasie- und Wahrnehmungsvorstellungen behandelten. Besonders im Hinblick auf die tiefe Verständigung der ausgezeichneten Bedeutung, die der uneigentlichen Vorstellung innerhalb unseres ganzen psychischen Leben zukommt, fühlte sich Husserl Brentano gegenüber verpflichtet. Diese "Uneigentlichkeit" macht nun den Kern der Husserlschen Frage nach dem Bilde aus, indem sie den psychologischen Standpunkt aufzuheben ermöglicht, demnach es zwischen Wahrnehmung und Phantasie nichts als Stufenunterschiede gibt. In Husserls Worten:

> Der vielverhandelte Streit über das Verhältnis zwischen Wahrnehmungs- und Phantasievorstellung konnte bei dem Mangel an klaren Begriffen und Fragestellungen zu keinen rechten Ergebnis führen. Daß die Aktcharaktere beiderseits verschieden sind, daß damit mit der Bildlichkeit eine wesentlich neue Weise der Intention Erlebnis wird, glaube ich zweifellos nachgewiesen zu können.[3]

Der Besonderheit der Aktcharaktere entspricht die Besonderheit des Objekts; dem Objektsinn des Fiktums aber, den man nur im Vergleich zum Objektsinn des Perzeptums sowie zu anderen unter die Bildesphänomenologie fallenden Modalitäten bestimmen kann, widmet Husserl sein Hauptaugenmerk. Es scheint hier besonders schwierig, sich an die streng phänomenologische Ebene zu halten; da man die "theoretischen Termini" anzuwenden imstande sein muß, die den phänomenologischen Charakter der Untersuchung versichern können. Einer dieser Termini ist die obengenannte Uneigentlichkeit, deren eigener phänomenologischer Zustand noch zu erforschen bleibt. Es ist nämlich zu untersuchen, ob sie aus der phänomenologische *Reflexion* über "Wahrnehmung und Bild" entsteht, oder einen *immanenten* Charakter des mit der Wahrnehmung zusammengehörigen Bildes als solches ausmacht.

Anders gesagt, es wäre zu fragen, ob die (phänomenologische) Irrealität des Bildes aus dem äußeren Vergleich mit der Wahrnehmung entsteht oder eben im Sinn des Bildseins eingeschlossen ist. Diese Frage durchdringt ganz die Husserlschen Betrachtungen, ohne aber zu einem entscheidenden Schluß zu kommen. Die hier zugehörige geschichtliche Hauptfrage betrifft den Sinn und den Wert des von Husserl über Brentano gemachten Fortschrittes, der paradoxerweise in der Vermehrung von Schwierigkeitn und im Ausbruch von Problemen und Aufgaben besteht, die bei dem Denken des Meisters noch im Verborgenen blieben. In der Tat wird durch die neueren phänomenologischen Perspektiven das Bild den Schwierigkeiten ausgesetzt, die anderen phänomenologischen Feldern wie der Ontologie (im Husserlschen Sinn), dem inneren Zeitbewußtsein, dem doxischen Charakter der intentionalen Akte oder dem Verhaltnis zwischen Erfahrung und Urteil usw. anhaften. Husserl selbst verheimlicht die Schwierigkeiten nicht:

> Meine eigene Studien, in die ich mich zumal ein Jahrzehnt später immer mehr verwickelte ... lehrten mich, daß die Probleme noch sehr viel verwickelter und schwieriger liegen, als Brentano sie damals gesucht hatte ... Eine systematische, vollständige Erledigung dieser Probleme konnte mir damals ... noch nicht gelingen. Es hängt mit der innigen Verflechtung und wohl auch mit der Eigenart der phänomenologischen Probleme zusammen, daß sie nicht isoliert zur Lösung kommen können.[4]

Der phänomenologischen Untersuchung ist eine Netzstruktur eigen, die der Darstellung wegen in eine Linearstruktur übersetzt werden muß, vermöge deren die Wechselimplikationen der verschiedenen Themen die logischen Zirkel vermeiden, bei denen die phänomenologische Rede scheitern würde. Auf die hier auftretenden methodologischen Fragen könnte nur dann einige Antwort gegeben werden, wenn erst der logische Zirkel in einem phänomenologischen umgeformt werden kann, dem die obengenannte beschreibende Oeffnung verpflichtet bleibt.

Zur Genesis der Husserlschen Phänomenologie gehört nun eine vom Verfasser so räsumierte Grundkritik:

> Selbst die bedeutendsten Forscher verwechseln konstant die sinnliche Inhalte der Wahrnehmung und der Gegenstand der Wahrnehmung. Durch metaphysische Vorurteile verwirrt wird als Gegenstand der Wahrnehmung ein unanschauliches Ding an sich gesetzt, während der wirklich angeschaute Gegenstand in der theoretischen Betrachtung übersehen und mit dem Empfindungsinhalt identifiziert wird.[5]

Es ist hier zu beobachten, daß der beklagte erkenntnistheoretische Psychologismus, dem die "phänomenologische Blindheit" unmittelbar folgt, jeden Unterscheidungsgrund des Bildes aufhebt, da dieses erst im Gegensatz zur Wahrnehmung bestimmt werden kann, die ihrerseits nur aufgrund des prinzipiell von jedem Inhalt unterschiedenen intentionalen Objektwert bestimmt wird.

Die hier enthaltene Verwirrung ist nach Husserl Brentano schuld, der zwar Gegenstand und Inhalt unterscheidet, unter Gegenstand aber das metaphysische Ding-an-sich versteht und sich so den Gang der phänomenologischen Untersuchung versperrt. Es sollte vielmehr die Analyse der verschiedenen Auffassungsweisen tiefer nachgehen. Dagegen,

> da diese fehlen (und sie fehlen bei Brentano, aber ebensosehr bei den übrigen Psychologen ...), so fehlt auch die Möglichkeit, die Streitfrage über das Verhältnis von Wahrnehmungsvorstellung und Phantasievorstellung methodisch richtig aufzufassen.[6]

Diese kritische Einstellung könnte den falschen Eindruck erwecken, daß die an diese grundlegende Unterscheidung gebundene Bildesphänomenologie als eine rein transzendental-subjektive Untersuchung, d.h. auf die Aktmodi zugewandte fortzuführen sei. Dagegen, wie wir sehen werden, entwickelt sie sich als Analyse des Bildobjekts bis sie die höhere Bildesgegenständlichkeit erreicht. In der Tat, indem Husserl seine Brentanokritik fortsetzt, klagt er darüber, daß sein Meister die

> wesentlichen Unterschiede zwischen Empfindungen und Phantasmen[6a]

nicht anerkennt, da er sie als Gegenstände und nicht als Akte begreift. Natürlich hängt der Unterschied *a parte objecti* von demjenigen *a parte subjecti* (Auffassungsweisen) ab; aber der Bezug der beiden Momente kann nicht zu Gunsten des einen oder des anderen angebracht werden. Dieser Bezug, wie wir bemerkten, muß an der phänomenologischen Ebene bleiben. Besser noch, wenn der Bildtheorie, wie es gewöhnlich geschieht, ein inhaltlicher Sinnunterschied, d.h. ein Lebendigkeits-Intensitätsunterschied, zugrunde gelegt wird, so scheitert die ganze Phänomenologie. Es taucht hier die *Gegenstandsinstanz* auf als Bestimmungsgrund zugleich der Bildesphänomenologie als auch einer Phänomenologie überhaupt. Dieser Instanz hätte eine weitere und tiefergehende Auseinandersetzung mit dem hier angedeuteten geschichtlichen Ursprung zu folgen.

3. BILD ALS OBJEKT

Dem Begriff des Bildes sollen wir nun den weitesten Sinn beilegen, der in sich sowohl das der real-anwesenden Wahrnehmung verknüpfte Bild als auch das durch einen psychischen Gegenstand konstituierte Bild umfaßt; seinerseits nachbildet derartige Gegenstand ein reelles Ding und umfaßt endlich unsere Seh- und Wortdarstellung. Gewiß impliziert die Wahrnehmung ein Bild (das auch Empfindung gennant wird), das letzte impliziert aber für sich allein kein Realanwesendes. Im Gegensatz also zur Wahrnehmung, die uns einer Gegenstand in Person darstellt,

erscheint in der Phantasie der Objekt zwar insofern selbst, als eben er es ist, der da erscheint, aber er erscheint nicht als gegenwärtig, er ist gleichsam so, als wäre es da, aber nur gleichsam, er erscheint *im Bilde*.[7]

Man bemerke hier zunächst, daß nicht vom Bild an sich, sondern vom Im-Bild-Erscheinen die Rede ist, die sich am Anfang als *indirekt* erweist. Dem Bild Objektität zuzuschreiben, bedeutet, es zu einem Wesen zu machen, dessen konstitutive Bestimmungen zu entfalten sind: Das "Wesen" des Bildes ist nur in seinem Bezug zu etwas anderem zu beschreiben. Da es von Anfang an als "sekundäre" Realität erscheint, insofern nicht etwas Anwesendes sondern Vergegenwärtigtes ist, so muß sich seine eigentümliche Unbeständigkeit in dem ihm angehörigen Redeweise widerspiegeln. Eben darum spricht Husserl von "Bildvorstellungen", "Phantasiebilder", usw. anstatt von Bildern schlechthin.

Die Bildesobjektität wird in der Differenz zwischen Gegenwart (bzw. "Gegenwärtigung") und *Ver*gegenwärtigung bestimmt. Nun können wir diese Dififerenz entweder als "Nebeneinanderschaltung" oder als "Reihenschaltung" begreifen, indem wir beide Weisen des Bildes als gleichursprünglich, oder die eine als von der anderen abhängig nehmen. Das Bildhafte ist für sich außerzeitlich und -räumlich; darum spricht Husserl lieber in Ansehung des Objektbildes von *Gleichsam*-Zeit und -Raum. Der Mangel an zeit-räumlicher Realität kann dann entweder als außerliche Beschaffenheit oder als innerlicher *negativer Gehalt* des Bildsinnes bzw. der Bildesintentionalität verstanden werden. Besteht nun der echte Sinn des Bildhaften im Abstrahieren von jeder reellen Ortbestimmung, oder vielmehr in der Spannung, wodurch jene Realität ferngehalten wird? Ist für das Bild die Tat der Irrealität etwas Wesentliches oder nur Zufälliges?

Wenn ich einer Erzählung folge, oder ein Gemälde betrachte, so kann man wohl sagen, daß ich mich nicht darum kümmere, ob der bezüglichen Darstellung etwas Reelles entspricht oder nicht, indem vorausgesetzt wird, daß dem Bildgehalt und dem Bildsein die Realität ganz gleichgültig bleibt.

Es ist auch zu sagen, daß das bildhafte Aesthetische seinen eigenen Wert darin besitzt, daß es im Akte des ästhetischen Genusses ein spezifisch "Irreelles" erzeugt. Es ist dann nicht nur, daß ich der Realität des Dargestellten bzw. Vorgestellten gleichgültig bleibe, sondern daß ich *positiv* seine Irrealität *erlebe*, während der Gedanke eines reellen *referents* den ästhetischen Wert eines geachteten Kunstwerkes verderben würde. Ich *verlange* Realitätslosigkeit.

Andererseits ist es auch denkbar, daß die reelle Setzung von der imaginären abhängt: Die Realitätswahrnehmung würde aus einem reellen und einem hinzugefügten Bilde zusammengesetzt hervorgehen, wobei lezteres auch motivierte doxische Momente enthielte. Die wirkliche Landschaft, die ich z.B. von meinem Fenster aus betrachte, würde *zuerst* ein Fictum und *dann* das Ergebnis einer Wirklichkeitsannahme sein, wodurch das Bild in eine Wahrnehmung umgeformt wird.

Der in Bezug zur Gegenwart gedachte Begriff von Vergegenwärtigung sagt aber noch mehr: Der Gegenwartssinn des Imbildanwesenden setzt eine aktuelle Anwesenheitsmodifikation voraus, d.h. den Akt, wodurch eine *verlorene* oder irgendwie negierte Anwesenheit als Bild (*in Person* aber als-ob) wiederhergestellt wird.

Es gäbe dann einen das bedeutungsvolle Abwesen des ursprünglichen Gegenstands einschließenden Anwesenseindruck, so daß zu jedem Bild gehörte eine im Begriff der Vorstellung als Wieder-herstellung eingezeichnete *Geschichte*. Was immer ich "im Bild" sehen, betrachten oder ästhetisch schätzen kann, das gilt für mich, nur insofern es *von einem verlorenen Anwesen spricht*, auf einer *abwesenden* Gegenstand hindeutet, der *künstlerisch* als abwesend mittels des Prinzips von der *künstlerischer Form* gehorchender Um-und Verformung dargestellt wird. In solchem Fall übersetzt sich die reelle Abwesenheit in die "intensive" Anwesenheit der künstlerichen Darstellung. Eben als ein von Realität *Fernes* hat das gemachte Bild, als anwesendes Ding-Bild, eine höhere Fähigkeit, sein bedeutungsvolles Anwesen auszudrücken.

Vom phänomenologischen Standpunkt ist es nun wesentlich, die Vergegenwärtigung als Eigenschaft des Fiktums nicht mit einer Art der durch passive Realitätsverdrängung oder Neutralisierung gewonnenen "Entgegenwärtigung" zu verwechseln. Diese kann zwar durch das wiedererinnerte Bild *denotiert* werden, wie es bei Erinnerungen geschieht, die auf Wahrheit Anspruch erheben, aber das Erinnerte als solches kann nicht Wirklichkeit enthalten, als ob es ein aktuelles Perzeptum wäre.

All das verursacht eine ontologische Grundveränderung des gesunden Verstandes, der jedes beliebige Seiende entweder für wirklich oder für unwirklich (imaginär) hält. Damit wird stillschweigend vorausgesetzt, daß das Wort "Gegenstand" einen eindeutigen, alle Gattung-Art-Unterschiede umfassenden Sinn besitzt. Dagegen führt die Bildesphänomenologie zu einer unüberbrückbaren Verschiedenheit innerhalb des Gegenstandsinnes. Wie Husserl behauptet,

drei Objekte haben wir: 1) Das physische Bild, das Ding aus Leinwand etc.; 2) das repräsentierende oder abbildende Objekt, und 3) das repräsentierte oder abgebildete Objekt... *Bildsujet.*[8]

Hier wird etwa der Fall eines Gemäldes oder einer Photographie erwägt, während dem bedeutungsvollen Fall der inneren Vorstellung eine besonderte Betrachtung gewidmet wird. Auch der letzteren mußte aber eine gewisse Objektität eingeräumt werden, in Anbetracht dessen, daß sie nicht nur als physisches Ding, sondern auch insofern dies etwas *darstellt* (s. Nr. 3) und erst durch das in ihm enthaltene Bild seine Darstellungstätigkeit entfalten kann.

Unsere Grundbetrachtungen können auch dazu dienen, das Objekt überhaupt phänomenologisch zu durchdenken, welches, insofern es zu einem nachbildenden im Bild-Ding fixierten Bild Anlaß gibt, in seinem Erscheinen den eigenen Objektsinn gewinnt.

Das Bild-Objekt dieses Bildlichkeitsbewußtseins besitzt seine alleinige Identität, die wir "eikonische Eidetizität" benennen möchten, und die dem Objekt im Sinn-2 entspricht. Das *Eigene* solch eines Objektes besteht nun in seinem vom Wahrnehmungsobjekt Verschiedensein, auf dem noch in der Darstellungsintention angewisen wird. Der Unterschied zwischen Bild- und Wahrnehmungphänomen ist von Natur aus phänomenologisch, was zuweil einen intentionalen Wechselbezug der Unterschiedenen impliziert. Das eikonische Eidetische ist daher weder ein Selbständiges noch bloße etwas sinnlich Einziges. Das Objekt-Bild hat seinen eigenen Bestand, der intentionalen Bezug in sich einschließt, ohne sich aber als bloßes Symbol in hinweisender Funktion aufzulösen. Dieses Objekt, wie Husserl sagt,

ist ein Erscheinendes, das nie existiert hat und nie existieren wird,[9]

was aber nicht heißt, es sei etwas Psychisches, wie irrtümlich der Gemeinsinn glaubt.

Die naive Auffassung irrt aber vor allem darin, daß sie das geistige Bild sich als ein dem Geiste reell immanierendes Objekt denkt. Sie denkt sich das Bild geradeso im Geiste vorhanden, wie in der Wirklichkeit ein Ding ... Ebenso existiert das Phantasiebild in

Wahrheit überhaupt nicht; das Phantasiebild hat nicht etwa eine psychologische Existenz.[10]

Das Objekt im phänomenologischen Sinn ist zwar kein metaphysisches Ding-an-sich, es ist aber auch kein Phänomen im phänomenalistischen Sinn, dem die Gliederung der drei oben dargelegten Bedeutungen fehlt. Der Phänomenalismus vermischt Bild und intendendiertes Objekt. Wie Husserl bemerkt,

wäre das erscheinende Bild phänomenal absolut identisch mit dem gemeinten Objekt, oder besser, unterschiede sich die Bilderscheinung in nichts von der Wahrnehmungserscheinung des Gegenstands selbst, so könnte es kaum noch zu einem Bildlichkeitsbewußtsein kommen.[11]

Da das Bild nichts anders als Objekt*erscheinung* ist, so steht es weder innerhalb noch außerhalb des Bewußtseins. Eben darum gewinnt die Phantasievorstellung eine *zweifache Objektität*, wodurch sie eine höhere Objektfülle in sich aufweist, die sich als unmittelbares Wahrnemen konstituiert. Husserl schreibt:

In der Wahrnehmungsvorstellung haben wir *einen* aufgefassten Gegenstand, und dieser ist in der vollständigen Wahrnehmung der gemeinte. In der Phantasievorstellung haben wir aber zwei Auffassungen aufeinander gebaut, zwei Gegenstände konstituierend, nämlich das Phantasiebild, das erscheint, und das bildlich dargestellte Objekt, das Bildsujet, welches durch das Bild dargestellt ist.[12]

Merken wir zunächst, daß hier das Objekt im Sinne-1 nicht mehr betrachtet wird: Das Objekt-Bild ist das "geistige Bild", dem aber das Im-Gesicht-Sein entnommen ist. Diese Weglassung beeinträchtigt, wie wir sehen werden, zu einem echten Verstehen des künstlerischen Objekts zu kommen, dessen Bildlichkeit nicht von der Materialität des Kunstwerks abstrahiert werden kann.

Außerdem setzt Husserl zu drastisch der mit Wahrnehmungsfähigkeit versehenen Phantasie, die phantasiemäßigen Dimensionen der zur Selbstgegebenheit zugerichteten Wahrnehmung vernachlässigend, die schlichte Wahrnehmung entgegen. Aber das, worauf für uns alles zukommt, ist hier die behauptete Duplizität des Objekts.

Der Gegenstand, den ich als mein bzw. als eines Anderen Bild wahrnehme, besitzt neben derjenigen des Bildsujets in demselben intentionalen Bezug seine eigene Intentionalität. Das Bildbewußtsein ist daher etwas mehr als das Wahrnehmungsbewußtsein. Wie Husserl sagt, ich vollziehe also nicht eine bloße Wahrnehmung; die Wahrnehmungserscheinung verbildlicht einen nichtwahrgenommenen Gegenstand.[13]

Die Möglichkeit einer vollkommenen Objektauffassung hängt daher von demjenigem Nicht-aktuell-Wahrgenommenen, das dem Bild- bzw. Phantasiefeld urprünglich gehört, d.h. von der zweifächen Objektität, die sich als Modifikation des Sinnes des Objekt erweist:

Das Gegenständliche gilt nicht als Gegenstand für sich, es gründet sich darauf eine Aehnlichkeitsrepräsentation als eine neue Auffassungsweise, welche die Beziehung zum Bildsujet gibt.[14]

Das Bildobjekt ist zunächst ein substantieller Gegenstand; eben darum kann es mit dem Bildsujet "im Doppel" sein, wie wir sagen würden. Das Doppelverhältnis heißt "Aenlichkeit", die jedoch ihrerseits einen eigenen Auffassungsunterschied verlangt. Es handelt sich nicht um eine objektive Aehnlichkeit zweier Figuren, sondern um das Aenlichsein selbst, das als Intentionalität vom Bildobjekt zum Bildsujet übergeht. So gewinnt die Aehnlichkeitsvorstellung eine die Bildobjektivität konstituierende Geltung.

Daß es sich um echte Objektität und nicht um bloßen Inhalt handelt, wird durch den phänomenologischen Gebrauch von "Erscheinung" ausgedrückt, die das von sich aus Sichzeigen der Sache selbst meint. Dem Bildsujet muß nun die Erscheinung zugeschriben werden, nicht aber indem ihm ein Erscheinendes entspricht, sondern indem er in und mit dem Bild erscheint, eben dadurch, daß die Bildrepräsentation erwächst.[15]

4. DAS OBJEKTMÄSSIGE ALS-OB

Der Objektsinn des Bildes wird im Ausdruck "Als-ob" festgestellt. In seiner vollständigen phänomenologischen Positivität verstanden, zeigt dieser Ausdruck die Nebensächlichkeit des Objekt-Bildes im Vergleich mit der Hauptsächlichkeit der wahrnehmenden Fassung, die kraft seines doxischen Charakters als "unmodifiziertes Bewußtsein selbst"[15] gilt. Die entsprechende Modifikation wohnt dem Phantasieobjekt als dessen Duplikat oder Nachbildung inne. Wie Husserl sagt,

zu jeder Reproduktion gehört es "Modifikation" zu sein, und das Modifizierte ist ein Als-ob.[16]

Die Phantasiemodifikation hebt bei dem unmodifizierten Bewußtsein den Doxamoment auf, wodurch dem Objekt selbst etwas Neues, nämlich der Möglichkeitscharakter, *hinzugefügt* wird. Man soll daher das Plus des Bildes über die Realität hinaus, in welchem der Objektwert des Als-ob besteht, genau untersuchen. Dieses Plus ist das Positive der Erfahrung.

Das Sein der Phantasiegegenständlichkeit — schreibt Husserl — ist "erfahrbares" Sein, erfahrbar sind die Phantasiegegenständlicheiten als solche . . . Unbestimmte Horizonte gehören wesensmäßig zu jedem individuellen Gegenstand . . . Ein Gegenstand ist in sich möglich, sofern er anschaulich vorstellbar ist, und ein gedachter Gegenstand, sofern er frei anschaulich phantasierbar ist.[17]

Vom Objektwert ausgegangen, sind wir nun zu einer allgemeinen *ontologischen* These gekommen: Zu jedem reellen oder wirklichen Gegenstand gehört ein Möglichkeits- oder Unbestimmtheits*hof*, der sich durch Phantasie in der Als-ob-Modalitäten als anschauende Darstellbarkeit konstituiert. Dies bedeutet daß in der Frage nach der Bildobjektität der Ursinn des Gegenstandes als solcher damit eingefügt ist, daß in der Bildesphänomenologie Sinn und Möglichkeit einer Phänomenologie überhaupt in Frage stehen.

Wieweit aber erstreckt sich die Herrschaft der Als-ob-Modifikation? Wie kann die Ausbreitung dieser Modifikation mit der eigenphänomenologischen Objektität und Konstitution übereinstimmen? Husserl behauptet:

Fikta bilden eine Gegenstandsregion, die Als-ob-Gegenstück ist der Welt und aller möglichen Welten. Fikta sind aber nur seiend als Korrelate meiner Phantasie und sind transzendente und Dauergegenstände nur aus Willkür meiner In-Geltung-Setzung und meiner In-Geltung-Haltung.[18]

Zunächst ist der quasi-positionale Akt, wodurch sich die Als-ob-Objektität konstituiert, kein Als-ob-Akt sondern eben ein *wirklicher*:

Sowie ich als aktuelles Ich . . . die Setzung des Phantasierten des Gegenstandes als ob (mache), habe ich nicht mehr Phantasie im ersten Sinne, sondern eben Erfahrung, und was ich erfahre ist ein Fiktum.[19]

Der Als-ob-Charakter wird so auf die Objektität verbannt, er wird nämlich zu einer Qualität irgendeiner Seinssphäre. Bilder sind nichts anders als Korrelaten der Einbildungskraft, die die wirkliche und objektiv mögliche Welt im Ganzen ausmachen. Als Objekte sind sie nicht minder als Wahrnehmungsobjekte "transzendente Gegenstände". Die Als-ob-konstitution ist daher *wirkliche* Möglichkeitskonstitution. Während wir früher bei Husserl sahen, wie er das Bewußtsein der Phantasie entgegensetzte, so hören wir ihn behaupten:

Auch Phantasie ist Bewußtsein und konstituiert also, aber reine Möglichkeiten, immanente oder transzendente. Transzendente aber in freier Gestaltung der weiteren Einstimmigkeiten oder Unstimmigkeiten.[20]

Dieser Gedanke verlangt eine Betrachtung der Objektität des Als-ob, dessen bewußtseinsmäßige Stellung nicht genauso ist wie die der reellen Wahrnehmung. Die hier hervorgehobene Beschränkung deckt sich aber mit der objektiven Verwurzlung des Faktums. Von letzterer hängt nach Husserl derdem Sein des Objekts angehörige Möglichkeitssinn ab:

Eine reine Möglichkeit ist undenkbar, es sei denn als Korrelat eines phantasierenden und daraus Möglichkeit gestaltenden Subjekts.[21]

Diese ontologische Stellung wird später verwandelt und feiner gemacht. Der volle Objektsinn ist an die Intentionalität und dann an die Subjekt*distanz* gebunden, die in der durch Glaubensmodifikationen erzeugten Ichabtrennung dargestellt wird. All das bringt neben dem Gedanken eines doppelten Objekts den Gedanken eines *zweiten Ich* (wie Husserl sagt) mit sich:

Indem ich schlechthin positional lebe, urteile, werte usw., bin ich in gewisser Weise selbstverloren, "unbewußtes", an die Gegenstände, Sachverhalten, theoretische Allgemeinheiten, Werte, Zwecke, verlorenes Ich . . . Ich vollziehe nun Reflexion, ich etabliere ein zweites positionales Ich, das in Beziehung auf das, was ich selbstverloren setzte, keine Setzung mitmacht . . . Ich setzte positional das Scheinbild als Wirklichkeit, als "Gegebenes" dieses *quasi*-Sehens. Ich etabliere ein zweites Ich, das den *quasi*-Glauben . . . nicht mittut, sondern es reflexiv betrachtet und darin das "Noema", das "Bild".[22]

Dazu glauben wir unsererseits einige Erklärungen hinzufügen zu können. Zunächst: Wie sich das Ich in der unmittelbaren oder erstrangigen Setzung verliert, so geschieht dem Gegenstand seinem Objektsinne nach dasselbe. Die Identifizierung von Subjekt und Objekt in der primären Positionalität hebt dem Objekt sein Ansichsein auf und läßt nicht das dem Subjekt Gegenübersein stehen. Diese Identifizierung findet beim erkenntnishaften Erleben statt, der Erkenntnisprozeß aber zeugt ständige *Wiederaufstellung* der schon verbrauchten primären Setzung, wodurch das "wiederaufgenommene" Objekt eine Seinsweise als An-sich-reell erlangt. Dies aber geschieht nur sofern die primäre

Setzung nicht mehr wirkend ist. Andererseits, nur durch die auf das Objekt ausgewirkte primäre Setzung erlangt sich das Subjekt wieder und erhält sich aus seiner primären Verlorenheit. All das, wie gesagt, war sowohl bei der unmittelbaren Setzung als auch bei der imaginären *quasi*-Setzung, durch die das Objekt im Als-ob-Modus erscheinte, ausgeblieben.

Zweitens: Daraus ergibt sich, daß, um Gegenstände in ihrer vollen Seinssinn gelten zu lassen und den phänomenologischen *Spielsetzung* auszuführen, wir uns erst des Irrealitätsmoments des Bildes bedienen müssen.

Durch die hier in Frage stehende Reflexion wird das Objekt sozusagen von der Als-ob-Modifikation befreit, aber nur insofern diese zur phänomenologischen Beschaffenheit des Objekts selbst wird.

5. ONTO-PHÄNOMENOLOGISCHE REFLEXION

Die zweifach Setzungstätigkeit erschließt den Reflexionsbereich als allgemeine phänomenologische Struktur. Sie taucht im Bild wieder auf, indem das Bildobjekt auf das Bildsujet verweist und sich in der Abbildung als inneres Bildeswesen anerkennt, dessen Gegenwart in sich das Moment der Vergegenwärtigung enthält, das seinerseits das Korrelat der zweitrangigen Setzungstätigkeit ausmacht.

Beide sowohl aneinander- als auch auseinanderziehende Momente drücken sich in der zweifachen Bedeutung von Phantasie aus. Wie Husserl sagt,

> das Wort Phantasie und Fiktion hat ... zwei Bedeutungsrichtungen. 1) Die eine geht auf Reproduktion (und Vergegenwärtigung) überhaupt, und dann heißt auch jede Erinnerung Phantasie ... 2) Die andere auf die Art des Vollzuges, wonach dann von perzeptiver Fiktion die Rede sein kann und andererseits die Erinnerung keine Fiktion, keine Phantasie.[23]

Phänomenologisch entscheidend erscheint hier der Husserlische Unterschied, oder besser sein Schicksal in dem Husserlischen Text. Im Falle, wo der Unterschied auf der ontischen bzw. psychischen Stufe stehenbleibt, wird die Phänomenologie selbst aufs Spiel gesetzt. Die vorliegende Texte bestehen aus den Schwierigkeiten einer Bildesphänomenologie, die um den obengenannten Unterschied kreisen. Es ist nun klar, daß beide Bedeutungen von Phantasie nicht ohne Wechselbeziehung bleiben können; wenn das reine Fiktum, unter der zweiten Bedeutung der Neutralisierung gedacht, als aktuelle Reproduktion fungieren muß, wie es die erste Bedeutung verlangt, so muß es in sich eine gewiße Anlage zur Reproduktion besitzen. Streng phänomenologische betrachtet, die Als-ob-Modifikation folgt dem Urmodus der direkten Setzung; sie ist eben Modifikation und das impliziert innere Reproduktionsmäßigkeit. Das Reproduktion-von-B-Sein von A gehört A selbst und ist kein äußeren Bedeutungsverhältnis. Damit aber das A in Reproduktion-von-B umschlägt, muß es schon neutralisiert und von der erstrangigen präsentierend-positionalen Unmittelbarkeit befreiet worden sein.

In der Tat tritt Husserl den phänomenologisch ausgezeichneten Weg der Vermittlung von beiden Bedeutungen an, von reflexiver Setzung höherer Ordnung abgesehen, die am Fiktum und an dem ihm entsprechenden Fingieren haftet, wonach das Als-ob-Urteilen die Stellung des unbeteiligten Zuschauers transzendiert.[24] Husserl hebt das Anwesen einer gewißen Verwandschaft beider unterschiedenen Sinne hervor:

> Insofern ist also das Als-ob wie jedes Erinnerung-als-ob nah verwandt mit der ... bloße Phantasie und andererseits doch verschieden.[25]

Im Fall des "bloßen Phantasie" waltet (nur einmal oder immer wieder?) ein *Setzungshindernis*, wodurch das Interesse vom dargestellten Ding (Bildsujet) zum Bildobjekt verschoben wird. Daraus folgt der Zusammenhang zwischen Bild und *Inhibition.* Husserl schreibt:

> Lenke ich das Interesse auf das Bild, auf das Wahrgenommene als solches, Erinnerte als solches, Gedachte als solches, so habe ich die Setzung des Identischen inhibiert ... Ich kann willkürlich einen Vollzug meines positionalen Aktes aufheben.[26]

Inhibition und bloße Bildhaftigkeit gehören zusammen; damit aber der Zusammenhang besteht, muß er *Vollzogen* oder durchgeführt werden. Dies ist nun der Ort, wo das *Wirken der Kunst* eintritt. Die ausgezeichnete Funktion des Kunstwerkes besteht nämlich darin, *den Inhibitionsakt bei dem Dinge aufzuhalten.*

Husserl selbst aber bleibt innerhalb einer bloßen Phänomenologie der Weisenunterscheidung stehen, und deswegen stößt er immer wieder auf die Schwierigkeiten der Setzungsmodi. Er neigt dazu, die Phantasie einerseits der Neutralisierung, andererseits dem ästhetischen Bereich zu entreißen:

> Der Ausdruck "Neutralitätsmodifikation" paßt auf die thematische Aenderung nicht aber auf die Phantasie ... Das Wesentliche ist für die ästhetische Einstellung also nicht

die Phantasie, sondern die Einstellung, auf das, was ästhetisch interessiert, Gegenständlichkeit im wie.[27]

Der echt phänomenologische Zug dieses Gedankens besteht u.E. darin, daß das Eigentümliche der phantasiemäßigen *quasi*-Setzung anerkannt wird. Dieser ist gewiß der von Husserl eingeschlagene Weg:

Und wenn ich ein Fiktum habe und mich nicht um die Nichtigkeit kümmere, bloß betrachte: ist das nicht ein eigener Modus, parallel dem Aktcharakter der Axiose?[28]

Die Eigenschaften des Bildobjektes sind nicht jene des Bildsujets, da diesen ein Widerstreit anhaftet, der sich in der Nichtigkeitsqualität des Bildobjektes selbst auflöst. Fügt nun Husserl hinzu:

Aber voll eigentlich ist es als nichtig charakterisiert, d.i. ich vollziehe das Nichtigkeitsbewußtsein in voll eigentlicher Weise, wenn ich von Wirklichkeits- zum Bildobjekt-Bewußtsein übergehe und eben in der "Vernichtung", der "Aufhebung" der Auffassungstendenz des letzteren Bewußtseins lebe. Nur dann, nur in diesem Uebergangsbewußtsein gewinnt das Bildobjekt den "eigentlichen" Charakter des Nichtig.[29]

So wird dem Bild als solchem eine *tätige* Neutralität zurückgegeben. Das Bildobjekt stützt den Uebergang zur Irrealität, es ist ein aktuelles Nichtigmachen, das nicht dasselbe wie eine vermeinte Nichtigkeit meint. Dadurch erweist sich das Bildobjekt als die Geschichte der Realitätsausschaltung bzw. der Irrealisierung des "bloßen Betrachtens".

Es ist nun zu untersuchen, wie der Fall im bildhaften Ontischen, der z.B. empiristisch dargestellt werden kann, zugleich die dem Bild innewohnende Reflexionstätigkeit als Bildbewußtsein zu verstehenist. All das wird von Husserl so formuliert:

Wir finden bei der reproduktive Phantasie, daß die Erscheinungen nicht etwa bloß perzipierende Erscheinungen, sondern imaginierende sind, d.h. in der Erscheinung stellt der Empfindungsinhalt etwas dar, und die Erscheinung selbst stellt Erscheinung dar, die Auffassung ist nicht einfach Auffassung, sondern Darstellung von Auffassung.[30]

Die reproduktive Phantasie wird hier der produktiven oder ursprünglichen und dem unmittelbar in der Wahrnehmung sichselbstgebenden Objekt schlechthin entgegengesetzt. Grundlegend wird aber der Streit zwischen Wirklichkeit und Täuschung, den zu explizieren es schon immer eine besondere phänomenologische Aufgaben gewesen ist. In jedem Fall muß all das als streng und echt phänomenologisches Thema erhalten bleiben. Außerdem veranlaßt uns die letzte angeführte Stelle

Husserls zu der Vermutung, daß dank seiner Fähigkeit, irgend etwas reproduktiv bzw. bildhaft darzustellen, das Phänomen in sich und von sich aus schon von wesentlich bildhafter Natur sei. Die konsequente Entwicklung dieser Hypothese könnte uns in den Stand setzten, den Keim der schon vor- und dargelegten Identifizierung der Bildesphänomenologie mit der Phänomenologie überhaupt zu erahnen.

6. TRENNUNG DES BILDES

Gegen die Setzungsfähigkeit der Wahrnehmung steht die Als-ob-Setzung der Phantasie, deren Erzeugnisse (Bilder im weitesten Sinn, Fikta)

> ohne Zusammenhang mit den Wahrnehmungen sind; das heißt: während sich alle Wahrnehmungen hinsichtlich der in ihnen vermeinten Gegenständlichkeiten zu einer Einheit zusammenschließen, ... fallen die Phantasiegegenständlkichkeiten aus dieser Einheit heraus, sie schließen sich nicht ... zur Einheit einer vermeinten Welt zusammen.[31]

Die Phantasiezusammenhänge laufen in isolierte Phantasieeinheiten aus, die ohne Zeit- und Raumzusammenhängen bestehen. Dadurch werden Zeit und Raum zu eingebildeten Als-ob-Zeit und Als-ob-Raum.

> Zeit — sagt Husserl — ist zwar vorgestellt, sogar anschaulich vorgestellt, aber es ist eine Zeit ohne wirkliche und eigentliche Oertlichkeit der Lage — eben eine Quasi-Zeit.[32]

Der angehörigen Fikta fehlt die eigentliche Einzelheit und Identität, da ihre Seinsweise in *Zusammenhangslosigkeit* besteht, die aber positiv, indem

> jede auß Zusammenhang stehende Phantasie ihre Einbildungszeit hat,[33]

verstanden werden muß. Es ist ferner eine Aufgabe der thematischen Wiedererinnerung, die Als-ob-Einheiten als *getrennte* Phantasieobjekte zu konstituieren. Derartige Einheiten scheinen aber von den reellen weltlichen Einheiten entliehen zu sein.

> "Einheit einer Phantasie" — schreibt Husserl — ist offenbar nichts anders als Einheit einer möglichen Erfahrung oder Neutralitätsmodifikation einer Erfahrungseinheit.[34]

Bilder sind "Objekte ohne Identität" und dieses ganz eigentümlichen Modus wegen sind sie von der Wirklichkeit getrennt. Soll aber eine derartige Irrealität als eine reale Tatsache oder als eine phänomenologische Eigenschaft genommen werden? Gehört zu der Als-ob-Phantasiewelt eine besondere (bildhafte, ästhetische) *Irrealitätsintention*? Die Frage bleibt noch in der Schwebe; Husserl selbst scheint mit den beschreibend-phänomenologischen Unterschieden zufrieden zu sein, die zwischen beiden Welten hervorgehoben werden können. Denn er bemerkt:

In der wirklichen Welt bleibt nichts offen, sie ist, wie sie ist. Die Phantasiewelt "ist" und ist so und so, soweit sie von Gnaden der Phantasie phantasiert worden ist.[35]

Der Phantasiebereich wird durch in Anführungszeichen gesetztes "ist" angedeutet. Dasselbe gilt aber überhaupt für den Gesamtbereich des "Als-ob". Es ist damit nicht gesagt, daß "Phantasie" und "Als-ob-Bereich" zusammentreffen müssen. Nach Husserl scheint vielmehr der phänomenologischen Bestimmung der Phantasie die des Als-ob voranzugehen, obwohl im Ist-Fall die Phantasie beanspruchen kann, als Prinzip zu gelten. Könnten wir diese Vorbehalte aufheben, so könnten wir das Bild als echten Urort des Seins*sinnes* denken, da in Betracht zu ziehen ist, daß für die Phänomenologie *Sinn und Anführungszeichen schließlich dasselbe bedeuten.* Statt als Einzelfall, Anwendung, Art usw. der allgemeinen Als-ob-Modifikation zu erscheinen, wird sich denn die fiktionale Trennung als die Epoché selbst erweisen, worauf die Phänomenologie errichtet ist.

Dasjenige nun, wonach sich diese Trennung auf der Ebene der phänomenologischen Konstitution aufhält, ist die immer wieder von Husserl behauptete Tat, daß dem fiktionalen bzw. ästhetischen Bewußtsein eine ihm eigene Intentionalität gehört. Durch eine Reflexion über diese Intentionalität *sui generis* werden wir zu jener Realität zurückgeführt, von der aus das fiktionale Bewußtsein eine Urtrennung durchgeführt hat. Als Husserl sagt,

Jetzt schließen wir das Wirklichkeitsbewußtsein aus: Es handle sich um bloße schlichte Phantasie oder bloße eikonische Phantasie, oder bloßes symbolische Vorstellen und Denken.[36]

Um die Phänomenologie überhaupt zur Bildesphänomenologie zurückzubringen, würde es nun genügen, der genannten Abschließung die Bedeutung von phänomenologischer Ausschaltung als Sinngebung wiederzugeben. Es handelt genau darum, dasjenige zu entdecken, was die fiktionale Intentionalität ausmacht, d.h. die besondere, der Wirklichkeitsausschaltung unterliegenden Stellungnahme. Die daraufbezogene Husserlsche These wird so ausgedrückt:

Die Erscheinungsweise ist trägerin von ästhetischen Grundcharakteren . . . Die Erscheinung ist Erscheinung vom Gegenstand, der Gegenstand Gegenstand in der Erscheinung. Vom Leben im Erscheinen muß ich zurückgehen auf die Erscheinung und umgekehrt . . . Der Gegenstand erhält . . . eine ästhetische Farbung um Erscheinungsweisen willen und die Rückwendung auf die Erscheinung bringt das Ursprungsgefühl zum Leben.[37]

Das Aesthetische (bzw. Fiktionale, Bildhaftes usw.) konstituiert in der Weise einer Genesis den Seinsinn von der Gegenständlichkeit überhaupt. Der in Anführungszeichen sich darbietende Seinsinn, d. h. der Sinn des Phänomens als solches, enthüllt das echte "Wie", an dem die eigenfiktionale (ästhetische) Vorstellung lebt. Die Wirklichkeitsausschaltung zielt auf die "Erscheinung der Erscheinung" ab, wie wir mit Kant sagen möchten.

Es ist nicht richtig – sagt nun Husserl – daß das ästhetische Bewußtsein auf das Erscheinende und so zu Beschreibende unangesehen von Sein und Nichtsein gerichtet ist, sondern ist darauf in der betreffenden "Erscheinungsweise" gerichtet . . . Also nicht dies macht den Unterschied ästhetischer Gefühle von anderen, daß dieselbe auf bloß Vorgestellten gerichtet sind.[38]

Während jeder Wirklichkeitssinn (eine doxische Modalität, die nicht mehr mit dem Urmodus oder "Null-modus" der erstrangigen Stellungnahme identifizierbar scheint) durch die Erscheinung hindurch geht, um sich nach dem erscheinenden Gegenstand zu richten,

geht nicht das ästhetische Gefühl durch die Erscheinung hindurch, sondern auf sie hin und auf den Gegenstand nur "um der Erscheinung willen".[39]

Das Gefühl verweilt bei dem Erscheinungsein der Erscheinung, erlebt ihre Phänomenalität. Es ist daher sowohl der thetische zum Gegenstand gerichteten Haltung als auch der bloßen *a-thetischen* Vorstellung entgegengesetzt, die nichts von Sein und Nichtsein weiß und die der ästhetischen Reflexivität entbehrt. Zur erstrangigen phänomenologischen Neutralisierung, die unter Ausschaltung der Seins- und Nichtseinssetzung zum Vorgestellten als solchen führt, ist nun etwas als eine *ästhetische Reduktion* hinzuzufügen, die zur Erscheinung *der* Erscheinung führt, dem einzigen Seinsmodus nämlich, der nie mit dem Ding an sich verwechselt werden kann.

In unserer Rede von ästhetischer Reduktion verstehen wir das Aesthetische nicht als eine der erkenntnistheoretischen Gegenständlichkeit übergeordnete Dimension, durch die, zu denen die ästhetischen Urteile zu zählen sind, als die den Seinsurteilen folgenden zu begreifen sind; vielmehr führt das ästhetische Gefühl *diesseits* der transzendentalen Setzungsintention, das Dieseits aber ist nich einmal die in such zurückgekommene Subjektivität.

Die hier angerissene Problematik bleibt noch nach vielen Seiten offen, vor allem nach der, die das Verhältnis zwischen Phantasie und ästhetischer Fiktion und demzufolge den Seinsmodus des bildenden Ichs betrifft. Die Trennung des gebildeten Bildes ist in der Modifikation der ästhetischen Trennung und dann in der Trennung des dadurch in seinem Seinsmodus bestimmten Subjekts wiederzufinden.

7. DIE ZUM BILD BESTIMMTE SUBJEKTIVITÄT

Das sich als unbeteiligter Zuschauer setztende Ich der transzendentalen Phänomenologie ist nicht dasselbe wie das in sich reflektierte, dem ästhetischen Gefühl eigene Bild. Indem sich Husserl der Bildesphänomenologie öffnet, hat er in der Tat die Selbstverständlichkeit des von Descartes eingeweihten sichselbstgewissen Ichs aufgehoben. Angesichts der Bildreflexivität ist vielmehr das "Ich bilde" der Grund der Reflexion als *Widerspiegelung*. So finden wir den Spiegelsein des Bewußtseins, der unter dem herrschenden Verstandestranszendentalismus verlorengegangen war. Um aber den verlorenen Sinn wiederzuerlangen, ist es zunächst notwendig, das Ich aus seiner selbst-bewußten Identität zu verdrängen und es vom Schicksal seiner Akte abhändig zu machen, die sich in verschiedenen Weisen und Dimensionen auf etwas *anderes* beziehen. Intentionalität ist Akt *des Ichs*, demnach der Seinsmodus, die Existenz, das *endliche* Leben des Daseienden entschieden wird. Diese Bestimmung als Ichwesen ist nicht auf dem cartesischen, dem von Husserl selbst betretenen Weg zu erreichen. Die transzendentale Reduktion kann andererseits in Richtung auf den unbeteiligten Zuschauer ausgeführt werden, der endlich, in seiner verzweifelten Suche nach dem anderen in ihm und außer ihm, in Angst der (sonst unmöglichen) "Einfühlung" geraten muß. Die auf dem cartesischen Weg thematisierte theoretische Haltung veranlaßt das Ich nicht dazu, etwas von sich selbst zugunsten der objektiven Konstitution aufzugeben, die ihn nicht wesentlich angreifen.

Wir können nun in der Entwicklung der Bildesphänomenologie eine veränderte Haltung bemerken. Das Ich, das sich in der Stellung eines bildenden Subjektes setzt, erweist sich nun als ein *Gebildetes*, das von sich aus etwas objektmäßiges ist, insofern es der Als-ob-Setzung und Seins- und Nichtseinstrennung, die im Bereich des Bildes stattfindet, verpflichtet ist. Das im Bild sich erzeugende Weder-Sein-noch-Nichtsein macht die onto-phänomenologische Macht des Bildes selbst aus und wirkt auf das Ich zurück.

In diesem Zusammenhang sei die folgende Husserlische Stelle angeführt:

> *Ich* phantasiere, ich vollziehe Erinnerung, sofern ich als Reproduktion in vollziehender Weise erlebe: Das kann aber so verstanden sein, daß das aktuelle, nämlich in den aktuellen Impressionen seiner selbst bewußte Ich sich mit des Phantasie-Ich eins weiß . . . Lebe Ich in der Phantasie, so ist das Ich . . . das Phantasie-Ich also ein Reproduziertes (gegenstand einer Reproduktion).[40]

Es wird so ein "bildhaftes Selbstbewußtsein" enthüllt und die These einer radikalen Bildhaftigkeit desjenigen Selbstbewußtseins angedeutet, das auf den cartesischen Wege auf der *anonymen* Vernunft allein gegründet werden konnte. Was ich in der erkenntnistheoretischen Haltung denke, nimmt zwar meine "konstitutive Kategorien" nicht aber *mich selbst* in Anspruch, setzt mich nicht als lebendes Subjekt. Meine Intentionalität selbst *gehört mir nicht*. Die in sich selbst unreduzierbare Phantasie vermag sogar im *bildenden* Ich eine unvereinbare Scheidung hervorzubringen. Schreibt nun Husserl:

> Lebe ich aber zugleich in der Wirklichkeit und der Phantasie, so ist abstraktiv gesprochen *ein doppeltes Ich* da, das wirkliche Ich und das reproduktive.[41]

Es ist eben dank dieses ganz eigenene Doppeltsein, in dem wir eine Grundmöglichkeit des Ichs selbst erkennen, daß es zu sich *phänomenologischerweise* zurückkommen und so als "Selbstbewußtsein" existieren kann. Der unreduzierbaren Verschiedenheit der Subjektsmodi wegen kann es keine vorsausgesetzte Ichsubstanz geben, da das echte daseinsmäßige Selbstsein nur auf der Weise des Doppelt- oder Vielfachseins verwirklicht werden. Das bildende Ich ist kein bloß denkendes Ich; es muß sich unter Bild-Modifikationen setzten, sich selbst bilden, ohne sich vom reflexiven Bildwesen im Bildbewußtsein befreien zu können. Als Husserl sagt,

Ich werde im allgemein und vielleicht notwendig selbst in der Phantasiewelt eine Platz haben als phantasierte Ich, die Phantasiewelt von seinem Standpunkt aus *quasi* setzten.[42]

Dieses existentiale Ergebnis übersteigt den Rahmen der "alltäglichen" Interpretation der Phänomenologie und eröffnet neuere Untersuchungsmöglichkeiten, die in erwogenen, vielleicht gequälten Texten, die dank dem wertvollen Werk von E. Marbach aus dem Haufen der unveröffentlichten Handschriften Husserls wiederhergestellten worden sind, gerade angedeutet scheinen.

8. DIE PHÄNOMENOLOGISCHE BESTIMMUNG DER MENSCHHEIT

Unsere Deutung einiger gewählter Texte Husserls scheint eben mit der Auffassung zu enden, nach der sich die fiktionalen Akte nur als Modifikationen der unmittelbar setztenden erweisen, die, da sie die Wirklichkeit als ihr Korrelat haben, "wahrheitsmäßig" zu nennen sind: Erst komme die unmittelbare Setzung und dann die mittelbare phantasiemäßig modifizierte.

Aber wie kann ein setztender Akt echt phänomenologisch analysiert werden? In der Idee solch einer noch lediglich "vermeinten" Analyse wird eine Darstellung der Aktstruktur und des gesamten Erlebens vorgezeichnet, die sich unter dem allgemeinen Titel "Positionalität" einordnen. Dieselbe Idee macht vor allem deutlich, daß es keine einheitliche und eindeutige "Position" gibt, sondern nur verschiedene Abwandlungen eines Begriffs, der nur "analogisch" gelten kann.

Es ist ferner zu fragen, ob die Positionalität ihrerseits eine spezifische phänomenologische Qualifizierung als Modifikation eines unmittelbaren, ontologisch begründeten Vernehmens konstituiert, die aber für sich allein keinen spezifischen Wert und eben keine objektive Gültigkeit besitzen kann. Husserl scheint auch hier bei der bloßen Unterscheidung der setztenden und nichtsetztenden Akte stehen zu bleiben und die Frage nach ihrer wechselseitigen Ordnung offen zu lassen. Diese Frage war in Wahrheit keine neue; wir finden sie bei Kant, für den die Objektivität (als objektive Geltung) nichts als die Leistung einer objektivierenden Tätigkeit ist.

An diesem entscheidenden Punkt eröffnen sich zwei theoretische Möglichkeiten: Entweder das Positionale erweist sich als Qualifizierung und Modalisierung eines vorauszusetzenden Vorpositionalen, das für sich neutral, ohne besondere Neutralisierungsakte besteht, oder dasselbe ist nichts als eine Nebenqualifizierung des Nicht-positionalen, dessen Verfassung als (kantischerweise) rein ästhetisch verstanden werden kann. Nach dieser Auffasung würde das Nicht-positionale dem Positionalen vorangehen. Kurz, ist die Wirklichkeit als Modifikation der Phantasie oder umgekehrt die Phantasie als Modifikation der Wirklichkeit anzusehen? Die "oberflächliche" Darstellung Husserls scheint die zweite Alternative nahezulegen, aber die "tiefe" Schicht des *Phänomens*-wesens und der gesammten Phänomenologie erfordert vielmehr, sich an die erste zu halten.

Auch diese tiefgehende Frage müßen wir hier offen lassen, da ihre angemessene Erörterung nur innerhalb einer — noch *in votis*! — Phänomenologie der *Kunst* geschehen kann. Diese letzte muß aber ihrerseits als der universale Boden des Bildes als Phänomen *sensu eminentiori* erörtet werden, derart, daß der Lauf der Beschreibung mit dem lebensweltlichen Verlauf, der das Dasein konstituiert, im Einklang steht.

Wir enden damit oder vielmehr wir brechen unsere Behandlung mit der Frage ab: In welchem Maße und welcher Weise, nach denen im positionalen Akt und im wissenschaftlichen Erleben überhaupt die Menschheit ihre Existenz übergegeben hat, ohne konkret eine endgültige Erlösungsperspektive zu haben, ein echt daseinsmäßige Menschsein realisiert werden kann? Aktuelle Aufgabe der Phänomenologie wäre ferner die Auseinandersetzung mit derartigen Grundproblemen, um darin den für sich unbestimmbaren Entschluß des Daseins für immer einzusetzten.

Zum Schluß sei es uns erlaubt aus den erwogenen Texten Husserls eine plausible Reform der "Analytik des Daseins" vorschlagsweise abzuleiten, indem wir den Satz aufstellen: Die Welt (das Worin des Daseins) ist nur als *Quasi*-Welt der echte Boden des menschlichen Daseins.

Universität Palermo

ANMERKUNGEN

[1] Husserliana, B. XXIII, S. XXVII.
[2] S. *op. cit.* S. XXX.

[3] *Op. cit.* S. LV.
[4] *Op. cit.* S. XLIX.
[5] *Op. cit.* S. 7.
[6] *Op. cit.* S. 10.
[6a] *Op. cit.* S. 92.
[7] *Op. cit.* S. 16.
[8] *Op. cit.* S. 19.
[9] *Op. cit.* S. 19.
[10] *Op. cit.* SS. 21–22.
[11] *Op. cit.* S. 20.
[12] *Op. cit.* S. 24.
[13] *Op. cit.* S. 26.
[14] *Op. cit.* S. 27.
[15] *Op. cit.* S. 28.
[16] *Op. cit.* S. 560.
[17] *Op. cit.* S. 564.
[18] *Op. cit.* S. 564.
[19] *Op. cit.* S. 564.
[20] *Op. cit.* S. 566.
[21] *Op. cit.* S. 566.
[22] *Op. cit.* S. 574.
[23] *Op. cit.* S. 575.
[24] S. *op. cit.* S. 579.
[25] *Op. cit.* S. 583.
[26] *Op. cit.* SS. 584–585.
[27] *Op. cit.* S. 591.
[28] *Op. cit.* S. 366.
[29] *Op. cit.* S. 367.
[30] *Op. cit.* S. 472.
[31] E. Husserl, *Erfahrung und Urteil* (Hamburg: 1954), S. 195.
[32] *Op. cit.* S. 198.
[33]
[34] *Op. cit.* S. 200.
[35] *Op. cit.* S. 202.
[36] Husserliana, B. XXIII, S. 388.
[37] *Op. cit.* S. 389.
[38] *Op. cit.* S. 391.
[39] *Op. cit.* S. 392.
[40] *Op. cit.* S. 349.
[41] *Op. cit.* S. 350.
[42] *Op. cit.* S. 467.

MIHAI PĂSTRĂGUŞ

AESTHETIC CONCEPTS OF A PHENOMENOLOGICAL ORIGIN

Though Edmund Husserl, father of modern phenomenology, is known not to have written an aesthetics in the proper sense of the word, his method proved to be very fruitful in this field of philosophical reflection. Noticing the virtues of this method even from the very beginning, Moritz Geiger, his colleagues for a generation not only dared to apply this method in the domain of aesthetics, but also thought that no aesthetics as an autonomous discipline can be conceived without combining it with the eidetic method. As a scientific approach in itself, phenomenological aesthetics aimed at studying aesthetic and artistic values, examining the structures of the aesthetic object, as well as determining their value. Still, such an aesthetics also implied imperiously some special theory on the essence of art, a reshaping of the theory on the psychological mechanism involved in the subjective and affective states of both the creator and the receiver.

An axiological perspective matching the new doctrine appeared just as necessarily with the personality of Max Scheler. He achieves what aesthetics really needs: abolishing the primacy of pure logic and demonstrating that the spirit does not contain only rational acts, but affective-volitional ones too, all of them cognitively available. For Scheler the *ordo amoris* characaterizes not only *ethos*, but *aesthesis* as well — it opens the way to the phenomenology of the emotional life, to the realm of collective-affective acts, including, in fact, the objects of art, the works. The aesthetic constructs that would fall within the framework of phenomenology were founded on the phenomenology of the emotional life, on this axiological metaphysics. Here scholars could come to an intuition of aesthetic essence, i.e., of value, beginning from an individual instance, with no need of abstracting it out of a multitude of previous experiences. Though at first sight it seems to be an aesthetics of feeling, the object of which is an analysis of aesthetic experiencing, it has been proved in fact that entering this still unexplored region — intentional objects and intentional acts (intentional consciousness) — represented a natural step in the logic of human knowledge.

In building up phenomenological aesthetics within the configuration

A-T. Tymieniecka (ed.), Analecta Husserliana, Vol. XXXVII, 147–165.

anticipated by Husserl and Scheler, several prestigious personalities beside Geiger made their contributions: W. Conrad, Roman Ingarden, Mikel Dufrenne, Guido Morpurgo-Tagliabue, Cesare Brandi, etc. Though each of them provided explanations and opened new vistas, the themes they approached were about the same: the structure of the aesthetic object, the essence of the art work, intentionality in the arts, the analysis of aesthetic experience, langauge, etc. In the perspectives of these thinkers — masters of nuances in logico-aesthetic thought — the universe of perceptible beauty, as a specific dimension of being, gets revealed to us in a series of concrete occurrences or subjective phenomena. In fact, phenomenology does not contradict the appropriations of classical aesthetics, but it differs in its manner of reaching the essence of beauty. In the new view, beauty is a positive possibility of noticing the appearance of subjective phenomena of a certain category, partly anticipated by the sense of what is really perceived — an aesthetics of sense and intentionality. But aesthetic intentionality — which is none other than the tendency toward beauty — views harmony itself, the essence structured proportionately to its inner relationships. Aesthetic intuition does not sense material forms and structures, but rather intentional structures and meanings. Phenomenological beauty no longer appears as a naturally given thing as in empiricism, but as a possibility of consciousness. Such a perspective makes possible the passage from a contemplative aesthetics to a generative productive one where symbols are but direction pointers of intentional consciousness. Ingarden pointed out here that works of art represent real critical structures of intentional existence, that the intentional object in the field of art work — painting, music, literature, cinema — has a twofold critical foundation: a real existence and an ideal concept.

Cesare Brandi, a fine expert in art history, uses the logical methods of phenomenology in the study of painting and music. He obtained famous results in the study of music through eidetic analysis, with specific reference to the Romanian-born composer and conductor Sergiu Celibidache.

Mikel Dufrenne, one of the most prominent contemporary aestheticians, who achieved spectacular results with the phenomenological method, develops a complex aesthetic analysis with the concepts of aesthetic object and aesthetic experience, both of them having various connotations, even among the phenomenological aestheticians. One of these connotations, drawn on by Scheler in his analysis of emotional life, particularly in his lengthy plea for the function of feeling, almost led him into the danger of irrationalism, introducing the idea that we could do without concepts in the ontological analysis of the affective life. In fact, this is how the empirical content of reality came to be more elastically and realistically interpreted. The analysis of emotional life from the angle of the concept of intentionality revealed more poignantly the idea of an axiological specificity of affective spheres, the relationship between the affective and the rational, the possibility of their intelligibility, and the working out of some authentic judgments with the participation of the whole "I." The use of the concept became obvious as soon as the method of intentionality had revealed new significances and meanings comprised by the emotional a priori. An eidetic analysis, if performed within the frameworks delineated by Husserl and Scheler, can lead neither into irrationalism, nor into pure intellectualism or formalism. At the same time, the fidelity and originality that Dufrenne uses in his famous *Phenomenology of the Aesthetic Experience* proves the need for concepts: aesthetic object, aesthetic experience, etc.

In a certain sense, an aesthetic object is in fact the substantive work of art. The epistemological object-subject correlation is meant to open the way toward to study of the object "for us," but the work of art becomes an aesthetic object only insofar as it is experienced and perceived as a functioning piece of art. A piece of music is music only insofar as it is performed and perceived affectively. Therefore, we are supposed to discriminate from the very beginning: *a sensible datum, a representation*, and *an expression*. Somewhat similar distinctions were also proposed by Etienne Souriau when he classed the artistic entity as physical or phenomenal existence, "reific" existence and transcendental existence. For Dufrenne, the aesthetic object is the sensible thing grown into an end and not a simple means to an end (but only insofar as it is perceived and it connects to the real world and the world of consciousness in its own time and space). Its truth is a mode of being consisting in its occurring in the form of signifying expressions (as Lucien Blaga also put it in his "axiological emphases"). In Dufrenne's view, the aesthetic object can appear with different specificities: as nature, as language (expression), as an art object and, naturally, in its ideal form, as an object in the world, i.e., a functioning artistic entity.

The aesthetic object can be perceived in representations, imaginations, intuitions, significances, and as a work of art. In order to be produced and to function in the world, the aesthetic must move both the life of the author and that of the receiver. Its correlation to the subject creates the aesthetic experience. However, it is here that the vicious circle also occurs: we define the aesthetic object by means of the aesthetic experience and the aesthetic experience by means of the aesthetic object. But this vicious circle is also motivated; it has the obvious anthropological aim of introducing the salutary concept of the *affective a priori*. Within aesthetic experience, as we have seen, one can distinguish the aesthetic object from perception, a distinction that seems valid when we notice that the unity of object and subject is of a historical character, that the intentionality expressing this unity does not eliminate a certain realism. But if we start from aesthetic perception, taken as a fragment of human experience, then we subordinate the aesthetic object to this perception and come to the point of having to take as an aesthetic object any object aestheticized by any aesthetic experience (see kitsch), or the ideal image that artists build about some project that has not yet been materialized.

In the phenomenological view, the *object of art* may become an aesthetic object only when it confronts a receiver that perceives its meaning and message. Still, one should not jump to the conclusion that the work of art is accepted within the order of objective realities and that the aesthetic object is ideal, or that the work of art is an object in the world and the aesthetic object is an object in itself, or a significance within consciousness; they have, rather, an essence-phenomenon relationship, and a particular-general one as well. Hence, one infers that the aesthetic object does not explain, but indicates and then invites us to linger on the experienced datum in order to decipher its intention through intuition, to decode the significance of an artistic gesture and transfer it through the material datum from the creator to the receiver.

The perception of this universe of meanings through an aesthetic experience is explained by a consubstantial solidarity by means of the affective a priori, which is, in fact, of a subjective origin and has the virtue of subjectivizing the objective. Sensibility, as an *affective a priori*, is the fundamental form by which the subject creates, discovers, and experiences the values of art. Hence, perception holds a primordial place in the artistic experience of a phenomenological type and functions as a model for the initial aesthetic experience through which the thing can be changed from a neutral object into an aesthetic one.

The aesthetic object amounts to the phenomenon in a phenomenological perspective — *a hypostasis of consciousness, and it introduces a new aesthetic category* side by side with other categories such as: the beautiful, the comic, the tragic, the sublime, etc. At the Second Congress of Aesthetics (1928), aesthetician Geiger uses the term of *aesthetic object* in its sense of a work as a phenomenon. His aesthetic view also emphasizes that the main interest lies in the structure of the aesthetic object, "... the laws of aesthetic values, the principal mode in which they find their groundwork is the aesthetic object, while value or non-value reside not in the real structure of an object, but in the phenomenal one."[1] Sartre in his turn thought "that there never are in the work itself sufficient reasons for the appearance of the aesthetic object, but only requirements for it."[2]

The points of view briefly mentioned thus far, reveal the fact that the aesthetic object is nothing but a concept defining the phenomenal structure of a work manifest in an aesthetic experience. From the available data we can risk the statement that most of the representatives of present-day phenomenology take *imaginative consciousness* to be the shaping domain of the aesthetic object, and hence assume a certain symmetry between the act of consciousness and the configuration of the aesthetic object. Imagination is thought to be an act envisaging an absent or non-existent object in its concreteness, by means of physical or psychic content, an object that does not yield itself as such, but rather as an analogical representation of the intended object.

Therefore, it becomes obvious that imaginative consciousness searches for its object in the field of perception and aims at its sensible elements, and that its intentional object may be almost non-existent or nullified. Any work has an intentional character and, according to the Polish phenomenologist Ingarden, this can be rendered obvious by eidetic reduction, by excluding work from the realm of real and ideal objects. Still, in this case one encounters the drawback that the work can no longer have an absolutely objective character. In this case one has to resort to using the concept of existential autonomy, independent of the act of knowledge but oriented toward it. *Any work behaves like a double existence: a material and an ideal one.* This heterogeneous character is taken as an essential characteristic of any work or of any aesthetic object. The eidetic description envisages the ontic foundation of the mixed work. The ideal in the aesthetic object is expressed by so-called *ideal concepts*, the role of which is that of yielding significances in the process of actualizing meaning.

The *act of signifying* — a particularly important process in converting the work of art into, or transcending it in the direction of, an aesthetic object becomes manifest as a subjective operation of consciousness, which, by actualizing the ideal meaning, makes up a unitary significant image.

Regarding the way the aesthetic object gets shaped and expressed, Ingarden develops a specific attitude in orthodox phenomenology. In his view, in building up the aesthetic object, the work of art co-operates with the aesthetic experience of the subject, and this co-operation is no simple encounter, but a process of determining a qualitatively new phenomenon. Generally, in phenomenology, positions raised upon common sense are rejected and this is why the reality of an object that does not arouse an aesthetic emotion is thought of as unnecessary, even if its very presence often triggers in us an aesthetic experience. Here Ingarden brings forth a series of arguments such as: (1) The object of aesthetic perception can belong to the realm of the imaginary or of fiction. An imagined tragic situation can create in us feelings of terror or sympathy of the same intensity that a real situation would provoke; (2) The specific domain of aesthetic experience is circumscribed by our response to artistic works, which belong to the category of intentional objects and not to that of psycho-physical ones; and (3) An analysis of the first stage of aesthetic experience — the one correlated with the perception of the sensorial object — points to the auxiliary role of the presence of a real object in shaping the aesthetic object. Ingarden tries to exemplify the idea with two different experiences in front of the Venus of Melos at the Louvre. The first one records only unmediated features, obviously a function of the existing subjective and objective conditions; the second perception endows the object of observation only with those qualities that are immanent to it and are structurally contained in the object.

The *aesthetic experience* is another way of regarding the aesthetic object. The aesthetic perception shapes or rediscovers, for instance, the image of a woman embodying the ancient goddess, and the fact that all we have is a simple block of marble full of impurities is overlooked. The aesthetic attitude is initially accepted and it uses a psychic type of effect — the play of the imagination — which fills up the missing parts and corrects the deformed ones, making the object of art become an aesthetic object on the basis of such features as: harmony, integrity, etc. But in the process of aesthetic experience there is a stage when the subject transported into the imaginary world, blocks out the other activities of a practical nature — a phenomenon called neutralization. With Husserl, this phenomenon allows the normal or exaggeratedly normal functioning of visual or auditory perception. Neutralization, besides being characterized only by certain phases of the subjective activity, manifests itself as a freedom to "believe" in the existence of the represented ideal object, and this is so especially when the aesthetic object contains a layer of represented objects sending us to a real referent. The aesthetic experience is, in this case, euphoria, a moment of revelation disclosing something new — an aesthetic qualitative pattern — having the force of reality, though it practically belongs to intentional existence, i.e., to the ideal one. Ingarden traces a common feature between the process of knowing as such and the process of aesthetic valuation. Discussing the specificity of the literary work, Ingarden also assures us that

> the same is true for objects represented in the literary work. These objects, though varied and multiple they may be, are independent, interrelate with one another within a more or less homogeneous whole . . . . And this is called the objective phase or the represented world. The words and sentences occurring in some work define not only isolated objects and people, but also the various relationships, processes and states they are in. In a normal instance, all these represent the components of a single whole.[3]

From the above, we can see that the Polish aesthetician meant to go beyond the exclusivism of objectivist and subjectivist tendencies. The same drive is noticeable with Dufrenne, who is aware of the existence of a vicious circle created by the fact that the aesthetic experience is defined through the aesthetic object, and this returns us again to the aesthetic experience. And Dufrenne observes that two contradictory situations appear now: if one starts from the aesthetic experience, one runs the risk of subordinating the aesthetic object to the aesthetic perception, which, as a matter of principle, can aestheticize any object; if one begins with the aesthetic object, then he discovers that it cannot exist without an aesthetic experience, basically without an aesthetic perception. Dufrenne thus proceeds in his *Phenomenology of the Aesthetic Experience* to have the concept of aesthetic object acquire such a function as to overcome the noted opposition, i.e., as to be a moment of reconciliation, a meeting place for the subjectivity of the aesthetic perception and the objectivity of the art work's structure. In Dufrenne's view, the aesthetic perception lays the foundation for the aesthetic object, and the latter is, in fact, the real existence of the work of art; the object fulfills the resources, the function, the social aim of the work.

In the context proposed by Dufrenne we can speak about a work of art only when it has also become an aesthetic object.

> What the artist has created is not wholly an aesthetic object yet, but only the possibility to be one offered to this object when the sensible is recognized as such by means of our sight. The object created by the artist gains its own specific existence only through the co-operation of the spectator, the receiver of the art work. In order to finish his work, the artist has got to become a spectator as well.[4]

This explanatory discourse prompts us to believe that the work of art and the aesthetic object are reciprocally dependent concepts and that they can be understood one through the other. The work is taken as directly linked with reflection — the reflection of the spectator who tries to find out where it comes from, how it is made, and what effects it produces. But the work will appear to the spectator as an aesthetic object only insofar as he decides to belong to it completely.

A phenomenological distinction similar to the one proposed by Dufrenne is also made by Souriau in his *Le correspondence des arts*,[5] where we find that the work of art enjoys two types of existence; a physical ("reifi") one, in virtue of which the work has a body, and a phenomenal one, by which it confronts our senses. According to Souriau, the study of the aesthetic object should follow three stages of the existence: phenomenal, reific, and transcendental. Aesthetic perception is meant to unify the three existence levels.

One thus notices that when the aesthetic object is an appearance signifying more than the work and the sensible is the whole of the work changed into an aesthetic object stripped down to its phenomenality — phenomenalized essence — then its components, among which are form, substance, significance and expression, become extremely elastic. In relation with a signifying object, the aesthetic object is characterized by the immanence of the signified in the signifier. In G. Morpurgo-Tagliabue's view, the main problem of phenomenology was the *relationship between the perceptive and the ideal aspect* of the knowledge and existence of the work of art, and it is on this problem that the attempts to use Husserlean concepts in aesthetics concentrated.[6] Another remarkable representative of phenomenology, W. Conrad was one of the first to draw attention to the relationships between our perception, where the art work practically exists, and the aesthetic object — the aim toward which tend all types of analysis and all the various moments of aesthetic perception. Along the same lines evolved the thinking of Cesare Brandi also, who, by adopting the principle of phenomenological reduction, was attempting to neutralize existence and perform some kind of artistic stabilization. Starting from the real, from the relationship between existence and image (like Sartre), Brandi maintains that by neutralization the artist squeezes the object out of the world, reducing it to a phenomenon — an optical one in painting or an accoustic one in music, and thus a fully made, constructed, and fixed image obtains shape in his consciousness. "Once the image gets fixed, it will have to be sensibly expressed so as to be thus turned out, conveyed, communicated . . . ; this way the mental image is conceived as a form."[7]

These briefly commented sources point to the fact that perception is highly instrumental in communicating the image and that it thus determines the technical process of creation, the elaboration of the work, and, implicitly, the preparation of the aesthetic object. *In this a view, the phenomenological reduction is the method isolating the aesthetic object from the other (real, natural) objects*, and the *apriorism* of consciousness helps in recognizing the value of the art object, i.e., in turning it into an aesthetic object. In his studies, Ingarden performs an eidetic type of analysis of the literary work. Still, his analysis remains all the time at the level of interconsciousness, and the structure of the literary work as he explained them "concern the work in the pure immanence of consciousness, as an autarchic reality existing outside any contact with external reality."[8] Ingarden's statements seem to be well-founded. His reasoning follows a pattern developed in Husserl's initial phenomenological stage, and Ingarden is known to have got into a very serious argument with his master concerning these very issues — the meanings of aesthetic experience.

Applying phenomenology in aesthetics, Ingarden particularly emphasizes: (a) the ontology of the work, (b) the phenomenology of aesthetic experience, and (c) the shaping of the aesthetic object. Even if he does not use a vocabulary similar to ours, Ingarden lays stress upon the "subject-work encounter as a special, decisive moment" for both components of this circuitry. Transcendental idealism of a phenomenological type does not exclude, but rather brings into focus (like other contemporary aesthetics) the working of the correlation of art — experience — aesthetic value. Even though in this aesthetic conception the systematic and praxiologico-materialist construction is not taken into account (a position we have got used to, but which does not always give satisfactory results and does not provide a genetic reconstruction of the work), still, phenomenological aesthetics in all of its variants is primarily concerned with the dimensions of aesthetic experience, with its relationships to practical life, with the axiological content of aesthetic values. It also has the virtue of having reacted consistently against empiricist psychologism and naked realism by analyzing in detail the aesthetic perception, which, if appreciated objectively, amounts to the concept of social practice. In fact, if we think of the social milieu when it was born, phenomenological aesthetics does nothing else but claim the rights of sensibility (not sensitiveness), of depositing any piece of information onto existence in its pre-analytical form, in the form of perception. And this, as N. Vanina has also emphasized in his essay, "... urges us to return to the objective world, to give back to the thing its concrete physiognomy, to organisms their own way of tackling the world, to subjectivity its historical inherence in finding phenomena, the layer of lived experience by means of which things are presented to us in a state of being born.[9]

Perception holds a very important place in phenomenology, maybe one approximating that of social practice in Marxist philosophy, which explains its having been explored by several thinkers. Among those who have pondered the concept from an aesthetic point of view too, Merleau-Ponty is one.[10] The phenomenological analysis of perception that he carried out is controlled by two principles. (a) Perceiving means seeing how an immanent meaning is born out of the cluster of data. In this type of thinking, significance is only a perceptive area that has not become an objective world yet — some determined existence. (b) The perceiving spirit is a spirit in its body. With his second principle, Merleau-Ponty goes against conceptions which approach perception as a simple result of the action of external things upon our body, and also against those pleading for an autonomy of consciousness.

On the other hand, these principles lead us to the conclusion that an analysis of perception implies an analysis of perceptive behaviour, an analysis of the relationship between the subject, its body and its world — a question that falls outside the limited psychological confines of these investigations.

Perceptive experience, particularly the aesthetic one — also taken as an experiencing of the world — gets identified with the sense of an open wholeness whose synthesis cannot be enclosed, whereas the experience of the "I" appears as something continually closing and opening up in time.

Phenomenology, through illustrious representatives in aesthetics — even though they sometimes may have deviated into transcendental subjectivism (through *affective apriorism*, e.g., Scheler and Dufrenne) — has pointed to the fact that aesthetic experience, a highly substantive variant of human experience, is the most authentic echo of human action and that against its background the most humane consistency takes shape, because art means humanity expressed in its fundamentally felt and idealized values. Artistic consciousness and practice may be thought of as a cauldron of intense heat where various values get distilled together to become a new reality — the reality of man.

On the other hand phenomenology also attracted attention to the fact that the creative and receptive acts are not only inseparable from the subject, but that, as acts of artistic experience, they are fully human, creative, and revelatory of essences. Consequently, the phenomenological method does not forsake the human by using the eidetic reduction; it only gives up, justifiably, the fetish of the material as a physical objective datum and sparks reaction in an age threatened ever more aggressively by technocracy by claiming its right to judge through feeling experience, sensation-perception, by and large. Phenomenology shows or means to show that through artistic experience life is represented in its most concentrated form, resulting from exercising one principle — that of the indivisibility of the "I," with the aim of rendering existence homogenous and meaningful even as it is threatened by anonymity. Art and artistic experience turn existence into valuable existence; art wrenches reality out of the anonymity of dissolution and confusion.

As the cardinal point of phenomenology, perception has the role of determining and prefiguring the destiny of the thing, of the work of art, of the aesthetic object. The thing — whatever it might be — when perceived, acquires a structure inherent to the human body, and the latter, as a figure ordered within existence, represents a synergetic system whose functions are taken over and integrated within the general movement of existence. Merleau-Ponty was trying to prove by this paradigm the value of auditory and visual experiences, their role in the creation of the arts. With the help of this idea, the ante-predicative unity of the perceived world is given foundation and through it — the verbal expression and intellectual significance — the question of art's language, the principal means of conveying any human message receives clarification.

If we were, however, to decide which of the two concepts has priority, then any aestheticism would favor the concept of creation, as it occurs as far back as the ancients, more particularly in Aristotle's *Poetics* or in Plato's *Dialogues*, but also because man's disturbing creative activity seems to be sufficient in itself. Still, the theoretical category of experience should be suspected of neither one-sidedness and incompletness, nor of the tendency to share the well-deserved glory enjoyed by the category of creation. Experience associated with experiment is meant mainly to reveal that the creative act as such has a social character, that the subject acquires his creative powers gradually, by a succession of stages, that there is nothing mysterious even in the activity of aesthetic creation.

The association of these two aesthetic categories in a non-contradictory pairing is of a comparatively recent date; more exactly, it came about as a result of phenomenological research (that of Geiger, Ingarden, Dufrenne, and more recently, H. R. Jauss). Although viewed from various perspectives, the concept of experience also shares common features: almost all interpreters agree that it is undoubtedly one of the springs of knowledge, too. From the position of subjective idealism, it is limited to a complex of the subject's experiences, where there is no such thing as an object independent from the subject. Apriorism allows for the possibility of experiences or knowledge independent of any type of experience. From the position of phenomenological aesthetics — to which, in fact, we owe much in the way of contributions to the understanding of these categories — the aesthetic experience means the act by which the audience of viewers contemplate the artistic object. In this perspective the value generating artistic experience signifies the totality of experiences making up the creative act. It is thought, as Dufrenne persuades us to believe, that it reaches a climax in feeling as a component of expression, as a form of communication. In this hypostasis, feeling indicates the presence of some affective qualities which we should take as the soul of the expressed world; it is only through experience and its affective quality that the whole universe of the work can receive unity.

Errors in interpreting the creative process could be looked for on several levels: first of all on the philosophical one, where, in contrast to the present, the habit of making diachronic analyses through probing the dynamics of connections was absent. Such an approach was also justified by the fact that art and its values once were the preserve of a restricted circle, that studies in aesthetics and art psychology, in the theory of art and culture were not elaborate enough. All these were also closely associated with a cluster of embarassing caste prejudices regarding creation and experience. The present age, with the enlarging of the field of influence that artistic values have, with the greater circulation of values, with the growth of a wider and more competent audience has fostered conditions for a more realistic interpretation of creation and reception, of the aesthetic experience. The aesthetic experience brings into play fundamental *a priori* categories of affectivity in the sense which Kant himself has in mind when he speaks of the *a priori* categories of sensibility and intellect. If the aesthetic experience is an act of discovery (as we tend to believe), is it not possible that the effect of art is in ourselves? We have to part with this subjective conclusion which takes into account effect rather than cause. It should not be taken into account as experience alone, but just as much as a cumulative action.

In order to come into being and make itself felt in people's lives the aesthetic must set into motion both the artistic life of the creator — which means action in the field of creation — and the aesthetic experience of the audience. Both species, though quite different in their content, have had a long common history and can consequently be conceived in terms of what we call productivity (creativity), perception, and communication, or, in H. R. Jauss' terms, *poiesis, aisthesis,* and *katharsis.*[11] These concepts are only partial derivatives of the artistic and aesthetic experience; they should not be taken as the expression of some value hierarchy, but rather as a set of autonomous functions grown out of the aesthetic experience, the more so as these are neither mutually replaceable nor reducible one to the other. However, the creative artist can naturally examine his own work as a contemplator or reader when he moves from *poiesis* to *aisthesis*, just as he can both paint and admire at the same time.

The creative hypostasis in art as a productive component of the artistic experience — liable to become aesthetic, too — when it finds its echo in the sensibility of the audience is the most complex of all stages since it involves the creative subject with the whole of the axiological universe. Both historically and currently, we notice that the productive side of the artistic experience is and develops under the sign of both social praxis and the artistic climate as the most original component of human experience. It has come to acquire a certain autonomy, too, through its specificity and function. This emphasis upon autonomy does not point to a break between artistic and aesthetic experience, on the one hand, and social practice as a whole, on the other, but only to an understanding of their structural and functional specificity. The artistic experience being essentially both *poiesis* and *praxis* does not purposefully bring about — like science — certain phenomena whose laws can then be studied or whose validation could be accomplished in some kind of laboratory. If in the sciences the experimental procedure moves from the unknown to the known, in the arts the beginning is in some given reality and the movement is toward some unforeseeable construct enfolding a wide range of possibilities and variants tending to attain the attribute of verisimilitude. Hence, the somewhat nasty joke of the British physicist Paul Dirac on hearing that his famous colleague J. Robert Oppenheimer wrote verses. Shocked at the news, Dirac said that "the man of science strives to express clearly something unknown until then, whereas the modern poet expresses unclearly something that has been known for a long time . . . ." To return to our problem, we can note that along his path into the creation of his work, the artist goes through all kinds of experiences ranging from scholarly to psycho-social and artistic ones. For the creator of art, experimenting means the validation of the investigative procedures, imagination, and poetic construction, and, hence, both uniqueness and originality, on the one hand, and art's renovating contribution, on the other. The artistic experiment does not follow prescribed rules, just as creation does not obey any canons. Contemporary literature, functioning in an age of profound questioning of human creation, cannot help being influenced in its constructive methods by the characteristics of contemporary creation: lucidity, quantification, anticipation, serializing, probability, etc. The frequently used phrase "poetic craft" means only the artist's craft: "While the theoretical craft means building real sentences and the practical one judging actions as functions of the criteria of Good and Evil, the poetic craft shows what can be done . . ."; it goes beyond the limited horizon of the *imitatio naturae* and proves that more can always be done in the realm of artistic reality than is revealed at a certain moment by landmarks established by theory and/or practice.

Poetry as the second nature created by man gathers together in the whole it builds the perfection scattered about the world and creates an illusion of things being other than they are in reality — more beautiful and more perfect. By this very idealizing function, the poetic function challenges the real-objective world and compels it to be redefined through new spatio-temporal features. In this space the artistic artifice parts with nature through its anthropomorphic character, through its capacity to create something new, something unknown in nature.

Creation and experience, therefore, are not mutually exclusive, but rather complementary and immanent to the artistic process. The essential condition for the creative act's becoming correlated with the artistic experience is made up of experiments occurring as functional entities which gradually regulate the creative act. The aim of experiment is exploration, prospecting and testing with a view to obtaining a certain result, which in the realm of art has a creative value. In this sense, experiment amounts to a moment of creation or a test for a new formula of expression. However, there are here certain situations that need to be mentioned: creators of art have always felt a need to dispense with or leave old fashioned forms and to propose novel formal variants, usually ones suited to their personalities. Some of them can even abandon their own style in the process; they either wander off from an unsatisfactory style and find their own, or adopt a fashionable mode that they find to be more appropriate to the requirements of the period, even at the risk of their personality's remaining unexpressed. In this case the experiment is of a semi-lucid kind and is thus a false experiment. Experiment may become fruitful only when conceived rationally, when it stimulates the creative impulse, when it changes it from a fact of consciousness into one of existence; otherwise it is a pseudo-experiment.

In a different hypostasis experiment amounts to the achievement of an artistic project that gradually goes through several stages or variants. Now, the experiment as an act of creation is also a refresher stage, though at other times it may be a terminal point, too. We find in experiment, *volens nolens*, a qualitative variation as well. That is why the experiment is sometimes the equivalent of approximation, of relative value, and at other times success or failure. The great creators, both classical and contemporary (in literature, the fine arts, theatre, music, etc.) have never achieved what they attempted, and not everything they thought to be finite was perfect. Here is Picasso, to give an illustration: "Of all the charges levelled against me, none is more untrue than the statement that my essential object is the search. When I paint I want to show what I have found already, not what I am looking for." Approaching the specificity of experiment in the act of musical creation (composition), one can distinguish between how the scheme of a musical form is realized and the perfect, live form. The discovery or imitation of a compositional pattern engages the intellect, but its becoming functional involves an affective-emotional component. The genesis of the pattern does not require the composer to feel the development or flux of the musical content, since it is only a matrix devoid of sound, incapable of suggestiveness. The pattern is not music, but a general project, some manifest strategy, which is why it has no capacity for self-communication, for carrying some artistic message. Isolated from its sonorous substance, the pattern is atemporal; it is only an experiment making up the composer's experience, it is an attempt.

By analyzing experiment we come across the existence of other traps: one of them has its roots in the idea that since the essence of experiment is a perpetual leaving open, an externally unfinished process, then it would be meaningless to speak of any stability, style, work, etc. on the grounds that adopting a formula leads to mannerism, stagnation, artistic death. In this instance, however, one ignores the fact that styles also go through a succession of stages before completion, and the outgrowing of one style may be taken as the development of another one, which, in its turn, implies a succession of other experiences and creations meant to express and buttress the new direction of the creative process. Style is, in fact, the locus where tradition and innovation organically fuse; it is the form in which what has been perfected through an endless line of experiments and through constant experience is fixed; it is the phenomenalized state of the consciousness of an epoch, it is the unity of expressive means and of language burdened with ideas that is embodied in the works — a unity which appears as the fruit of spontaneity and lucidity.

Style denies anyone's right to and possibility of making new experiments, of innovating; still, however many and however different styles may be, the existence of a closed style — like the Wagnerian one in music or the classical one in architecture — is no longer feasible. After Wagner, who took the style of opera music to its limit, the only choices left were either an extension and amplification of dissolving tonalities as in the achievements of Mahler, Bruckner, and taken to the limit by Schönberg, or an opening toward new horizons in the manner of

Debussy. The unoriginal utilization of a style leads to mannerism, and mannerism lacks any kind of challenge; it neither experiments, nor creates, it only repeats, freezes, and occasionally finishes or emphasizes shades and details.

Another dimension or functional component of the artistic experience is the perceptive one, which lends itself to a pertinent epistemological approach, especially through the modern methods of semantics and semiotics. Semantics and semiotics have become necessary in the horizon of art and have the particular aim of ordering and clarifying the host of information and knowledge delivered by a multitude of stimuli; otherwise, it seems, the comprehensive development of the significant values of contemporary art would be very endangered. In the new circumstances of the contemporary scientific and technological revolution which influences both artistic creation and its reception as experiential facts, wide avenues have been opened toward new and unexpected fields of knowledge and interpretation. We are in the age of a new emancipation of the senses; a new nature is being explored (according to other principles) — one that is different from the one seen with the naked eye but, still, one engendered and prepared by the latter. Photography and film, by immortalizing the moment and revealing the accidental have challenged the optical mechanism of the subconscious. We are well beyond the stage of seeing a natural detail through feeling alone. In the twentieth century art of the novel and drama, artistic experience developed new levels of creation and reception, reaching a climax in antinomous positions (witness Beckett, Proust, or Brecht). By liberating literature and the arts from narrative realism and by adopting methods experimenting with live discourse, the creator and reader are already familiarized with the dependence of experience on logical and linguistic clichés; by phenomenalizing action, the hidden dimension of things, the causal transparency of the artistic universe is revealed.

In the contemporary experience of literary creation — vast in both its range and methodology — a new type of writing is being inaugurated, one that seeks to possess the poetry of things unalienated from or indistinguishable from the beauty of aims while also preserving the possibilities of verisimilitude.

The third moment derived from aesthetic experience is its communicative dimension. If the productive and receptive function of the artistic experience prepares the creative act up to a certain stage, we could say that the communicative function completes it. Catharsis as a communicative moment both contains and has to take into account certain assumptions: the initial Aristotelian meaning of artistic enjoyment, the role of curiosity in triggering artistic communication, the influence of feelings on the peruasive power of a poetic discourse. Communication as a functional fragment of artistic experience — differing in time and from one art form to another — requires that the creative personality and the receiving subject enter into a process of identifying the art work with a set of values related to some ideal and some aesthetic taste. But the identification can be either a normative-conformist or a sympathetic one. In the first instance it generates and sustains the illusion of empty forms, it is simple ideological or religious ritual, which, in fact, hinders the development of art. In the second instance, sympathetic identification is productively enjoyable and stimulates affective-rational energy as it proposes new models and values, renews artistic expression and refurbishes content. There are cases, however, when a new system of artistic values faces strong opposition from dubious, conservative tastes and ideals; identification here involves the fiction of a real or imaginary object, but only through it does identification lead to significant satisfactions.

Describing the instances and moments in which artistic creation and experience take shape and function has revealed the complex character of the social status of art and the need for a special epistemology that could take more closely into account the logic of things in this realm of human existence. This survey has also sought to bring into focus the idea that, however strong its natural foundation and however many affective-rational resources the creator of artistic values might have, the act of artistic creation and the artistic and aesthetic experience are inseparable from reception. Apart from it neither the act of sensitizing nor that of affectivizing — both fundamental in the educative act — can occur. It is also worth revealing that the domain of the aesthetic constitutes a framework for any artistic experience, wherein are rooted all the other types of human experience. Artistic experience, in its restricted sense of an artistic attitude toward reality, grows out of the aesthetic quality of reality fashioned as a fiction or as a reality in which each process reaches its climax and each experience its completion.

However, the aesthetic experience is fulfilled only to the extent that it becomes an aesthetic taste and ideal. We will never entirely agree with Kant's view that the judgment of taste is to be referred exclusively to the gnoseological model. The taste conditioning sensibility is the fruit of a cooperation between imagination and fancy, and between intellect and feeling, with none of them subordinated to an other. Aesthetic taste and sensibility cannot be postulated, but can be educated; they result from permanent cooperation among a succession of experimental acts of feeling and judgment. The one who judges is not a priori judged only by his psychic peculiarities, but also by both social and individual experience, by the conditions society creates for him. The aesthetic experience and aesthetic feeling at the same time signify a correct apprehension of the work of art, and this perception is conditioned by the existence and the quality of aesthetic taste. Once created through the mechanism of the judgment of taste or through that of creation, sensibility becomes the existence of the aesthetic object. The object of art in general and of music in particular can only exist in the medium of the affective-rational state. Through feeling as the main instrument of sensibility, the subject transcends it uniqueness and becomes available for the axiological universe. That is why the perceived aesthetic object has the quality of joining people together at a higher level, and aesthetic contemplation becomes an act of social integration and unity. Hence, the conclusion that the existence of artistic value gains its right to existence through aesthetic sensibility, whereas the audience becomes a relational category setting up and conditioning not only the creative act, but also the perennial condition of values, their very existence.

Iasi

NOTES

[1] Moritz Geiger, *Zugange zur Aesthetik* (Leipzig: 1928), p. 143.
[2] J. P. Sartre, *Qu'est ce que c'est la literature* (N. R. F., 1948), p. 59.
[3] Roman Ingarden, *Studii de estetică* (Bucharest: Ed. Univers. Bucuresti, 1978), p. 38.
[4] Mikel Dufrenne, *Fenomenologia experientei estetice* (Bucharest: Ed. Meridiane, 1976), Vol. 1, II, pp. 64—65.
[5] Etienne Souriau, *La correspondance des arts* (Paris: 1947), pp. 44—70.
[6] G. Morpurgo-Tagliabue, *Estetica contemporană*, Vol. II (Bucharest: Ed. Meridiane, Bucharest, 1976), p. 205.
[7] Cf. G. Morpurgo-Tagliabue, *op. cit.*, p. 206.
[8] Ingarden, *op. cit.*, p. 147.
[9] Nicolae Vanina, *Tendinte actuala in estetica tenomenologica* (Bucharest: Ed. Stŭntică, 1974), p. 52.
[10] M. Merleau-Ponty, *Phenomenologie de la perception* (N. R. F., 1945).
[11] H. R. Jauss, *Experientă estetică şi hermeneutică literară* (Bucharest: Ed. Univers, Bucharest, 1983).

ASBJØRN AARNES

A POET'S LIFE AND WORK IN THE PERSPECTIVE OF PHENOMENOLOGY

This paper will draw your attention away from the main currents and problems of phenomenology to a poet who was not a philosopher and of whose work only a few specimens have been translated into other languages. Of Emil Boyson's ouevre, two poems have been translated into English by Martin Allwood, another by the poet himself, and some have been translated into French by Pierre Emmanuel and myself.[1]

Emil Boyson, was born in Bergen in 1897, and he died in Oslo in 1979. He published six volumes of poetry in his lifetime and three novels or, better, tales of a particular kind, namely, epic-lyrical prose tales. He wrote besides a number of essays, among them one on Coleridge and the concept of imagination and another on the Norwegian scientist and philosopher of the Romantic period, Henrich Steffens. He published a translation of selected writings of the French painter Delacroix, and also wrote newspaper articles from Berlin just after the Second World War. He was co-editor of a dictionary of literary and philosophical terms, and he made several justly famous translations of European poetry, published in a volume entitled *European Poetry*. He took the initiative to publish a large anthology of Norwegian poetry and another one of French poetry translated into different Scandinavian languages.[2]

Today Emil Boyson is recognized as one of the most outstanding poets in Scandinavia in this century. He belongs to the late Symbolist tradition, his artistic roots being in France, in the movement instigated by Baudelaire, Verlaine, and Mallarmé. He regarded his own work as a continuation — or rather as a shadowy imitation — of those masters. Once when he was asked about his favorites, he answered: "Give me a handful of poems by Baudelaire, Verlaine, and Mallarmé, and that will be enough for me." By critics he has often been cited as a "Norwegian Mallarmé" — a formula that he himself found pretentious, ridiculous, and impossible.

Boyson's poetry had, according to his own statements and those of his critics, two main sources of inspiration: on the one hand, Husserlian phenomenology; on the other, French Symbolism. The Husserlian

A-T. Tymieniecka (ed.), Analecta Husserliana, Vol. XXXVII, 167–179.

phenomenology will be the essential topic of this paper. Boyson was engaged by it, especially in its early expression, as it was worked out in the Göttingen period by Husserl and his young scholars, Adolf Reinach, Hans Lipps, Edith Stein, Jean Hering, Hedwig Conrad-Martius, and Wilhelm Schapp. The latter's *Beiträge zur Phänomenolgie der Wahrnehmung* (Contributions to the Phenomenology of Perception) was his favorite reading, from the beginning of his career to the end of his life.

It was the *realist* trend of phenomenology — or what was expected to be a new realism in philosophy — that captured his mind. Because he was not a regular scholar, he was spared the deception discovered, with the publication of *Idéen* I in 1913, "that we were working on the foundations of a new idealism," as Roman Ingarden puts it.[3]

Boyson's philosophical formation was thus linked to the early version of phenomenology. That was "enough" for him, as he had said about the handful of poems by Baudelaire, Verlaine, and Mallarmé.

His studies in philosophy did not stop there, however. He followed with eager attention any new signs of budding realism, in the writings of Dietrich von Hildebrand, Samuel Alexander, Gabriel Marcel; and of course any new interpretation of the early phenomenologists. He followed the work and fate of Edith Stein with a particular interest and empathy.

An intellectual and human surprise that confirmed his life long interest in phenomenology was reserved for Emil Boyson's old age: in 1966 the Polish philosopher Roman Ingarden came to Norway.

Ingarden was one of Husserl's first scholars. He attended his lectures as early as 1912. In his major work, *Der Streit um die Existenz der Welt* (The Controversy over the Existence of the World), he developed a consistent realist philosophy within the scope of phenomenology. Ingarden had been faced with an ultimatum by the authorities in Poland: either teach Marxist logic or retire from office. He chose the latter and in 1967 he came to Norway for six months. At the University of Oslo he gave two series of lectures: one on Problems of Aesthetics and the other titled: Introduction to the Phenomenology of Edmund Husserl. The latter series was published in Norwegian in 1970, a highly important book of about 350 pages.[4]

Ingarden's visit was for Boyson a crowning event. He attended all of his lectures, and was particularly interested in those on Husserl. A warm friendship was established. When Boyson later participated in the translation from German of Ingarden's lectures, he had, as he said, the satisfaction of paying off some of his debts and of sustaining a "mood of thought" that had been "the light of my youth."

I asked Boyson once why he did not pay much attention to Ingarden's work on aesthetics. He replied by telling me an anecdote: "A traveler came once to an inn where he wanted a room for the night. They had only one room, and that one, they warned him, was haunted. The traveler took no notice of that, but before he went to bed, he put a card on the door: 'Phantoms, no admission. I phantomize myself.'" So Boyson said, "I don't worry much about aesthetics, I aestheticize a little on my own."

It was Boyson's opinion that artists — especially painters and poets — have, from the time of Descartes, been haunted by the idea of nothingness.

Since Descartes, poets and painters have been condemned to live and work on second-hand reality, shadows of reality. Feelings, colors, sounds do not belong in the world; they have no other root in being than in the poet's sensorium. A poet is a person who works with and within himself, on his own ideas of sensation, ideas produced within himself. Savoring aesthetic dignity, he may be proud of his autonomy, calling himself a king, or, like Baudelaire, he may see imagination as the Queen of Faculties, but here were a king or queen without land, a John Lackland — or as Ibsen puts it: ". . . interpreters on the foundation of the self" — *fortolkere . . . på selvets fundament.*[5]

When Boyson started to write during the First World War, the spiritual climate in Scandinavia was dominated by the psychological associationism (*psychologisme associationiste*) in which the poet is bound to the endless refinement of impressions, which circle the ego, and, like the Lady of Shalott, he sees nothing but shadows of the world.

It was while caught up in this mood that phenomenology happened to open Boyson's eyes. In his own words: "The reading of *Logische Untersuchugen* (Logical Investigations) and some articles in the *Jahrbuch*, confirmed a quite natural belief for me: 'that this tree in the garden is a real thing and not my idea of sensation nor an association of such idea.' There *is* something in the garden, something inexhaustible that *gives* ideas of sensation. It is only by an act of violence that this experience may be dissolved into elements of sensation: it is only by deliberately closing my eyes that these ideas cease to compose things, bring things to my knowledge — and so disappear in ultimate constituents of sensation. Or, from my point of view: I am on my way towards a source of being, towards a profusion of things, without running dry. I felt as if I had awakened from a dream, and the phantoms of the ego evaporated."

The problem of the existence of the Real was not solved, but there was a *beginning* that annulled all prior hestitations: a faith in reality was achieved, a conviction of belonging to the world. It was indeed no "naive" realism, for Boyson returned to this starting point over and over again, asking how this world and our relation to it may be understood. He found that the Real is not given once and for all. Using the terminology of Bergson and Marcel, he preferred to speak of the *giving*, instead of the *given*: *non datum sed dans.*

For this contact with the Real he often used the word *meeting.* To *meet* reality on the poetic level is not to be struck by something already constituted — things *met* being of a great variety. He described it in this way: "There are several relations of transcedence, aspects, colors, sounds coming forth in an abundance that overwhelms my capacity for perception, in a process of confirmation and deception. I experience a plenitude that I am not able to retain or express, my *locus* being in the middle of transcendences."[6]

The poet, as Boyson saw him, is a person who wanders around real things which persuade and dissuade him. He is outside the scope of a definite victory or definite defeat. This was what he meant by *meeting,* to experience by *meeting.* The generosity of being means that the world is richer than any of our ideas of sensation or the sum of our ideas. And seen from the subject's side, we are in the modalities of our sensibility richer than we ever can tell.

The creation of a poem supposes a certain influence of grace — when the artist happens to meet and transpose into words and rhythms a vision of determined configurations of appearances, as when a sad face turns to light, and like the inevitable sincerity of a child who has discovered something evil and who asks you for an explanation . . . . And not only in these traditional human lineaments does the poet look for the essence. Let us listen to Boyson's vision of the miracle of spring, of the beginning of spring with the melting of the snows, the bursting out of water and sun, the birds singing and the heart hesitating towards unspoken words. Here is the translation of it by Pierre Emmanuel and myself; the poem is entitled "The Melting of the Snows":

La fonte des neiges

A travers l'espace nu
déchiré par la morsure du soleil
un vent du sud se précipite,
sourdement sauvage et lourd.
L'odeur du terreau monte des prés,
le bleu lointain
scintille comme de la glace,
l'esprit balance, puissant,
entre le froid et le feu.

Le pouls argenté du ruisseau
au soleil,
le sifflement blanc et clair
des mésanges,
l'air saluant haut
lance des étincelles
jusqu' au sang qui frissonne.
Plus sombrement, il sanglote,
sans bruit,
sous le pont des neiges . . .
Le coeur, alerte comme un aveugle,
tâtonne vers des paroles *inédites*.[7]

With this, I have tried to give some glimpses of one source of inspiration in the life and work of Emil Boyson: his discovery of a realist ontology. I think that further investigation into his writings will clarify the sense of this realism.

Another Norwegian poet, Nordahl Grieg, has declared that: "Botany is what the flowers detest more than anything else." Boyson, being in profound disagreement with the dogmas of modernity, contested the pretense of the natural sciences that they have a monopoly on knowledge of the visible world. I think that in his meditation there was a knowledge that flowers find pleasant. Several times he spoke of an intuition of essence that may come to artists, and surely not only to artists — an intuition that comes as a call or a consent from the things themselves. In notes in books of his favorite poets — Stefan George, Trakl, Claudel, Saint-John Perse — we find expressions like the following: "Here the phenomenon lies bare"; "here the essence of waves splintered by the rocks in unveiled"; "this is the pure phenomenon of *anemona nemorosa* — the white anemone."

His way of "saving the phenomena" was *Theoria* — theory in the etymological sense: a concentrated attention of vision — or as Rilke puts it in the *Aufzeichnungen des Malte Laudrids Brigge*: endless patience in learning to see.

We must admit that Boyson's sympathy went a step further back than phenomenology claims or allows — back to the Thomist tradition of the *philosophia perennis* — in which not only the poet is pleased to know the flowers, but where the flowers too "enjoy" entering into the mind of man, fulfilling thus the purpose of their existence.

Boyson's realism was not "naive," because it was not unquestioning; but there was a trace of dogmatism in it. He was so fascinated by reality that he sometimes confounded it with normality. He had no time for eccentricities. To the great eccentrics for instance in the Russian novels he preferred the normal *dramatis personae*, even if they were only bystanders. The universe of Kafka was closed to him. "I am willing to read Kafka once more," he said, "if somebody can convince me that the labyrinths of his novels have a *fundamentum in re*, as for instance the Austrian Habsburgian bureaucracy."[8]

The second source of inspiration in Boyson's life and work — the second half of the same love — was, as I have already mentioned, French Symbolism. In the works of Baudelaire and his school he had learnt how to transpose the intuition of essence into words, into metre, breathe, rhythm. But that is a topic for another paper.

In the last part of this lecture I want to consider in some greater detail Boyson's relationship to phenomenology.

What you have heard so far about this poet is not spelled out explicitly in his books. Most of it I have gathered from conservations with him during our friendship which lasted for more than twenty-five years — and from his numerous letters.

When Boyson died in 1979, a new phase in our acquaintance with his work began. Papers that were found after his death confirmed to a certain degree what we knew about him in his lifetime, but we also became aware of new lines, new perspectives. Boyson was very shy, quite like the Spanish master Ramón Jiménez whom he admired so much. In conversation he spoke mainly of what could be seen, heard, or felt in the present; he never talked about himself. When we examined his notebooks and his excerpts from the books he had studied, it became clear that the aim of his studies was neither the gathering of information nor a logical *divertissement* in solving metaphysical problems. It had been a meditative exercise in reshaping perception and in achieving understanding in accord with reality.[9]

I asked him once how a poet works: Does he gather material like a scientist, formulate hypotheses to see them confirmed or refuted by facts? Or is he simply a writer, an author being, as Roland Barthes has it, "somebody for whom language has become problematic"? Is he an "*arrangeur de mots et de syllabes*" as Malherbe understood him — a conception of which Mallarmé and Valéry are the heirs.

While Boyson respected all these questions and positions — except the linguistic point of view — which he considered to be the linguistic fallacy of our time — his own answer was different: "My task is to clarify my eyes and keep my senses alert to the spirit of reality."

In his younger days he would probably not have used the word "spirit" — *reality* would have been "enough" for him. In some of his early poems there is indeed such an abundance of descriptions and such an attention to the multitude of signs that they "often become difficult to understand, and appear fragmentary," as Martin Allwood, his translator, says.[10]

His prose is also loaded with the description of details, and it is not easy to bring one's reading to a conclusion: Sometimes you have the feeling that the human eye cannot sustain so much reality! He was, however, aware of this himself; in the opening paragraph of one of his novels he wrote: "In this room there is nothing to see for the person who asks and look around in order to run further . . . . The person who possesses the patience of vision may learn who (the author) was."[11]

A critic once wrote to him: "Is this fiction? Is it not rather a phenomenological description of physical and human space?" From a literary critic's point of view, this was a rather dubious compliment, but Boyson was pleased by the statement and remembered it with gratitude.

In his later poems the images are clarified and reduced to their essential features. We have a proverb, saying: "You cannot see the forest for the trees." Boyson, discovering the "spirit" of reality, did not lose sight of the infinity of trees, but he got a better view of the forest — the distinct parts are fused into a whole, still keeping their singularities, their nuances, their aspects.

We witness here a sort of "eidetic reduction," and perhaps this change may be seen in the light of what I have already called the last step in his life and work. That step took him in the direction of the Thomist tradition but did not take him away from phenomenology; I think he saw it himself rather as a fulfillment of the phenomenological approach to reality.

He never became a Thomist or neo-Thomist. He wanted openness to "life" from philosophy. Philosophy in which the conceptual apparatus dominates intimidated him. When he preferred Gabriel Marcel and Etienne Gilson to Jacques Maritain it was probably because of the former's "empiricism," openness to experience.

His meditations on the realist position made him aware of the dialectics of noesis and the noema, the reality of thinking and the reality of the world. Their "mutual penetration" — to use one of Bergson's expressions — seemed to him a mystery. If we want to go beyond experience, beyond the moments of evidence, to understand this union as a substantial truth, only faith can testify to it.

The *genius malignum* of Cartesianism had caused a dislocation of the regions of the Real, inflicting a wound on both man and nature. The result was the suffering that Marcel has described as *le mal de l'avoir*,[12] the illness of possession: man reduced to a phantom, the epistemological subject — and nature reduced to the market place, a reservoir of provisions used by man. As Boyson saw it, relief from this suffering could only be found in a philosophy open to the mystery of the Christian faith.

Earlier in his life he had entered the Roman Catholic Church, and although he never became a defender of the faith, he always asserted his conviction that Catholicism is more open to reality than the Protestant tradition of our country. He saw Protestantism as a consequence of idealistic subjectivism wherein the way of *Je pense* — I think — was replaced by the dogma of *I believe*.

I think that the problem of culture and the problem of praxis played a part in his moving beyond Husserl. Spiritualism had always been prominent in his research, but over the years this spiritualism grew more and more concrete, and finally drove him away from the philosophy, the culture, and the everyday life of modernity. A regressive tendency came to dominate him — hostility to science, to technology, to the machine age. He even felt animosity towards the Roman Church because of her efforts at modernization — her attempts at bringing up to date the form and content of her ancient legacy.

Teilhard de Chardin, although a pupil of Bergson, was not Boyson's man; and that was precisely because of Teilhard's striving to combine faith and science and to, on the premises of science, abolish the gap between the natural and the sacred.

One of Boyson's satirical poems against modernity has been translated into English; it is difficult to explain, but the intention in clear:

Contact Useless

In the Magnito-Gorsk of the Giant's Mill
I met (and this is one of the lesser dreams)

The beast we have reared, the machine-ape-phantom,
The anthill king, killer of embryos,
The concentration camp experimenter,
God of the strongest battalions,
Professor of Man-of-flesh-and-calculation.
"All that you know" — was my word to the beast —
"I know too;

But you do not know the world I love."
The beast supposedly replied: "I will destroy the world
you love."

I answered: "You will, but you cannot know it."
The conversation ended with no result."[13]

Between Boyson and Ingarden there was a community of mind, both in philosophical and religious terms. Boyson appreciated not only his realist ontology, but admired also his methodological seriousness, his attention to vision, and his untiring descriptive patience. One of his last wishes before he died was to have translated into Norwegian what he called Ingarden's little masterpiece: *Ueber die Verantwortung* (1970), (On Responsibility). "I know no other book" he said, "that teaches so eminently how to work in philosophy and how to be a philosopher."

Ingarden was not optimistic about the future of phenomenology. Phenomenology was destroyed, he said, by the two H's: Hitler and Heidegger.[14] Boyson repeated this *calembour* with a sarcastic smile. Their hopes for the future of phenomenology were pinned on the USA — and on Latin America, Boyson used to add.

When I consider the source and the initiative of this Congress, I understand what both of them meant.

It is time now to take leave of this poet. If unfortunately, you have heretofore had little opportunity to learn of him, I am glad to have enabled you to make his acquaintance.

University of Oslo

NOTES

[1] Martin Allwood, *20th Century Scandinavian Poetry* (Marston Hill, Mullsjö, Sweden: 1950). In 1979 I gave a lecture at the Sorbonne, "Tendances actuelles de la poésie norvégienne," and on that occasion, Pierre Emmanuel and I translated several Norwegian poems, among them, two of Boyson's. For Allwood's anthology, Boyson translated one of his own poems, "Hær-tatt land," titled in English, "Norway, 1941."

[2] Here are the titles of Boyson's major poetic works: *Skumring mellem søiler* (Dusk between Pillars) 1932; *Varsler og møter* (Forebodings and Meetings) 1934; *Tegn og tydning* (Sign and Interpretation) 1935; *Gjemt i mørket* (Hidden in the Dark) 1939; *Sjelen og udyret* (The Soul and the Beast) 1946; *Gjenkjennelse* (Recognition) 1957; *Utvalgte dikt* (Selected Poems) 1959; *70 dikt* (70 Poems) 1974. His novels are: *Sommertørst* (Summer Thirst) 1927; *Yngre herre på besøk* (Youngish Man on a Visit) 1936; *Vandring mot havet* (Wandering towards the Sea) 1937, new edition in 1974. His article on Coleridge is published in *Litteraturforståelse. Diktning og kritikk* (Understanding Literature, Poetry and Criticism), Idé og Tanke, 1961; *Eugène Delacroix Dagbøker* (Diaries) 1954; *Henrich Steffens. Forelæsninger og fragmenter* (Lectures and Fragments) in *Norsk Filosofi i det 19. århundre* (Norwegian Philosophy in the 19th Century) 1967; *Litterært Leksikon* (Dictionary of Literary Terms) 1964, 1967, 1976; *Europeisk poesi* (European Poetry) 1965; *Norsk Poesi* (Norwegian Poetry) 1961, new edition in 1983 by Asbjørn Aarnes; *Fransk poesi* (French Poetry in Scandinavian translation) 1963, new edition 1970, for Boyson's seventieth birthday in 1967 a *Festschrift* was published, *Poesi og virkelighet* (Poetry and Reality).

[3] A statement from one of Ingarden's lectures in Oslo; in his *Introduction to the Phenomenology of Edmund Husserl* he tells of the astonishment of Husserl's students when they read in *Idéen* statements like these: "If we take away the consciousness, we take away the world" or "If there is no pure consciousness, there will be no world"; see pp. 44ff.

[4] From the German manuscript, this book has been translated and published in Poland, 1971.

[5] Henrik Ibsen, *Peer Gynt,* Act IV, set at the Lunatic Asylum in Cairo where Peer meets the master of "hermeneutics," Mr. Begriffenfeldt.

[6] This statement summarizes a passage in Ingarden's introduction to Husserl's phenomenology.

[7] The poem is in Norwegian entitled *Sne-løsing*. It was published in *Varsler og møter* (Forebodings and Meetings); here is the original:

Gjennem det nakne rum,
flenget av solens bít,

velter fra syd en bris
ulmende vill og tung.
Muld-lukt slår opp fra mark,
blået blinker som is,
sindet vakler av kraft
midt mellem frost og ild.

Bekkens sølvpuls i sol,
meisenes blanke fløit,
luften vinkende høi
gnistrer mot sittrende blod.
Mørkere hulker det stilt
innunder sneens bro . . .
Hjertet, på vakt lik en blind,
famler mot usagte ord.

[8] In Boyson's correspondence with Alf Larsen (see *Edda*, 4, 1982), he sees the main difference between himself and Larsen in the latter's affinity for the Russian genius; what he, himself, prefers in Russian art and culture is the *ballet*: "the most Western and the least Russian of that culture." I think that nobody succeeded in convincing him of the existence of a *realist* background for the works of Kafka, so that this author remained forever a stranger to him.

[9] The *Nachlass* of Boyson was taken to Hamar in 1980, where Odd Solumsmoen organized a *Boyson Archives*; at the death of Solumsmoen in 1986, the Archives were taken over by Universitetsbiblioteket in Oslo.

[10] Allwood, *op. cit.*, p. 113.

[11] *Vandring mot havet* (Wandering towards the Sea), 1974, pp. 7 and 9.

[12] See, for instance, his *Etre et Avoir* (Paris: 1935), especially "Esquisse d'une phenoménologie de l'avoir," pp. 223—225.

[13] The title of the original is *Kontakt fruktesløs*; published in *Sjelen og udyret* (The Soul and the Beast):

I Grottekværnens Magnito-Gorsk
møtte jeg (dette er en drøm av de mindre)
Dyret vi har oppfødd, maskin-ape-spøkelset.
maurtuekongen, fosterdreperen,
koncentrationsleirexperimentatoren
gud for de sterkeste bataljoner,
professor i mennesket av kjød og regning
"Alt hvad du vet" — var mitt ord til Dyret — "vet også jeg;
men DU kjenner ikke den verden jeg elsker."
Dyret formodes at ha svart: "Jeg skal ødelegge den verden
du elsker."
Jeg svarte: "Det skal du, men du skal ikke kjenne den."
Samtalen endte uten resultat.

Allwood's translation is true to both form and content. It is something of an "occasional" poem that Boyson did not reprint in *Selected Poems*. The poem *Norway, 1941*, translated by himself and published in Allwood, *op. cit.*, is also an "occasional" poem (The Norwegian title: *Hær-tatt land* means "Occupied Country"), but that one is reprinted in *Selected Poems.* Here are the original and the poet's translation:

Norway, 1941

My country, mutely guarded by the stars:
Ere I resign myself to the mirage Sleep
Slowly turning to thee I summon my soul,
Scrutinize myself before thy tranquil face.

The brotherly silence of spirits that Death made pure
Looms in my mind, reminding me of thy fate;
Those who died for thee, did share in death
Their loneliness with thee, therefore I am lonely.
Empty and weaponless is my hand:
But this I know, that fathered by thee my will,
Mute and austere like the stars, desires
To serve thee, fighting in thy holy war.

Grief and patience are the portion of the living:
But though I be never elected, may at least
The icy clarity of these starry spaces
Strain, like the steel of a sword, my soul!

Soon my will, still watching thee, shall drown
Deedlessly in Sleep that wipes out all.
Fidelity be my last prayer! Soon only
Thy honor immaculate mirrors the guardian stars.

Hær-tatt land

"Fedreland under stjernenes stumhet!
Før jeg trer inn der søvnen hildrer,
stanser jeg langsomt, kaller min sjel,
gransker mig selv foran Ditt stille åsyn.

Broderlig taushet av ånder som døden adlet
demrer i minnet, maner mig med Din skjebne;
hver som døde for DIG, var i sin død
ensom med dig, derfor er jeg ensom.

Våpen eier min hånd ikke;
ett er likevel visst: Du avlet
vilje i mig, som disse stjerner
stum og rên, i din strid at tjene!

Sorg og tålmod er de levendes lodd;
men om jeg aldri til strid vigsles — la enda
stjernedybdenes kolde klarhet
spenne mitt sind lik en klinge av stål!

Snart skal vilijen, som ennu ser dig, i søvn
dådløst drukne. Der munner Alt ut.
Bønn om troskapp blir siste tanke.
Og urørt speiler din ære stjeernenes værn.

[14] A statement from Ingarden's lectures which he often repeated in conversation; it is a leading thought in the first two chapters of his book about the destiny of phenomenology.

WACLAW M. OSADNIK AND LUKASZ PLESNAR

ON THE QUASI-INTENTIONAL NATURE OF REPRESENTED OBJECTS IN A FILM WORK OF ART

In several previous studies on the theme of the phenomenology of the cinema, we indicated the advantages resulting from the formal analysis of a film work.[1] In this study, we want to refer to the article in which we proved the quasi-linguistic character of a film work (if we consider it as a system of signs).[2] The aim of this text, however, is the description of the quasi-intentional character of the objects represented in a film work. According to Ingarden's thesis, the represented objects forming the represented stratum of a film work possess an intentional nature. Let us begin with the explanation of the very notion of an intentional object. Psychology and philosophy differentiate "the act of thinking" from "the content of thinking." This refers to all sorts of representations, judgements, statements, etc. It is maintained that different acts may possess the same content. One of the simplest examples of such a case may be the situation in which two persons are thinking about the same event at the same time, for example, that it has started raining even though the sun was shining a few minutes earlier. We have to do with two acts of mind although the content of both of them is identical. In Latin, "the content" bears the name "intensio." There is, as well, the term "intentio" which is used to describe a certain act directed towards a certain object. In accordance with Brentano's terminology, this object bears the name "intentional" and may be identified with the content of thought.[3]

Kazimierz Twardowski put forward two important arguments against the identification of the notions "content" and "object." Firstly, the content of an act belongs to the mind, whereas the object does not. Secondly, the content is something real, existing always, whenever we either fancy or make statements, whereas the object does not exist in certain cases (e.g., when it is given contrary features). In speaking of the existence of objects, the word "exist" has to be understood in a different way, that is, as indicating either a phenomenological or intentional existence.[4] We would agree with Marciszewski's idea,[5] that the above differentiation between real existence and intentional existence avoids contradiction. This contradiction may occur when we

A-T. Tymieniecka (ed.), Analecta Husserliana, Vol. XXXVII, 181–192.

state, for example, that acts do not exist, as well as, that every act has an object (which implies that if it does, then the object has to exist). Taking into consideration the above formulated assumptions, we speak of non-existence, having in mind real situations. On the other hand, in attributing existence to a given act we mean intentional existence. A. Meinong expresses a similar viewpoint on the subject. He even maintains that we can examine acts which are directed at objects non-existent in reality. These are either objects which are possible in a logical sense, or ideal beings such as relations or numbers. Therefore, if an act possesses as its object an unreal, fictional product, then it is directed to a specific intentional object. Meinong calls it object-objective and adds that it is characterized by a weaker form of existence, namely, subsistence.[6]

Edmund Husserl approaches the problem of intentional objects in a slightly different way. He rejects the understanding of an act as an activity, and treats it as a certain state of mind. According to Husserl, (from the perspective of intention), it does not make any difference whether a given object exists or not. If it does exist in reality, then there exists its supposition as well. Then we are talking about derived intentional objects. If the object itself does not exist but its supposition does, we have to do with a so-called purely intentional object.[7]

Roman Ingarden refers to Husserl's views. He agrees with the thesis that an intentional object is the supposed production, and agrees to differentiate purely intentional objects from the objects which are called in his terminology, "also intentional." Let us quote the relevant fragment from his book *On the Literary Work of Art*:

> . . . the purely intentional object in its total existence and essence is dependent on the existence and essence of the appertaining act of consciousness. It arises from the execution of determinately conditioned acts of consciousness which lead to its "constitution" as a whole that transcends this act. But already in this "dependence" on the acts of "consciousness", in this "total submission to the sphere of influence of the 'I' of the consciousness" and, on the other hand, in the lack of genuinely creative power of the pure consciousness, of the type we humans are capable of realizing in our experiences, there lies the final argument for the fact that the purely intentional object as such is, itself, a "nothing" in terms of ontic autonomy, that in itself it can neither exist nor be capable of changing itself. We say that it is "projected", "created" immanently contains the determinations assigned to it by the act. They are merely assigned to it, and this assigning is not creative in the sense that the assigned determinations can be "incorporated" into the object — especially into the content of the purely intentional object. In the mere intentional state of "having" determinations "assigned" to it, the purely

intentional object contains in its content nothing that could give it its own ontic foundation. It is in a true sense ontically heteronomous. . . . One must only note that the range of these possibilities is different for purely intentional as compared to ontically autonomous, "also intentional" objectivities. Thus, for example, there is no real "wooden iron" or ideal "round square", whereas purely intentional objects with such contents are quite possible, even though they cannot be intuitively imagined.[8]

Summing up Ingarden's thesis, we can say that their nature is to some extent "metaphysical."

K. Ajdukiewicz on the other hand, put forward a different conception, trying to resolve his reflections on intentionality on the plane of linguistic (semiotic) analysis.[9] Ajdukiewicz bases his position on Lesniewski's calculus of names, called "ontology" in Polish logic. This is a logical theory based both on the primitive term "$\in$" ("is") and on the terms of the calculus of propositions and quantifiers. The term "$\in$" is introduced into the language of "ontology" by means of the axiom:

A1. $$a \in b \equiv \Pi x(x \in a \rightarrow x \in b) \wedge \exists x(x \in a) \wedge \wedge \Pi x, y(x \in a \wedge y \in a) \rightarrow x \in y$$

which is read "*a* is *b* if, when every *a* is *b* and there are *a*'s, and there is at least one *a*."

The definition of "object" is introduced as well:

D1. $$\text{ob}\, a \equiv \exists x(a \in x).$$

This expression reads: "*a* is an object iff for a certain *x*, *a* is *x*." D1. gives us the theorem iff: "*a* is an object if there exists *a* and exists at least one *a*":

T1. $$\text{ob}\, a \equiv \exists x(x \in a) \wedge \Pi x, y(x \in a \wedge y \in a \rightarrow x \in y)$$

From this theorem the following one issues:

T2. $$\text{ob}\, a \rightarrow \exists x(x \in a)$$

which states: "if *a* is an object, then *a* exists."

Since in the framework of "ontology" itself, one cannot construct any constant name, for which it would be possible to prove any sentence like "*a* is *b*" (having its name for a subject) or a sentence like "there exists *a*" (in which that name would never occur as an argument of the functor "there exists"), Ajdukiewicz creates two languages just for the needs of his study. The first of them (let us call it L_1) is made by joining to the sentences of "ontology" all those sentences: firstly, which may be obtained from sensible (in "ontology") propositional functions by substituting the relevant constants for independent variables; and secondly, which have propositions which are recognized on the basis of an experimental criterion. Here, we mean particularly all sorts of so-called perceptive propositions and other ones derivable from them according to the methods recognized by Science and the Arts. In this constructed language, the notions which are the meanings of the terms "there exists *a*" and "*a* is an object," are, Ajdukiewicz suggests, to be called "the notions of real existence" and "the notion of a real object" respectively.

Furthermore, by enriching the thesis of "ontology" with all the sentences (with those sentences having been made sensible in "ontology," thanks to the substitution of a certain constant for independent variables) registered in the Polish translation of Homer's *Iliad* and *Odyssey*, and widening it appropriately by an obligatory inserted rule of substitution, Ajdukiewicz obtains a second language (L_2), in which the following sentences occur as recognized theses:

"Zeus is an Olympian god."
"Polyphemus is a Cyclops."
"Zeus is an object."
"Olympian gods exist.", etc.

"Because of the fact," writes the author, "that the direction concerning the recognition of sentences like "*a* is *b*" in the language just now under discussion differs from that of the language discussed prior to this one, one should assume that the word "is" (likewise other words in the second language, consonant with the certain words of the first language) has in the second language a different meaning from the one it has in the first one.

Eventually, defined by means of the definition in the form taken from "ontology," the word "*ex*" (i.e., "exists") and the word "*ob*" (i.e., "object") will have different meanings in the second language than in the first one, since in those two languages the sense of the word "*is*," to which both of the words "*ex*" and "*ob*" are resolved by definition, is different.

In the language that is being discussed just now, we come to a different notion of "existence" and "object" which we had before. Here again, it seems to be the natural course of events to define the notions of object and existence from the second language by the notions of the intentional object of Homer's epics.[10]

Obviously, the notions of *intentional object* and *intentional existence* may be given a desirable generality by enlarging the language of "ontology" with all sentences (which are made into sentences that are sensible in "ontology," thanks to the substitution of certain constant names for independent variables) which are found not only in the *Iliad* and the *Odyssey*, but also in all texts, both in belles-lettres, scientific writing, or even common parlance, either written or spoken.

Witold Marciszewski[11] here is grammatically (syntactically) parallel to Ajdukiewicz's. He proceeds out from Lesniewski's ontological theses enriching them with sentences met in various texts either current or potential (that is to say only imagined), which have not been carried into effect either orally or in writing. Still, he imposes on these sentences certain conditions concerning their recognition. They serve to explain the difference in the approach to be taken to, e.g., the sentence "Odysseus is Telemachus' father," and sentences like "Odysseus is Penelope's father," or "Odysseus is a hundred and eighty centimeters tall." We are obliged to accept the first of these sentences, but as far as the second one is concerned, we accept its negation, whereas in the case of the third one, we accept neither the sentence itself, nor its negation. Here, we must emphasize the difference between the notions of recognition and truth. It may happen that we accept a false sentence and reject the true one. For example, we accept the sentence "Genii are very small," although it is, in the so-called strong interpretation, a false sentence. We do not accept, however, the sentence "Genii are very big," although, in weak interpretation it is true (given the theorem of the inclusion of an empty set in any other set).

The conditions which should be imposed on the activity of recognition were constructed by Witold Marciszewski in the so-called "system U," taken from "The Basis of the Logical Theory of Belief."[12] Here, the key word is the expression "*U x p*" ("*x* recognizes *p*"), which is characterized by this arranged system of three axioms:

U1. $U x p \rightarrow \sim U x (\sim p)$

U2. $U x (p \rightarrow q) \rightarrow (U x p \rightarrow \sim U x q)$

U3. p is a tautology $\rightarrow U x p$

Axiom U1. says, that if *x* recognizes *p*, then it is not true that *x* recognizes not-*p*. Axiom U2. says, respectively, that if *x* recognizes that if *p* then *q*, then if *x* recognizes *p* then *x* recognizes *q*. Finally, axiom U3. says, that if *p* is a tautology, then *x* recognizes *p*.

From axioms U1.—U3. it follows that:

U6. $U x p \wedge U x q \rightarrow U x (p \wedge q)$

U7. $U x (p \wedge q) \rightarrow \sim [U x (\sim p) \wedge U x (\sim q)]$

U8. $U x (p \rightarrow q) \wedge U x (\sim q) \rightarrow U x (\sim p)$

Theorem U6. says, that if *x* recongnizes *p* and *x* recognizes *q*, then *x* recognizes *p* and *q*. U7. says next, that if *x* recognizes *p* or *q*, then it is not true that *x* recognizes not-*p* and *x* recognizes not *q*.

Two other theorems are accepted as well:

U4. $\sim P x, p(U x p \rightarrow p)$

U5. $\sim P x, p[\sim U x p \rightarrow U x (\sim p)]$

The first says, that it is not true always that if *x* recognizes *p*, then *p*. The second states, that it is not true always that if *x* does not recognize *p*, then *x* recognizes not-*p*.

The notion of recognition so characterized renders well the common intuitions comprised in everyday language practice. Marciszewski then constructs the criterion of intentional existence, basing it on this characterization, formulating it in the following way:

"If in an actual or potential text of a language J, there occurs a sentence Z in a form '*a* is *b*,' and if the recognition of this sentence fulfills the conditions imposed by system U, then this sentence may be recognized as a true one in language J and consequently there may be recognized sentences like '*x* is an intentional object', there exists intentional *x*.' *The objects, whose existence is ascertained by means of this criterion, may be described as intentional objects.*

It should be stressed here that the limitation which can be found in the criterion of intentional existence should be accepted as a thesis for only those sentences that can be recognized in a sense of system U. that prevents us from creating intentional objects by using sentences involving contradiction. Thus, Marciszewski draws the following conclusion: "If the sentence '*a* is *b*' exists in J as an internally contradictory sentence or if its results are internally contradictory (e.g., *p* and not *p*), then there is no foundation for recognizing in J the sentence 'there exists intentional *a*.'" This conclusion is not, however, implied by the criterion of intentional existence given by Ajdukiewicz, since its only requirement is that the sentence belong to the actual text of the language having been formulated in *oratio recta*.

From the several short, and by necessity, incomplete characterizations presented so far, it is evident that the notion of intentional object is by no means indisputable and that is why various investigators often obtain different explanations.

Despite the differences, one feature of the intentional object is unanimously stressed in all conceptions: namely, that an intentional object is never a material one.

It has been just such a definition of the intentional object that has provoked objections for many years. The basic one was formulated by A. N. Prior in his book called *Objects of Thought*.[13] It is based on two theorems, which concern the situation in which somebody is thinking about something, and which seem to be discordant between themselves:

(1) If *y* is thinking about *y*, then *y* is always the element of this relation (of thinking) equally with *x*, so *y* exists.

(2) In some cases if *x* is thinking about *y*, there does not exist any *y*.

Theoretically, there are three possibilities, mentioned by Prior, which help us to avoid the above antinomy. The first would have us accept that thinking about a real object is an activity that differs from thinking about an unreal object, so, when thinking about something, we think about it either as about something existing, non-existent (e.g., fictional), or even neutral, as far as existence is concerned. The two former cases may be defined by means of the following conditions:

W1. *x* is thinking about *y* as about something existing, iff, when *x* recognizes that *y* exists.

W2. *x* is thinking about *y* as about something non-existent, iff, *x* recognizes that *y* does not exist.

W1. and W2. can be written symbolically by means of the terms "ME" and "MN" which signify the relations of thinking about something as existing and thinking about something as non-existent:

W1. $x \, ME \, y \equiv U x (\exists y)$

W2. $x \, MN \, y \equiv U x (\sim \exists y)$

If we differentiate between the two meanings of the word "think" as it occurs in expressions (1) and (2), then we obtain the following pair of sentences:

(3) $\Pi x\,(x\,\mathrm{ME}\,y) \rightarrow \exists_i y$

(4) $\exists x\,(x\,\mathrm{MN}\,y) \wedge \sim \exists_i y$

which are read:

(3) For every *x*, if *x* is thinking about *y* as about something existing, there exists an intentional *y*.
(4) For a certain *x*, *x* is thinking about *y* as non-existent object and there does not exist an intentional *y*.

Thanks to this distinction, sentences (1) and (2), if properly interpreted as (3) and (4), are not in danger of contradiction. Whereas the truth of sentences (3) and (4) is guaranteed by the criterion of intentional existence given by Marciszewski, which allows us to infer that if a given sentence is recognized, then there exists its subordinate intentional "fact." This solution, however, contains a certain incongruity since it refers only to the sphere of intentional objects. If, however, what we mean is concrete existence, then the real sentence (1) always appears to be false.

Marciszewski noticed that the pair of sentences (3) and (4) raises another difficulty. We can ascribe to sentence (4) a sense which would make the sentence false:

"If *x* is thinking about *y* as about something non-existent, then *y* is the subject of thought; that means that *y* is an intentional object, so *y* exists intentionally despite thesis (4)." It is Ingarden's differentiation between the structure and the content of the intentional object that protects it from this antinomy. To the content of certain constructions of thought may belong the fact that they do not exist if the sentence about non-existence belongs to the relevant text, and is recognized in its language. The intentional object, however, does not exist as far as its structure is concerned, when it has not been created by any of the acts of thought and thus there does not exist any such text that could be spoken of.

A second solution that could possibly allow us to avoid the contradiction indicated by Prior, is based on the assumption that if we are thinking about something, we are not relating to this object itself, but to its "idea." This "idea" would, in certain cases, represent a real thing, whereas in other cases, this kind of representation would not occur at all. This way of presenting the matter is rather controversial and involves considerable difficulties concerning the problem of so-called "indirect realism." Let us, therefore put aside this possibility and proceed directly to the next one, which has as its background a very powerful philosophical tradition. It demands that we differentiate between the stronger and the weaker manners of existence, suggesting at the same time that all objects of thought, without exception, are entitled to at least the weaker manner of existence, and that only some of them are characterized by the stronger one.

This suggestion, seemingly the most natural and sensible one, has many opponents, among them Bertrand Russell.[14] According to him,

> Existence is essentially a property of a propositional function. It means that that propositional function is true in at least one instance. If you say 'There are unicorns', that will mean that 'There is an x, such that x is a unicorn'. That is written in phrasing which is unduly approximated to ordinary language, but the proper way to put it would be '(x is a unicorn) is possible'. We have got to have some idea that we do not define, and one takes the idea of 'always true', or of 'sometimes true', as one's undefined idea in this matter, and then you can define the other one as the negative of that. In some ways it is better to take them both as undefined, for reasons which I shall not go into at present. It will be out of this notion of *sometimes*, which is the same as the notion of *possible*, that we get the notion of existence. To say that unicorns exist is simply to say that '(x is a unicorn) is possible.'
>
> It is perfectly clear that when you say 'Unicorns exist', you are not saying anything that would apply to any unicorns there might happen to be, because as a matter of fact there are not any, and therefore if what you say had any application to the actual individuals, it could not possibly be significant unless it were true. You can consider the proposition 'Unicorns exist' and can see that though it is false. It is not nonsense. Of course, if the proposition went through the general conception of the unicorn the individual, it could not be even significant unless there were unicorns. Therefore when you say 'Unicorns exist', you are not saying anything about any individual things, and the same applies when you say 'Men exist'. If you say that 'Men exist and Socrates is a man, therefore Socrates exists', that is exactly the same sort of fallacy as it would be if you said 'Men are numerous, Socrates is a man, therefore Socrates is numerous', because existence is a predicate of a propositional function, or derivatively of a class.[13]

Russell suggests, as can be seen in the second part of his statement, the specific reduction of supposed objects to mental phenomena. This operation would consist in resolving a sentence like "a exists intentively" to the sentence "there exists such a y that y is thinking about a"; and a sentence in the form "a is b" to a sentence proclaiming that "there exists such a y which is thinking that a is b."

The necessity of this reduction may be precisely demonstrated using the formal theory of the functor of "thought." But intuitively it seems to be correct. There is no room for doubt that if somebody is thinking that "*a* is *b*" then he is thinking about *a* at the same time.

At this stage the problem of intentional objects may appear to have already been solved. Such a conclusion, however, would be premature since the success of the reduction implies only the possibility of the elimination of the notion of "intentional object" and hence also of the notion of "intentional existence," which does not mean that such elimination would be advisable. All this is connected with the enormous utility that the notions of "existence" and "intentional object" prove to have in mathematics, in physics, in epistemological and semiotic investigations as has been demonstrated by Ajdukiewicz, or in research into works of art.

Nowadays, it is difficult to imagine any complete analysis of a film, a literary work, or any other work of art, which would not, at least in an indirect way, take advantage of the problematic connected with the issue of "intentionality." Our suggestion, which is, as it turns out, a continuation of Marciszewski's, is based on the acceptance of the notion of "intentional object" as a derived term introduced into the language of film studies on the basis of the definition of identity.

The intentional object itself would stand in connection with its purely theoretical construct, which probably has no reference in empirical reality. The argument in favor of this approach to the problem of intentional objects can be supported by the fact that even the most powerful scientific theory makes use of terms that do not denote (on a limited scale) previously introduced definitions. Those terms have no reference in reality, but nevertheless they are essential in the language of a given science.

We have here added a suggestion to Marcizewski's conditions for the recognition of sentences that refer to the notion of intentional object. Thanks to this, we avoid the danger of creating contradictory objects — which in turn allows us to construct so-called intentional models. This very fact is a matter of consequence, as far as research into a film work is concerned. Furthermore, it assumes something which is very important in our view, namely, the possibility of realizing and explaining the world in its cause-effect aspect.[15]

By means of another criterion, all events represented in a film work may be divided into events presented either objectively or subjectively. The former are related to present time narration and the purely behaviorist presentation of activities, behavior, and states of particular objects. Subjectively presented events, however, usually appear in visual form as the stories, memories, and dreams of the persons presented.

Summing up, we can say that intentional objects are the basis for the creation of the representational stratum of a film work. However, we suggest that it is preferable to speak of the *quasi-intentional* character of those objects just as we speak of the *quasi-linguistic* nature of film communication. The word "*quasi*" means that the represented objects have been treated as theoretical ones. We do not decide whether (and in what way) they exist in a film work, or not.

It is not up to us to examine the pure ontology of a film work. In other words, the represented objects (treated as theoretical ones) are brought into being by the observer (or researcher) in the very process of construction. Thus, we can speak of the constructive but quasi-intentional character of the representational stratum of a film work.

University of Alberta and The Jagiellonian University

NOTES

[1] W. Osadnik and L. Plesnar, "Filmologie, Semiologie et Linquistique," *Linquistica Silesiana*; Vol. 9, (1987).

[2] W. Osadnik and L. Plesnar, "Film Study and Linguistics," *Problemy semiotyczne filmu* (Katowice: 1980).

[3] F. Brentano, *Psychologie vom Empirischen Standfunkt* (Leipzig: 1924), p. 124.

[4] K. Twardowski, "O tresci i przedmiocie przedstawien," *Semiotyka Polska*, ed. J. Pelc (Warsaw: 1971).

[5] W. Marciszewski, "Problem istnienia przedmiotow intencjonalnycg," *Studia Semiotyczne* 1973.

[6] A. Meinong, *The Theory of Objects. Realism and the Background of Phenomenology* (Illinois, 1960).

[7] E. Husserl, *Logische Untersuchungen*, Vol. II (Halle: 1900—1901). See also: A. Poltawski, *Swiat, spostrzezenie, swiadomosc* (Warsaw: 1973), pp. 45—46.

[8] R. Ingarden, *The Literary Work of Art*. Translated, with an Introduction by G. G. Grabowicz (Evanston: Northwestern University Press, 1973), pp. 121—123.

[9] K. Ajdukiewicz, *W sprawie pojecia istnienia. Jezyk i poznanie*, Vol. II (Warsaw: 1965).

[10] *Ibid.*, p. 148.

[11] W. Marciszewski, *op. cit.*

[12] W. Mariciszewski, "Podstawy logicznej teorci przekoman," *The Basics of Logical Theory of Belief* (Warsaw: 1973), pp. 196—7.

[13] A. N. Prior, *Objects of Thought* (Oxford: 1971).
[14] B. Russell, *Logic and Knowledge* (London: 1956), pp. 232—233.
[15] W. Osadnik and L. Plesnar, *An Introduction to the Semiotics of Theatrical and Cinematographic Acting*, European Journal for Semiotic Studies, Vol. **2** (1991), 111—30.

PART THREE

THE LIFE-SIGNIFICANCE OF LITERATURE AND ITS INTERPRETATION

MARLIES KRONEGGER

TYMIENIECKA'S VINDICATION OF THE LIFE SIGNIFICANCE OF LITERATURE

Homo Ludens and Homo Creator: Scapino

INTRODUCTION

Imaginatio Creatrix at the Origin of Art

Anna-Teresa Tymieniecka's *Logos and Life, Creative Experience and the Critique of Reason*[1] is a feast for all the senses and the mind, for all the vital synergies that stream together, when "initial spontaneity ... springs forth from the Human Condition." (175) Once again she confirms that art is the sublime mission of man, since it is the exertion of the mind, trying to understand the world and the human condition, and to make the world and the human condition understood. She proposes that "the creative act of man is the Archimedean point from which to gain understanding of the life functioning, and that it is the only access to the exploration of the Human Condition." (3—16) In the creative orchestration of human existence and of everything-there-is-alive, is Tymieniecka's path and vision: she takes the path in order to become the path herself for us who will follow her vision in a new appreciation of Scapin, the leading character of *commedia dell'arte*, as he emerges in Callot's engravings, in the writings of Molière and Claudel, and as enacted on stage in the plays of Molière and J.-L. Barrault. Tymieniecka's treatise on the life-significance of literature has liberated it from the cultural and theoretical prisons to which it had been sentenced for many years. Her treatise finally retrieves the poetic sense, the smile of the mind, in the interplay of several modes of experience. The voices of drama, art, dance, music, poetry, sculpture, and opera are harmonized in Tymieniecka's *Logos and Life, Creative Experience and the Critique of Reason.*

Tymieniecka's life-world conception challenges Husserl in the evolution of its form and lays the cornerstones for a specially human self-interpretation-in-existence of which the creative act is the vortex. (121—317) Tymieniecka transcends Ingarden's interpretation of a

A-T. Tymieniecka (ed.), Analecta Husserliana, Vol. XXXVII, 195—212.

literary work of art as having value qualities of two kinds, artistic and aesthetic, in insisting on the generic nature of the real as revealed in "the relationship between the body and the mind" (97—99) when the creative function of consciousness comes into being. She radically challenges the so-called "de-construction" of our life-world into a "primordial life world" (132) and proposes a reality which "maintains an existential status of its own, emerging and poised between the workings of the network of stream of life and man's creative act." (132)

With Tymieniecka we are convinced that the only knowledge worth acquiring is creativity, the irreducible element of the human condition, the freedom to transcend our ties to Nature and the conditions of our own life-world, achieving the full potential of human personality in the creative orchestration of our own faculties and in relating to everything-there-is-alive. Her treatise, *Logos and Life*, opens for us our path for exploring the life significance of art, focusing on Scapin in various media. Scapin's *imaginatio creatrix* decomposes the life of his opponents into a mechanical ballet; he surrenders himself to his masters never, but only to the smile of his mind and so with body and soul.

With Callot's, Molière's, Claudel's, and J.-L. Barrault's recreations of Scapin, we focus on the "poetic sense" and "smile of the mind," following Tymieniecka's path from initial spontaneity and improvisation to creative vision in Scapin's performances with the *commedia dell'arte*, then and now.

I. SELF-INDIVIDUALIZATION-IN EXISTENCE: CALLOT'S SCAPIN

(1) *Callot's Life-World and the Human Condition*

Callot's stock characters are ahistorical[2] symbols of the theatrical idea itself since we can trace them from antiquity to the present day.[3] Their deep principles energize our imagination and illustrate the spiritual potentialities of human life in periods of repression, obedience, superstition, and restraint of all kinds. Of Greek and Roman origin[4], the *commedia dell'arte*, this popular form of character comedy, was revived in the fifteenth and sixteenth centuries, as illustrated by Callot. Its popularity reached phenomenal proportions from the seventeenth to the twentieth centuries. Callot's sketches show us what delighted Europe and France on the verge of world domination, in the only age when its theaters really lived and flourished. Perhaps only Callot had the energy of spirit to pull together the vast inchoate mass of real life, and re-create it in sketches giving a unified, organic vision of real life situations.

In Callot's engravings, art and drama are able to bring us back to the real life of his age, through dynamic, constantly changing scenes from whose depths meaning simply materializes. His vision appeals to our pre-logical faculties, pointing to the secret sources of being and the mysterious origins of the human condition.

I Balli di Sfessania were created in Nancy in 1622 from sketches Callot made of outdoor entertainments in Florence. It is a series of twenty-four etchings celebrating the typical behavior of lovers, musicians, captains, gamblers, in dances, in amorous incantations with gestures mimicking shouts or songs, fright or enjoyment, pain or passion. He incarnates the realities of life in concrete images of man's essential fears, hopes, anxieties, joys, pains, and beliefs in given situations imposed by authority, war, amorous encounters, etc. . . . , recreating the life-world he knew for the spectators of both his and our times.

These characters are certainly ugly in their pride, vanity, aggressiveness, frivolity, exhibitionism, cowardice; yet their beauty exists in that they have character and express joy or sadness, enthusiasm or despair, calm or rage, love or hate, power or impotence. Beauty is character and expression, because in the contortions of their physiognomy, in the scoring of a vicious mask, in every deformity, in every blemish, the inner truth brusts forth. Callot reveals truth from life which is movement rhythmed by gestures and time, knowing how to make his figures come to life. The physiognomic deformity of these characters reflects their spiritual deformity. This explains why they remain vital and visible in our own days, as they are truer than truth itself, and more real than persons that we encounter in daily life. Their movements are grotesque, their gesticulations obscene, their poses awkward, yet they are true to themselves.

(2) *From Gesture to "the True Reality behind the Screen of Life"*

The character's inner truth is an unveiling, a bursting forth of the originary being, as Callot seizes human essences in both gesture and dance. In his sketches, all of life's phenomena are expressed as in the spirit, as existentially, and as economically as possible. Callot's drama speaks the many theatrical languages that appeal to the senses, rejecting the language of words that appeal to the intellect. Language as a means of intellectual communication has no function in the drama depicted by Callot. Both Molière and J.-L. Barrault, as we shall see, also base the essential core of their acting on gesture.

The magical appeal of the *commedia dell'arte*, as illustrated by Callot has been to rediscover the magic of the performer, the magic of enacting human frailties and foibles in an honest unpretentious way. The performers are mimes whose plastic eloquence communicates by means of gesture symbols: gesture, smile, glance or reflective air, or physical action. Gestures do not duplicate, but, rather, complete the spoken word. All we see is gesture, the continuation of thought, and the minimum of gesture corresponds to the maximum of expression. The realities of the character's inner life can only be depicted by art in which mind and emotion are compounded — and not by intellect or feeling in isolation. With Callot's sketches, as with J.-L. Barrault's interpretation of Scapin, in plays by Molière and Claudel, every pose, every gesture of the performer has its inner justification. Their gestures provide a pretext for the emotion to expand indefinitely, making the spectator aware of inner truths that lie beneath the appearances since it is the beauty of life that plays in them.

(3) *The Equivalent of the Poetic Image in Movement and Dance: the "Life-Symphony Constructs Itself"*

Viewed from the stage, threater begins with silence prior to the performance. A dialogue of gesture and dance emerges out of silence. Gesture and silence come from opposite poles, yet they run to meet each other in Callot's sketches. Tymieniecka holds that dance is the art of silence, "silence of the perfect instant, delicious suspense of breath and heart." (109) From this quietude comes the most elevated and creative experience. The silence itself becomes an aspect of the total movement of gesture, channeling the dancer's and the performer's vital synergies into a symphony of being, expanding consciousness beyond mere sensitivity, time, and space. Callot's performers communicate this expansion clearly and with artistic skill.

The chain of movements is like a dialogue, a series of questions and answers, as Tymieniecka explains in her analysis of the aesthetics of movement in dance. (109) The meaning carried by the dancer's movement is not signified, it is expressed. Signification becomes expression of life. Callot reveals the inner world of his dancing figures in the use of gesture and movement. He gains access to what cannot be said or seen, to what can only be lived. In Tymieniecka's view, "when lived experience slips through the language of gesture and movement, it is capable of catching the most significant elements of life in its meshes, the body and the mind are revealed as analogous structures." (108) Art, then, is the language of the human spirit. When Tymieniecka refers to the mastered body of the dancer, she is, whether she realizes it or not, also describing the performance of such roles as the two Scapins created by Molière and J.-L. Barrault: "It is the universe of sensations, of emotions, which arises from a mastered body . . . ." (109) For the actor Barrault, the chain of movements is the art of pushing and pulling: "The whole life is either pushing towards oneself, or else pushing away from oneself, and oneself is the nave."[5] As we see in Callot's sketches of the performers, there is no physical action of the body which does not involve desires, aspirations, and objectives of feeling which justify the action.

Dance in Callot's sketches is musical self-expression, a gestural rendering of musical forms in spontaneous movements, when every fiber of Scapin and Captain Zerbino's bodies, eyes, fingers, neck, chest, wrists, hips responds to the music of incipient impulse with subtle inner orchestration.

In the generation of the dancer's movements, there is truth: who can tell the dancer from the dance? Look at Callot's Rinuciolina: "can she not make love visible by her beautiful acts . . . she is a wave." (110) Tymieniecka's view refers to her also as being the essence of dance. Truth is eternity made visible in a dancer, as Tymieniecka explains: ". . . we can see complete life as being comparable to a dancing woman . . . . Body and mind are revealed as analogous structures . . . the soul translates its own dimensions into those of space and matter . . . the body, transformed by art, becomes the place where the real, the unreal and the intelligible fuse and combine as the power of the Muse dictates." (107—109) Thus, sense is within the sensuous, since it is both perceived and felt.

All this bears witness to the intimate tie between physical action and all the inner elements of the creative state when Callot shows the deep-seated impulses of ordinary people in the imaginative expression of the sketches. We also *see* what the artist feels stirring within himself when indistinct impulses of his inner life are seeking expression. Dance is rhythmical, spontaneous movement and leaves to each spectator the task of imagining a dialogue which is ever true since it is always in accord with the emotions received.

II. POETRY IN SPACE

(1) *Scapino and Captain Zerbino: the Spirit of Movement and Moment*

Callot makes visible the passage of one pose of Scapino into another; he indicates how imperceptibly the first glides into the second. We see the gradual unfolding of movement. Time does not stand still. Callot succeeds in producing the impression of a gesture that is executed in several instants, bringing together different phases of an act in the same figure, and mirrored in his counterpart.

Here is the expression of life, happiness, and robust sensuality in the simple *lazzo*, the fight over a bottle of wine. The etching is essentially movement, as much as it is the intensity of something dark on something light. The way in which a black is set off against a white, the modulation of the transition from one to the other contribute to the suggestion of a specific rate of speed. The sketch exemplifies how Callot fixes very rapid gestures or gestures so transitory that there are no longer lines, but here is movement, here is life, here is drama. Callot conflates the action and depicts three different moments of the conflict simultaneously both visibly and invisibly. Callot does not proceed by reason or by artifice.

As to the *scenario*, it is visible here: Scapino's and Captain Zerbino's souls seem to be drawn first to a gesture of negotiation, yet very soon their argument unfolds as if they were on stage. Callot wants to represent not only passing gestures but also an extended action. We see first Scapino and Captain Zerbino beginning the action, then continuing it in the middle ground, and finally we perceive the spectators who are enjoying their performance in the background. They are a reflection of ourselves, the viewers. The *scenario*, as seen by Callot, encourages participation on various levels: as a theater of the people it was enjoyed by both the unlettered and intelligent audiences; it was an everyday experience and spoke to many levels of the human being and satisfied all their appetites: emotional, physical, spiritual, and intellectual. It is from the existential reality of the actor, of Scapino and Captain Zerbino that Italian Comedy draws its own life.

Scapino's and Captain Zerbino's fight over the bottle in the middle ground of the engraving is a *mise-en-abyme*, a play within a play, framed by their swords is the oversize bottle, whereas the church in the background barely approaches full size. We see how Captain Zerbino in an argument with Scapino seems to be in torpor in the foreground, and how Captain Zerbino changes equilibrium in the middle ground, where his right foot advances and takes control of the earth. At the same time, the left shoulder is raised as if to bring the weight of the torso to its side and help the back leg come forward. We see how Scapino raises both his sword and his bottle, a movement doubled and mirrored in Captain Zerbino's holding up the cape draped over his left arm — a fighting pose.

Scapino and Captain Zerbino seem to be twin sided mirrors of one another: Scapino's inner being seems to reveal itself in Captain Zerbino's actions, and Captain Zerbino's in Scapino. The utilization of identical figures is particular to Callot, Molière, and Claudel. Both the unexpectedness in behavior and repetitious description in different circumstances create the illusion that they are living beings.

(2) *Scapino's Unfolding from within His Life-World*

In Callot's sketches, performances of Italian Comedy involving Scapin seem to be an everyday experience; it is everywhere. Theater here is a unifying force in the life of the spectators. It seems to reflect some aspect of life, of truth, and to point to the secret sources of being, to the mysterious origins of the human condition: the play of masks seems to conceal the soul, and the soul conceals itself, yet with Scapin the truth shows forth in his acrobatic skill, promptness, flexibility, imagination, and unmatched vitality.

Scapin's mask represents the way Callot sees him, or how the audience responds to him, emerging from virtual space as an intangible image resulting from the formal relationships established within the given defined area. Tymieniecka's discussion of focused or polarized space as being rhythmically divided into spatial rhythms is illustrated here. (225) The pulsation of the etching discloses the vital rhythm that gave birth to it, brings us into contact with it, and communicates itself to the viewer, as a kind of *poetry in space*. We also see that dance and drama are not distinguished generically, and are central and organic to the community.

Callot's satirical masks on stage elevate the actors and distance the

Scapino.
Cap. Zerbino

audience. The sky is astonishingly empty, dominating groups of people by its great calmness. There are powerful contrasts of depth and value between the oversized dancing figures in a theatrical *mise-en-scène* which overpower spectators and nature. The figure of Scapin emerges in another sketch as he scrutinizes the world from an elevated vantage position, turning his back to the spectators in the distance, enjoying the panorama of the city square, contemplating his engagement in the action and in his life-world. As in Molière's *Les Fourberies de Scapin* and Claudel's *Le Ravissement de Scapin*, the play belongs to him, the actor, the stage manager, the master of improvisation who looks at us. In the following analyses we shall see that in the case of Scapin, Tymieniecka's view holds true when she declares: "Drama makes the actor not only the co-creator of the work, but simultaneously one who is co-created by it for this specific purpose. It is the network of real-life co-recreative relations which emerge and hence, carries that real-life event of creative transition." (280) Illusionistic virtuality has always been one of the dominant characteristics of the spectacle. Scapin's observation area from which the spectators are viewed develops in a spatial sequence from a plateau, framed by houses which act as the traditional corner motif forming an intermediary between the viewer and the middle and far views of Scapin. Scapin seems to mimic and mock those who sanction or impose restraints. His inner urge to misbehave and the joy of release is in direct proportion to the severity of the restraint, imposed in his case by both the mask and the scenario.

(3) *From Restraint to Improvisation: from Mask to Creative Inwardness and the New Functional Orchestration*

A mask is a metaphor for what lies behind the visible stereotyped image of Scapin. Masks are the full expression of character itself, and the play of the body is a supplement to the mask. The mask takes away the person we know. It invests the wearer with something we do not know, transforming a particular human personality into a dramatic statement. Scapin's mask reactivates in us forces of imagination and desire. It evokes possible worlds of Scapin's ingenuity. Jacques Copeau, a famous stagedesigner and actor, is convinced that the actor who performs under the mask receives from this *papier-maché* the reality of his part. While he is being controlled by it, a new being emerges, the existence of which he had before never suspected. It seems that the entire body of the performer becomes an instrument of the expression of the mask. It orchestrates the body and mind of Scapin, who emerges as a "testimony before the universe . . . it is the universe of sensations, of emotions, which arises from a mastered body." (109) The body always expresses the spirit for which it is the shell.

Molière's *Les Fourberies de Scapin* (1671) and Claudel's *Le Ravissement de Scapin* (1949) are the apotheosis of Scapin, the actor with a mask, the embodiment of his traditional face and of the imaginative life behind it. In both plays, Scapin's metamorphic world, its kaleidoscopic mobility, is an image of his infinite playfulness, flexibility, and resourceful ingenuity. His identities are interchangeable, since he manipulates the audience to adapt to his rogueries, to his solace of total imaginative freedom. The focus in both plays is not on the author, but on Scapin's creative mind. With Scapin, the theater is in the process of birth. He is the master of acting and his part is among the most demanding to play, whether for Molière in the seventeenth century or J.-L. Barrault in the twentieth century.

It is Scapin's creative imagination and creative orchestration of his whole being, of "the body, soul, the spirit" (392) that transforms the lyrical potential of text and world into theater poetry, restoring the freedom of the theater to illustrate the human condition: all of Scapin's inner faculties participate in each performance, conveying to the public things which are visible but ineffable: the spiritual content of the play institutes "new forms of sensibility through . . . [Scapin's] creative act." (72)

III. CREATIVE VERSUS CONSTITUTIVE PERCEPTION

(1) *From Scenario to a New Life Significance of the Text*

The *scenario* belongs to the constituted world of stock characters. From Callot to Molière and Claudel, from Mozart to Rossini, these are: two helpless lovers, two authoritarian fathers, two servants, and Scapin. Their *lazzi* handed down from generation to generation would impose a mechanical, well-known pattern of man, when life is reduced to mechanical forms or fixed sociological, psychological, and biological ways in which man is defined within an increasingly narrow frame.

Scapin, better than other figures of the Italian Comedy, knows how to break away from expected norms in his lapses from the imposed mechanical *scenario* pattern. Following his bliss, he realizes his true being. The basic plot is this: while the two fathers are away, the two sons fall in love with two girls and then appeal for help to their two servants. It is up the Scapin to find rescue. *Scenari* of this kind frequently recurred all over Europe, passing over all frontiers.[6] Scapino knows the line of action, the artistic kernel of the play. This he must transplant into his soul, and from that moment will begin the creative process, the process which is akin to all organic growth in all aspects of nature: Scapin confronts the world in order to transform those elements which constitute it and to realize himself in becoming who he is — "Werde der Du bist," as Goethe adjures us. Scapin is the image of metamorphosis itself: not being, but becoming, not essence, but existence. His is a world of labile emotions rather than of facts, rules and events, and one of textures, associations, and tonalities rather than clear-cut outlines; his is a world of the uncontrollable and unpredictable rather than of rational direction and completion.

(2) *Scapin's Poetic Sense in a World of Spontaneity and Improvisation: Claudel and J.-L. Barrault*

Tymieniecka refers to the sixth sense which transposes truth through poetry as the poetic sense. (405—422) To this poetic sense both Molière and Barrault were particularly sensitive. With them, Scapin's first appearances on stage follow the rules of tranposition. In transposition there is no apparent logic, there is metamorphosis. Scapin's love of improvisation, his wit for wit's sake, his desire to outdo himself at all occasions in any unexpected life situation, his ingenious tricks, dupery, games, and impersonation of fourberies bring this creative and resourceful actor to his apotheosis. As we have seen him in Callot's sketches, as in Molière's and Claudel's plays, he is the author, stage designer, actor, and performer of other parts all at once. His self-confidence and imagination are put to the service of the young lovers. Scapin improvises spontaneously all his tricks and plays, therefore, in a much livelier and more natural manner than the one who learns his role by heart.[7]

Barrault's improvised means for enacting Scapin's fourberies in 1949 were a delight to Claudel. His dream of *the theater in the process of birth* became with Barrault a reality on the stage. After having seen a misinterpretation of Molière's *Fourberies de Scapin* in February 1949,

Claudel suggests devices which were adopted by J.-L. Barrault: "il aurait fallu quelque chose de joué dans un cabaret par des amateurs dans une atmosphère de vin et de gros tabac, et les latrines ne sont pas loin. Faire jouer les vieillards par des jeunes gens et Scapin par un prêtre défroqué, Zerbinette, une vieille entremetteuse. Les jeunes gens, de vieux ivrognes, etc."[8]

For Claudel and Barrault, Scapin embodies one of Molière's most characteristic types whose virtuosity in deceit and trickery, whose love of challenge for challenge's sake must recapture the original ambiance of Italian Comedy. In "Rules of Acting," J.-L. Barrault asks himself four vital questions, when playing the part of Scapin himself on stage: "Whence do I come from? Where I am going? In what state am I? What am I doing here?"[9] Barrault recreates the hot ambiance of Scapin's Naples, where he is King of Siesta, eats, sleeps, makes love, and plays, and, without any ambition, trusts that things would come to him by themselves. "Sleepy, he will awaken progressively. His first tirade flows forth from the fumes of sleep and wine and is garlic flavored."[10] Claudel defines the mission of his theater as a process of birth, when the actors discuss their performance: "No need for an author — we are the author! Something is going to happen, and it matters to us what it is! . . . We have plenty of actors to draw on; they are in the pub, doing one thing and another — drinking, smoking, yawning, arguing, playing cards . . . . And suddenly someone brings on a large skip full of costumes and wigs. It's like an electric shock! Each immediately decides on his part. The play is taking shape all by itself before our eyes!"[11]

This basket of costumes makes actors of everyone in the tavern. When the curtain rises, Callot's scenes from *I Balli di Sfessania* are transfigured in the smoke laden atmosphere of a tavern where people are playing dice and *mora*, while one group of men are laboriously hoisting a barrel toward a trap door in the ceiling. In this Neapolitan tavern scenery, emerge the hideous stock characters of Callot's sketches; from here Scapin will be born with each performance anew, playing his part with sincerity, authenticity, and spontaneity. Always in the service of youth, love, and the society which will emerge from his interventions, he uses the full range of his imagination, orchestrating his presentation of the situation carefully as he takes on the challenge to extract large sums of money from two old suspicious misers by exploiting the psychological mainsprings of his victims.

With J.-L. Barrault's Scapin, we reach an apotheosis of the role player when the effect of his acting goes far beyond cause. With him we discover the power of theatrical illusion. Faced with a disastrous fall, Scapin rebounds victoriously and is carried off in triumph to a festive banquet at the end of Molière's and Claudel's plays, the traditional ending of comedies since Aristophanes, Plautus, and Terence.

In Claudel's preface to *Ravissement de Scapin*, first intended to be called *Impromptu de Scapin*, comedy is defined as an instrument of liberation. To a society with a compulsion for hierarchies, order, and obedience to rules, Claudel's Scapin responded with irreverence and laughter, with agility rigidity, pure ingenuity against unimaginativeness. The laughter which Claudel wishes to evoke brings Callot's and Molière's Scapin on stage in order to temporarily lift us from the weight of the world, of morality and mortality. Like to his predecessor Molière, toward the end of his life, Scapin's spontaneous improvisations of Claudel offer both rational and spiritual liberation from the restraints of law and morality, reaching an extreme point of lyricism in a theater in the process of birth.

(3) *The Theater in the Process of Birth*

Molière's play is for Claudel only a pretext. The play takes shape before the audience, not without discussion or argument. The artifice of repetition recalls impromptu plays of Corneille, Molière, Régnard, whose atomsophere passes via Marivaux, Musset, Goldoni, and Pirandello to Giraudoux, Ionesco, Claudel, and Beckett. Claudel acknowledges with them that the actor's part, especially in the case of Scapin, reduces the role of the author, and imbues the actor, the artist, with the thought and feeling of the person he is playing in order to bring a character to life in a given situation. J.-L. Barrault, in *Connaissance de Claudel* (1952) emphasizes this kind of creative imagination in a way congruent with Tymieniecka's view: "J'aime ce genre de spectacle. Il m'émeut comme m'émeut une source. C'est spécifiquement du théâtre à l'état naissant. Il y a là une sorte de nouveau départ qui pourrait nous apporter peu à peu la matière d'un style neuf, capable de nettoyer bien de choses."[12] Scapin's reality is flexible, illusory, and temporary, because it changes with each performance in contrast to the *scenario* or text which are fixed (in Claudel's case, the text of Molière). For Scapin, life is an endless pleasure, a perpetual rapture, a wild intoxication, since he can act spontaneously. He denounces the absurdity of life and endeavors to restore its equilibrium. Each new situation in Scapin's life means a struggle to survive, and in this very struggle the his very life emerges, creating his new reality.

Behind the given text, not reality but the spontaneity of life reveals itself to Scapin, the actor who knows to create with each performance a real event, endowing the text with new life significance on stage, just as Tymieniecka explains: "The creative process penetrates with the style of performance most intimately into the life-world's meaning bestowing channels, bringing about the direct transformation of its life-significance." (208) This same transformation occurs also in the *opera buffa* created by Mozart and Rossini! Cherubino, Figaro, Count Almaviva, Basilio, and Suzanne have form and mysterious roots in the *commedia dell'arte*. The difference is in the written *libretti* not in the words, but rather, it is the music of the composers that has changed the hearts of the known stock characters toward the tender and passionate in a sublime blend of wit and melancholy worthy of Callot.

All these transformations of texts, words, scenario, and libretto into gesture or music illustrate that texts are to be relegated to an unimportant place. The theatrical experience itself is the "useless necessity" of a country's well-being. With Callot's etchings, Molière's and Claudel's transformation of scripts, Mozart's and Rossini's versions of the *commedia dell'arte*, we realize that dramatic art is not fixed in the script: "the emphasis is shifted to the situation within the life-world in its existential fullness. It is from the existential reality of the actor that it has drawn its own life." (280) In short, a play is created in the theater, and aside from this creation, it has no true being and cannot be judged.

Therefore, the actor's voice must be heard again, since it is he who answers our search for authentic reality which is eternal and true. Tymieniecka's vision refers to the total human being who is in command on his physical, spiritual, emotional, and intellectual potential. She is proposing a way of life in which man is skillful, perceptive, creative, spontaneous, authentic, united, and generous as realized by actors of modern times who perform in communion with the audience. Let us listen to the voice of Stanislavski which is in harmony with that of Tymieniecka:

> Artistically, the actor can only get the fullest value of this *communion with an object* on the stage if he has trained himself in prolonged exercises, so that he is capable of wholly receiving and wholly giving himself up to this communion, and also of absorbing it with

a maximum of intensity. It is only then that the action on stage reaches its proper expressiveness; between the characters in the play a living bond is formed, such as is necessary in order to convey the drama to the fullest extent, while always observing the right rhythm and tempo for the over-all performance.

Yet, no matter what an actor's circle of attention may be — that is, whether it is lessened at times to put him into a state of "public solitude" or enlarged to contain all the characters on the stage — *creativeness on the stage, whether during the preparation of a part or during its repeated performance, demands complete concentration of all his physical and inner faculties.* This concentration includes his sight and hearing, all of his external senses; it stimulates not only the periphery but also the depths of his being; it challenges his memory, imagination, emotions, reason, will. The actor's entire spiritual and physical nature should be involved into what is happening to the character he has imagined. In moments of "inspiration", the spontaneous exaltation of all his faculties, this is what happens to an actor.[13]

The creative process of the actors is never finished: each successive performance requires that they be in a creative state, that all their inner faculties participate in their acting, adjusting the scores of their parts to the capricious mutations which are constantly taking place within them, since the mind must be conquered. "In this conquest," Tymieniecka says convincingly, "resides the very life of the mind that opposes itself to the natural life of the senses." (42)

Tymieniecka's treatise is a testimony to the creative imagination, and in the field of literature, in particular, a unique call to investigate further the creative process in an endless list of impromptu plays from Molière, Corneille, and of Régnard to Marivaux, Musset, Goldoni, and Pirandello to Giraudoux, Ionesco, Claudel, and Beckett.

Why is the emotional memory so attached to Scapin who comes to life not only in Italian Comedy, but in engravings and *impromptu* plays? In sum, we have seen that the spontaneous interplay of complex organic emotions is visible in each muscle of Callot's *commedia dell'arte* performers, audible in the numerous love duets of Italian Comedy and *opera buffa*, and when Scapin, the actor, interpreted by Molière or Barrault, etc., imparts by means of rhythm the lyrical state to the audience. With Scapin, we have discovered "the spontaneous unfolding of initial spontaneity which plays a crucial role within the creative orchestration of man's faculties" (107) as Tymieniecka observes. Her treatise aims at revealing a new significance of human reality, of life, concentrating on the "human functioning with its powers of body and mind . . . which in its making corresponds to the creative potential of the Human Condition." (234—235) The enactment of the text (*scenario, lazzo, libretto*, engraving) on stage is a real event which lives and dies with each performance, yet the reincarnation of Scapin or Rinuciulina, etc. in various media demonstrate Tymieniecka's "synergetic cohesion" of the mind in the creative orchestration and in the organic unity of everything-there-is-alive.

CONCLUSION

We have chosen Callot's *I Balli di Sfessania* to illustrate some key positions of Tymieniecka's *Logos and Life, Creative Experience and the Critique of Reason*, for we have found in Scapin an ideal example. His creative impulse rejects any pre-established pattern imposed by the rational organization of the constituted world. While he did not need to grow in wisdom, he knew how to live on and on from his birth in remote ages to the present day. Scapin's dynamic gestures, what the French would call "la dynamique du corps," has slipped into Molière's stylistic devices (into the play of word and gesture), into Mozart's and Rossini's *opera buffa*, into Claudel's theater in that progress of birth wherein each word becomes gesture and rhythm, a feast for all the senses. And this is also the meaning of Tymieniecka's treatise for us, since "the mind is a call toward eternity, the absolute, the true." (62) The smile of the mind of Tymieniecka's lyrical-musical prose or poetry might elude any philosopher or literary theorist who bases his findings on reason and logos alone.

Michigan State University

NOTES

[1] Anna-Teresa Tymieniecka, *Logos and Life. Creative Experience and the Critique of Reason*, Book I (Dordrecht: Kluwer Academic Publishers, 1988). All references in parentheses in the text refer to this edition.

[2] Gerald Kahan, *Jacques Callot Artist of the Theater* (Athens: The University of Georgia Press, 1976); Thomas Schröder, *Jacques Callot. Das gesamte Werk*. Handzeichnungen (Herrsching: Manfred Pawlak Verlagsgesellschaft MBH, 1968).

[3] Allardyce Nicoll, *The World of Arlequin. A critical study of the* Commedia dell'arte (Cambridge: Cambridge University Press, 1987).

[4] Pierre Louis Ducharte, *The Italian Comedy*, trans. Randolph T. Weaver (New York: Dover, 1966); Luciano Mariti, ed. *Alle origini del teatro moderno. La Commedia dell'arte* (Rome: Bulzoni, 1980); Odette Aslan and Denis Bablet eds., *Le Masque. Du rite au théâtre* (Paris: ECNRS, 1985).

[5] Jean-Louis Barrault, *The Theater of Jean-Louis Barrault*, trans. Joseph Chiari (New York: Hill and Wang, 1961), p. 60.
[6] Giuliana Cilajanni, *Les scenarios franco-italiens* (Rome: Edizione de Storia e Letteratura, 1970).
[7] Evaristo Gherardi, in *Le Théâtre Italien* (Paris: 1694) states in his introduction: "To say a 'good Italian actor' is as much as to say a man who has depth, who plays more from imagination than from memory, who composes, while acting, all that he says; who seconds and supports whoever is with him on stage, that is to say, fits his words and actions so perfectly to those of his comrade that he instantly enters into all the acting and movements which the other requires of him in such a manner as to make everyone believe that they were prearranged."
[8] Paul Claudel, *Journal II* (Paris: Pleïade, 1965), p. 673.
[9] Jean-Louis Barrault, *op. cit.*, pp. 32–59.
[10] Ibid., p. 37.
[11] Paul Claudel, *Claudel on the Theater*, ed. Jacques Petit and Jean-Pierre Kempf, trans. Christine Trollope (Coral Gables: University of Miami Press, 1972), p. 169.
[12] Quoted by J.-L. Barrault, in "Un Claudel vivant," *Cahiers Renaud-Barrault*, no. xxv, p. 9.
[13] Konstantin Stanislavsky, *Selected Works*, ed. Oksana Korneva (Moscow: Raduga Publishers, 1984), pp. 149–150.

SILVIA MANTEIGA POUSA

THE ENIGMA OF AVANT-GARDES

Notes for a Phenomenological Analysis of the Reception of Avant-garde Art

If in Husserl phenomena imply the facts as observed and interpreted by a cognizant subject, and if we agree with Ortega's theory about the *ordinary man* versus the uncommon *egregious man*, as analyzed in *La deshumanización del arte*,[1] then it is necessarily the case that avant-garde art is incomprehensible to the majority of its receivers, who are just not able to understand its function, or rather, not able to give it any meaning. This is all the more true if one takes into account the fact that many avant-garde movements take literature and painting as a starting-point but later depart from language and figuration, and then are no longer codifiable (e.g., trash art). Is it then necessary to state that avant-garde art means the death of art? Or that avant-garde art leads to non-language, and non-communication?

Calligraphs, a sum of poetry and figuration, would be the first step towards the primitive pictograph, a mere concept deprived of verbal expression. Does this lead us towards a hieroglyphic art which can only be decodified by an accustomed minority — that is, to a situation in which the power of the work of art is centered in the power of the receiver?

"Poetic language" establishes a set of preconceived rules. Is this true or is it a fallacy? If one considers the language crisis in the avant-garde, e.g., the negation of language in Dada and mechanical writing in Surrealism, this statement could not be further from truth since here language is only programmed to provoke an answer, to shock. Therefore, for as long as it is not yet perceived, an avant-garde work of art *is not*, does not exist, is just a ghost, a possibility. This is the reason why it is not appropriate for museums or libraries, where reception is collective and passive. In a photomontage exhibition in Cologne in 1920, Ernst invited the visitors to use an axe to destroy one of his sculptures. It is a question of shocking.

In their time, at the beginning of the century, the historic avant-gardes were considered to be a social and political phenomenon, an

A-T. Tymieniecka (ed.), Analecta Husserliana, Vol. XXXVII, 213—216.

ideological recoil, the expression of consciousness of a crisis. Nowadays it is our turn to explain the artistic value of the work of these movements, their aesthetic value. No doubt one of the main values of avant-garde art is catharsis, the power of liberation, but it is still art for a minority. How is it possible then that the apostles of compromise, of critical consciousness, express themselves with an art which is far from utility and the pragmatic, so far from life but at the same time so vital? Is this like Kafka's an art which prevents collective suicide and holocaust?

This art is an excuse, a pretext to avoid death, a way of evasion that goes back to the aesthetics of decadence; it is so related to agony and lucidity at the same time and witnesses the collapses of values. Is this why the avant-garde looks back to childhood and its artificial Eden which Bergamin[2] spoke so of, the lost paradise of non-figurativeness, of non-myth, of the pre-Babel babble? A virginal art, in short, which knows not art, which knows no tradition?

Avant-garde art is the testimony of the truth of an age through falsehood, through a series of impossibilities, of frustrations. Is it therefore an art bound to fail from the very beginning because it conveys incommunicability through non-communication?

If the artist is a prophet, or if he ever were one, it is now the receiver's turn to be one, since the message is to be ciphered. It is an enigma. In this way, avant-garde artistic manifestations become meaningful through an analysis similar to that of the Rorschach test. They project the obscure possibilities of the human secret onto the symbolic forms of language — by dissociating these forms from their own context. In avant-garde art it is not the principle of identification but the principle of alienation that holds true. A good example of this is the definition of beauty that Surrealism took from Isidoro Ducasse, the dark Count of Lautréamont, i.e., the accidental meeting of a sewing machine and an umbrella on an operating table.[3] Plato spoke of Art in terms of beauty and utility, but the avant-garde follows Aristotle in that they talk about art as being useless and gratuitous. On the other hand, almost all the "isms" fought for freedom, except for Italian Futurism, and even that movement espoused the "incumbency myth," to use Frie's terminology, because they believed Fascism was the revolution.

A writer so compromised as Louis Aragon said that the principle of utility will be alien to all those who practice the surrealist vice.[4]

Futurism looked at the progress of machines as the symbol of a new era, mankind's salvation. In contrast to this religion, which praised modernity, Expressionism, protects mankind against the evil of bureacracy and technocracy. Is its art dehumanized? Objects break down in Cubism, and Surrealism serves up putrefaction — after the scandalous rage of the iconoclastic Dada. Surrealism uncovers all the hidden sides of man. One must insist. Is this a dehumanized Art?

Let us put aside the intentionality of the artist and let us think about the role of the public in avant-garde art. Let us use a phenomenological analysis. The phenomenon is caught by the "intentional consciousness." It is the subject who makes it meaningful and gives form to the object giving it a *function*. The represented object functions autonomously in the work of art, but according to the meaning which the receiver assigns to it. It is in the *act of reception* and especially in the *reception context* where all this play obtains full meaning.

In Cubism, the word as object, the autonomy of the artistic object, has its origin in the *receiver's intention* to provide meaning. The objects built up are far from empirical reality. They are images, but not mirror images. The receiver's visions and memories give them the form necessary to make sense. Cubism, as Ortega[5] says, paints Ideas, ideal objects that are immanent to the subject — intrasubjective. It is a question of going back to the optic synthesis found in ancient Egyptian art.

Art only shows aspects, partialities. In the avant-garde movements relativism works full-time. Avant-garde artists do not judge, they reveal by summarizing the most meaningful features. That is why caricature is so relevant.

To fill in the blanks, Ingarden's *places of indeterminacy*, the receiver will need a peculiar competence. Therefore, only a minority is chosen to be co-creators of the work of art.

The avant-garde shatters the expectations with which traditional art is approached. That art sits stagnant in museums like monuments in a cemetery waiting to be passively consumed. The new art shocks the receiver. This is why all sorts of pseudo-artistic phenomena are so important.

Indeed, the task of objectivating the art piece — actualizing the maximum of aesthetically valid qualities — is limited to a sensitive and sensitized minority because this new art rejects the generic traditional codification and tends towards declassification and interferences. Should we say, echoing Schopenhauer's[6] comment on a symphony by Beethoven, that there is here a wonderful order hidden under the seeming disorder? In any case, the avant-garde was revolution, a rebellion against anxiety, and all new art will always be avant-garde – will show forth the heroism of launching into the unexplored. As Dali put it: "todo el arte directo, producido bajo una intensa presión de lo desconocido y de los instintos."[7]

University of Santiago de Compostela

NOTES

[1] This book, which treats the dehumanization of art was published in 1925.
[2] J. Brihuega, *Manifiestos, proclamas, panfletos, y textos doctrinales* (Madrid: Cátedra, 1982), p. 283.
[3] M. De Micheli, *Las vanguardias artísticas del siglo XX* (Madrid: Alianza Forma, 1987), p. 184.
[4] Brihuega, *op. cit.*, p. 281.
[5] *Ibidem*, p. 275.
[6] *Ibidem*, p. 434.
[7] *Ibidem*, p. 302. The English translation of this quotation would be: All direct art, produced under an intense pressure of the unknown and of the instincts.

DARIO VILLANUEVA

PHENOMENOLOGY AND THE PRAGMATICS OF LITERARY REALISM

Realism has not only shaped important schools and periods in the evolution of world literature but has also constituted a basic constant in all literary art since the formulation of the principle of *mimesis* in the *Poetics* of Aristotle. In this sense, it has become one of the central concepts of literary theory, one deserving clear definition of its conceptual limits. When theorists have risen to the task of correcting the imprecision, polysemy, and ambiguity with which the realist principle is applied, their efforts have often been opposed.

It is in no way an easy task given the diverse implications surrounding realism, all of them of great importance. The first of these implications is, of course, the philosophic. The very denomination *Realism* has its origins in the old dispute over the question of the universals, or the archetypal ideas to which Plato granted full existence. Consequently, to the idealist philosophers poetic imitation appeared conditioned; this is the argument made in the tenth book of *The Republic*, which W. J. Verdenius (1949: p. 16) has described as "his conception of a hierarchical structure of reality."

According to these Platonic suppositions, there are three levels of reality: that of ideal forms or archetypes, whose ontologic plenitude is not questioned; that of visible objects or Phenomena which are nothing more than images or reflections of the ideal forms; and a third composed of images proper which one finds in the mimetic arts in general and literature in particular, which take as their model a reality — the perceived reality — which is an imperfect copy of the genuine one. Consequently, the most direct of literary imitation is two steps below the essential nature of things. For art to truly strive for realism it should elevate itself from the ever precarious material world in order to approach the ideal reality, the essential nature of things, which is different from their visual appearances.

Confronted with this paradoxical Platonic identification of realism with idealism and a mimetic principle that completely transcends the restrictive realm of art, we refer to the basic relationship extant between the archetypal (*paradeigmatos eidos*) and the sensually perceptible

A-T. Tymieniecka (ed.), Analecta Husserliana, Vol. XXXVII, 217–235.

(*mimema paradeigmatos*). Aristotle's contributions are of extreme import. Without rejecting the concept of universals, Aristotle does not consider them as being outside the things themselves but rather embodied in them and only abstractable from them. Thus, perceptible reality is no longer an image of anything which transcends that reality and *mimesis* definitively confines itself to the specific boundaries of art and literature.

Consequently, for Aristotle the word *imitation* becomes a term specific to the arts, and "it distinguishes them from all else in the universe, liberating them from any rivalry with all other human activities." (Abrams, 1953: p. 25)

Given this ordering of things, an appreciable connection is clear between the ontological, epistemological, and philosophical foundations, in general, of what Jan Bruck (1982) terms the "bourgeois" realism of the eighteenth and nineteenth centuries and the very conception of reality and art formulated by Aristotle (although Bruck chooses to view that realism as something different from, and outside of Aristotelian *mimesis*).

Effectively, the tradition of rationalism, sensualism, and empiricism extending from Descartes, Locke, and Berkeley to Thomas Reid's *common sense* thinking and prolonged by nineteenth century positivism, does not represent any substantial break with Aristotelian philosophy, but tends rather to confirm the full reality of the perceptible objects themselves, outside of the perceiving mind.

Derived from this philosophical foundation is a clear aesthetic implication which allows for purely linguistic considerations when discussing literature.

What words certainly are not, save rare exceptions, are iconic signals. They are rather purely symbolic and their capacity to signify is based on the need to traffic directly with that which they signify. This is not sufficient reason to deny to all the undramatic genres their mimetic potential. However, this does introduce a great number of new elements which cannot be ignored. These complicate beyond measure the process of the imitative representation of reality which verbal art in fact accomplishes.

Very distant to us is the empirical theory of language of Locke's *Essay Concerning Human Understanding* (1690), according to which, words are direct images of reality, of the things perceived through the senses. However many other conceptions of language concluded like

Locke that language has substantiality right up through the "first" Wittgenstein (1921) who made language into a sort of scale map of the entire world. As we read in the *Tractatus Logico-Philosophicus* (5—6), "*Die Grenzen meiner Sprache* bedeuten die Grenzen meiner Welt." (1921: p. 163)

The distance between the conception of literary realism based on Platonic thought and that founded in the Aristotelian conception of the world is no greater than that between a *mimesis* based on linguistic proposals like that ones we just have cited, and still another one which, in the manner of the "second" Wittgenstein (for whom there exist no essential meanings, only relative meanings), excludes the conception of words as transparent images and grants them, on the contrary, full capacity to create the world of which they speak for themselves, to internally render and compose things within "language games."

In the end, all questions pertaining to language, among them that of its relation to reality, have been converted to one of the principle preoccupations of contemporary philosophy, if not the primary one. This is true in both the so-called analytical and the neo-positivist schools.

In the present situation, what is most important is finding a conception of realism that maintains an equilibrium between the principle of the autonomy of the literary work before the determinations of reality, and the relationship that doubtless exists between words and reality — a relationship without which literature could not sustain itself in the role of authentic social institution which it actually fulfills. Without that relationship the literature of a given time would lose interest for the readers of all successive generations.

Any loss of balance here will produce either of two fallacies in the estimation of literature which we could well name, respectively, the *mimetic* and *aesthetic* fallacies. To these are attached the least pertinent conceptions and practices of the two modalities of realism which dominate the present state of the question in literary criticism.

In genetic realism all trust is placed on the existence of a unitary reality, anterior to the text, before which is situated the perceptive consciousness of the author, and which works by means of long and efficacious observation. All that will yield as a result a true reproduction of the referent, thanks to the transparency or the thinning of the expressive medium of literature, language.

In the essay by Zola "Les Romanciers naturalistes" we read: "Je voulais bien une composition simple, une langue nette, quelque chose comme une maison de verre laissant voir les idées à l'intérieur . . . les documents humains donnés dans leur nudité sevère." (*Oeuvres complètes*, XI, Paris, 1968: p. 92) Zola's idea had an earlier expression in a letter to Valabrègne of the eighteenth of August 1864 in which he developed his theory of the three screens, the classical, the romantic, and the realist which is "un simple verre à vitre, très mince, très clair, et qui a la prétention d'être si parfaitement transparent que les images le traversent et se réproduisent ensuite dans leur réalité. L'écran réaliste nie sa prope existence." (See Alain de Lattre, 1975: p. 98ff.)

What is attempted with this realism is a maximal concealment of form in order that language be completely transparent. This inclines to that which Hayek called the "fallacy of conceptual realism." Consistent with this is the belief that behind every word one finds the designated object which corresponds to it; the more imperceptible the word may be the more the object gains presence and corporality. Here is, likewise, the "propositional phantom" of Wittgenstein's *Tractatus* which we have already mentioned, which can be related to a semantics of a neo-positivist ilk, influenced by that science which believes in the certain existence of objective truth and of a compact and unarguable world.

In light of the expression of genetic realism represented by the naturalist theory of Zola, it is necessary to expand on much of what is usually covered in treatments of the Marxist aesthetic of "reflection," especially in that treatment of his formulations by Georg Lukacs.

Lukacs places the reflection of objective reality at the very center of his aesthetics. Between that reality and the literary text which it represents, there is no direct relation, in the manner of the medieval "aliquid stat pro aliquo," but rather a relation in the manner of the *interpretant* of Peirce's semiotics which is situated between the referent object, or *designatum*, and the *representamen*, or sign. Between the two poles is inserted a third discourse which in the thesis of Lukacs is an ideology, an essentialist interpretation of reality: Marxism.

In summary, Socialist Realism is, paradoxically, a faithful reflection by artistic means of a world interpreted ideologically through Marxism. From the standpoint of concrete reality (in which would reside the genetic principle of the work of literature which intends to represent it), the work that would be more realistic in Lukacs's view is that work which passes along reality's reflection (the reflection which reality gives the text and which the text passes on) by means of the "third discourse" or "interpretant" of Marxist ideology.

There is nothing more logical, thus, than the identification proposed by Thomas E. Lewis (1979) between the concept of referent in the Marxist theory of reflection — expounded on more explicitly by Althusser than by Lukacs — and the semiotics of Umberto Eco, who is essentially faithful to the principles of Peirce. For both systems the relation established is not between the sign and the referent, but between the sign and a convention understood as a "cultural unit" or as an ideology ("a system of representation of images, concretized in specific practices" according Lewis) (1979: 470) — which is then projected onto the referent.

Thus, Althusser's Marxist epistemology, whose aesthetics will not be harvested much here given the condensed nature of this paper, "subscribes to the same complex of assertions about the relation of thought to the real that figures in the semiotic notion (at least as defined by Eco) of the relation of language to the real expressed in the concepts of the interpretant and meaning as a cultural unit." (Lewis, 1979: 469)

In contrast, the second way of understanding this literary phenomenon displaces the central axis of the same from a reality which precedes the text, to be scrupulously observed and reproduced with absolute sincerity by the author, to a world created autonomously within the work. This second realism, going beyond pure imitation, gives primacy to an imaginative creation that cleanses the objective material which could lie at the origin of the whole process and subjects it to a principle of immanent coherence which gives it meaning, and does so more via the path of estrangement than by that of the identification of factual reality itself.

We will speak here of *formal or immanent realism* as opposed to the previously discussed *genetic* realism. We will attribute to it, instead of a "hermeneutic of reconstruction" (in the mode of Schleiermacher), the principle, developed by Gadamer (1965: 125—128), of *aesthetic distinction.* It accomplishes the abstraction of everything constituting the root of a work such as its original vital context — all functions, religious or profane, in which its meaning could have dwelt; that is to say, the abstraction of the non-aesthetic moments which are inherent to them: object, function, and the meaning of the content. In this way, concludes Gadamer, in virtue of the "aesthetic distinction" by which the work of art is made to belong to the aesthetic consciousness it is dislodged from its place and from the world to which it belongs.

Gustave Flaubert's conception of the novel exemplifies full-blown this formal and immanent realism that magnifies, in the manner of the

Renaissance Neo-Platonists, the artist as a creator of worlds along with God.

According to Abrams (1953: pp. 482—505) such ambitious identification seems to have had its origin in the Florence of the fifteenth century, specifically in the Cristoforo Landino's 1481 commentaries on Dante. Thus, Tasso can contend that: "Non merita nome di creatore se non Iddio ed il Poeta" (Abrams, 1953: p. 483), and since then the Italian and English Neo-Platonists (from Sir Philip Sidney on) have maintained that the artist is, similar to God, creator of a second nature. The extremes of this view are to be found in the Goethe's dialogue *Über Wahrheit und Wahrscheinlichkeit der Kunstwerke* (1797).

We cannot pay much attention in a summary like this to the historical development of such demiurgic notions of the writer, which have very recently found fresh expression, for instance, in Andre Malraux's *Les Voix du Silence* ("les grands artistes ne sont pas les transcripteurs du monde, ils sont les rivaux"). It is enough here to cite the notion as one of the key points for a more solid theoretical characterization of immanent and formal realism. Jacques Derrida wrote about this years ago in a volume dedicated to tracing the theme of mimesis from a deconstructivist perspective:

"L'Artiste n'imite pas les choses dans la nature, ou si l'on vent dans la *natura naturata*, mais les actes de la *natura naturans*, les opérations de la *physis*. Mais puisq'une analogie a déjà fait de la *natura naturans* l'art d'un sujet auteur et, on peut même le dire, d'un dieu artiste, la *mimesis* déploie l'identification de l'acte humain à l'acte divin. D'une liberté à une autre." (Jacques Derrida *et. al.*, 1975: p. 67)

Thus for Derrida the true mimesis is established between two producing subjects and two operations of producing, not between two produced realities with one of these (that of the literary work) being a reflection of the other (that of nature). The writer does not imitate anything but the productive liberty of God, and the analogy established between both "n'est pas seulement un rapport de proportionnalité ou un rapport entre deux — deux sujets, deux origines, deux proportions. Le procès analogique est aussi remontée vers le *logos*. L'origine est le *logos*. L'origine de l'analogie, ce dont procède et vers quoi fait retour l'analogie c'est le *logos*, raison et parole, source comme bouche et embouchure." (Derrida *et al.*, 1975: p. 74)

As Guy de Maupassant said in a famous prologue (that of *Pierre et Jean*), according to the standpoint of this second realism, narrative art is characterized above all by the sensation of an illusion of truth, an idea which has been applied to painting by Ernst N. Gombrich in his conception of a wholly antigenetic realism, one completely formal and conventionalist, a conception which (through Nelson Goodman and others) has influenced literary theory and criticism beyond measure.

This second appraisal of reality in literature avoids the dangers of genetic mechanism and interprets the work from specifically artistic standards which are more in accord with its essential nature than other external references would be. But it carries in its breast the opening of a total fissure between the created world and reality. This is exacerbated by the most radical immanentism, which drives us to yet another dead end with a warning sign, that called the "intentional fallacy" (or "genetic fallacy"), which involves a confusion of the literary work with its own origins.

Warnings against this last excess have been widespread in the work of theoreticians and literary critics. But the preventatives they offer are only palliatives. This is precisely the stimulus that has prompted us to approach this theme in our research and to develop it in the terms which we now explain.

We will adopt here the perspective of literary reception as a third path that can be taken in the consideration of a problem which has not been perfectly resolved by the other two paths already elaborated in the analysis of the literary communicative process.

In effect, genetic realism bases everything on the relation of the writer to the world of his orbit, which he conceives through observation and reproduces mimetically as faithfully as possible. Conversely, in formal realism everything is centered on literariness: it is the work itself which institutes a reality disconnected from the referent, a "textual reality."

We will arrive then, at a Phenomenology and a Pragmatics of realism because both methodological pillars today support all understanding of literary fact including the missing perspective of the reader, which is currently growing in importance.

Concretely, following Roman Ingarden and Wolfgang Iser, there is included in the dialectic established between the work as a schematic structure and its concretization as an aesthetic object the fundamental problem of the theory and criticism of literature. Consequently, we also have here the fundamental frame of reference for the balanced comprehension of realism which we defend.

For Ingarden literature has its origin in creative acts of the intentional consciousness of the author. Its ontic base resides in a physical foundation — printed paper or manuscript, magnetic band, computer disk, etc. — which permits its prolonged existence. Its internal structure is a multi-layered one, in which there operates a stratum of sounds and verbal formations, another of semantic unities, a third of represented objectivities, intentional correlations of sentences, and lastly, the stratum of the schematicized perspectives (*ansichten*) in which those objectivities appear. The artistic intention creates a solid unity among all these, thus, arranging the polyphonic harmony of the work. Thanks to its double linguistic layer (phonic and semantic) the work is intersubjectively accessible and reproducible, such that it becomes an intentional intersubjective object that refers to an open community, spatial and temporal, of readers. Precisely for this reason the work of literary art is not a mere psychological phenomenon; it transcends the experience of consciousness, as much those of the author as those of the reader.

The work of literature leaves many elements of its own ontological constitution in a potential state; thus, it is its own fundamentally schematic entity. The active actualization of this by the reader fills these blanks of indeterminacy (*Umbestimmtheitsstellen*) or brings out latent elements, and if this actualization is realized with a positive aesthetic attitude, the artistic object (the work) is converted into an aesthetic object.

In this order of things, each one of the strata that ontologically composes the work demands different actualizations. But, without detriment to the concretization of the whole set as a unity, this releases in particular the process of meaning lending which the reader engages in, beginning with the semantic units and the represented objectivities, a task in which the schematicism of which we speak requires the bringing in of those absent or undetermined elements without which the work would not reach true existence.

This leaves open a margin of variability between the inherent artistic values of the work itself and the aesthetic values reached in the concretization or concretizations which provides it its total ontological plenitude.

The fundamental difference between a work of literary art and its actualizations is that in these are concentrated the potential elements and therein the blanks or spaces of indeterminacy are filled in. The artistic values, which belong to the diverse strata, comprise some of those potential elements. Their aesthetic productivity depends in great part on the system of relations established between them, that is, on the qualitative harmony comparable to a *Gestalt*, or, in other terms, to a structure.

Concretely, besides the idea of literature as a schematic formation and stratified structure in which is integrated, together with other levels, a level devoted to the representation of objectivities, and the idea of actualization as the key element for the ontological fullness of the same, we will borrow from the phenomenology of literature for the futherance of our investigation the following two theoretical tools:

1) The principle of intentionality, which is deeply rooted in phenomenological philosophy and from whence it has been applied to literature; intentionality according to Husserl, is the activity which the cognizant *I*, extends toward the transcendental phenomenon in order to endow it with meaning.

As John D. Boyd (1968: p. 63) reminds us in his book on the function of mimesis, the intentionality that Husserl takes from his teacher Brentano, which is also to be found in Heidegger, Sartre, Merleau-Ponty, etc., has a long tradition in Thomist thought, up through Jacques Maritain. However, the author of *Logische Untersuchungen* placed it in a clearly Cartesian matrix. Thus, in Husserl's *Ideen* (1976: p. 98) we read: "*Der Thesis der Welt, die eine 'zufällige' ist, steht also gegenüber die Thesis meines reinen Ich und Ichlebens, die eine 'notwendige',* schlechtin zweifellose ist. *Alles leibhaft gegebene Dingliche kann auch nicht sein, kein leibhaft gegebenes Erlebnis kann auch nicht sein*: das ist das Wesensgesetz, das diese Notwendigkeit und jene Zufälligkeit definiert."

Thus, for Husserl, Kant's distinction between *noumenon* (that which is in the subject) and *phenomenon* (that which is not therein) is not valid. There are no things unto themselves: there is no being other than the knowable, a thesis whose value for a theory of literary realism does not escape us. In an enlightening monograph on the development of intentionality within Husserlian phenomenology D. Souche-Dagues clearly explains:

> ce n'est désormais ni le *Je suis* ni *le monde est* que peuvent réprésenter le point de départ authentique d'une ontologie phénomenologique, mais plutôt le: *Je suis le monde est*, c'est-à-dire le thème de la corrélation intentionnelle au sein de toute expérience. C'est bien donc l'intention qui devienne le fil directeur de la pensée, c'est-à-dire de la pensée de l'être. (1972: p. 240)

In effect, phenomenology studies this *erlebnis* which intentionally points toward an object, real or otherwise, which (in any case) forms a constitutive part, as an inexhaustable reference, of the intentional entity of life. It is for this reason, consequently, that we study it. As we read in the fourth lecture in Husserl's *Die Idee der Phänomenologie*:

> Die Erkenntniserlebnisse, das gehört zu ihrem Wesen, haben eine *intentio*, sie meinen etwas, sie beziehen sich in der oder jener Art auf eine Gegenständlichkeit. Das sich auf eine Gegenständlichkeit Beziehen gehört ihnen zu, wenn auch die Gegenständlichkeit ihnen nich zugehört. (1958: p. 55)

In this way, intentionality, according to Husserl, is the activity in which the knowing *I* departs towards the transcendental phenomenon to give it a meaning. This phenomenon to which we so often refer could well be a reality like a simulacum of itself, in the same way that, as Gilbert Durand says in *L'imagination symbolique* (1964: p. 8), consciousness makes use of the two ways of representing the world. One is direct, when the thing seems to present itself to the spirit; and the other is indirect: "Dans tous ces cas de conscience indirecte, l'object absent est ré-présenté à la conscience par une *image*, au sens très large de ce terme." He concludes:

> A vrai dire la diffèrence entre pensée directe et pensèe indirecte n'est pas aussi tranchée que nous venons, par souci de clarté, de exposer. Il vaudrait mieux écrire que la conscience dispose de différents degrés de l'image . . . dont les deux extrèmes seraient constitués par l'adéquation totale, la présence perceptive ou l'indéquation la plus poussée.

Morse Peckham in an article about realism with a flashy title, "Is the Problem of Realism a Pseudo-Problem?" (1970: p. 98), considers things to be immediate signals which upon being imitated artistically make room for other mediate symbols.

2) In a very obvious way, the concept of *epoché*, or phenomenological reduction, consists in suspending belief in the reality of the natural world and consequently placing in parenthesis the inductions which it creates — in order to stay exclusively by the given, according to Husserl's assertions.

This fundamentally phenomenological attitude is the same one which convinces the reader of literature to accept the pact of "fictionality" as it has been called or, in the appropriate words of Coleridge, "the willing suspension of disbelief."

Toward this particular literary *epoché* are directed three types of questions, each of which deserves separate attention. In the first place, the very pact of fiction in which one suspends disbelief implies the notion of play, by convention, and also the hermeneutic repercussions of the one notion on the other. Then there is the pragmatic question of which are the "speech acts" belonging to literary communication. Finally, and very directly related to this last question, is that of the logico-semantic status of the ficticious or fictional (derived from "fictionality").

The sceptical and phenomenological origins of this *epoché* are not in contradiction to the ludic aspect to which we will soon refer. Since Plato (*Republic*, 602b; *Sophist*, 234ab; *Politics*, 288c; *Law*, 796b, 889de), play has been considered to be fundamental for art and, particularly, for imitation. In any case, by play is not meant arbitrariness and futility, but a transcendental exercise, for the imitative artistic game expresses something different, something more profound, and more compelling than that which it appears to represent (W. J. Verdenius, 1949: pp. 21—27).

This epistemological rendering of play is what justifies its being of growing interest for philosophy and science since the second half of the nineteenth century, a development traced by Mihai Spariosu (1982).

Previously, play (in its serving double duty as an attitude without transcendence and as a theoretical tool for mediating between inapprehensible reality and reason) had occupied a notable place in Kant's philosophy. On specifically artistic grounds, Schiller, in his letters on the aesthetic education (*Über die ästhetische Erziehung des Menschen*, 1795), completely identifies art and sport.

The question of intentionality, for its part, involves matters of great importance for the literary phenomenon, like imagination, symbol, and meaning. Reality gains meaning through an act of understanding or intentional experience, to which may be compared, in the communicative process of literature, the apprehension of the world by the writer, the production of the text, and the reading of it by its intended audience. Like all intentional acts writing constructs intentional objects. It can be said that these objects are the reality perceived by the author, the work of literature created by him and the world which that work creates, projected by the reader.

For Jean Paul Sartre also, "l'intention est au centre de la conscience: c'est elle qui vise l'object, c'est-à-dire qui le constitue pour ce qu'il est."

(Sartre, 1940: p. 27) "L'imagination n'est pas un pouvoir empirique et surajouté de la conscience, c'est la conscience tout entière en tant qu'elle réalise." (1940: p. 358) "Il n'y a pas un monde des images et un monde des objets," "les deux mondes, l'imaginaire et le réel, sont constitués par les mêmes objets; seuls le groupement et l'interprétation de ces objets varient. Ce qui définit le monde imaginaire comme l'univers réel, c'est une attitude de la conscience." (1940: pp. 45—46) The value of this last argument for a theory of realism is unquestionable.

This is the pragmatic emphasis which we encounter on one of his pages: "la conscience imagineante que nous produisons devant une photographie est un acte . . . . Nous avons conscience, en quelque sorte, d'*animer* la photo, de lui prêter sa vie pour en faire une image." (1940: pp. 44—45) We can substitute for the photograph a work of literature, and we will have the same schema for a process identical to that which we considered in the framework of realism, which is the object of our investigation. Sartre remembers that for Husserl the image was a "remplissement" (*Erfüllung*) of the meaning, and adds: "ce qui constitue l'image et suplée à toutes les défaillances de la perception, c'est l'intention." (1940: p. 64)

The semiotic implications of phenomenology are fundamental and to be gathered not only in Peirce but also in Husserl, as Jacques Derrida has shown in *La Voix et le Phénomène* (1967), where he insists on something which is a key piece in the functioning of realism from the perspective we are applying: co-intentionality.

In the first of his *Logische Untersuchungen*, which is dedicated to expression and meaning, Husserl (1984: p. 39) asserts that the communicative process is accomplished because he who listens comprehends the intention of the speaker, which makes possible, above all else, spiritual intercourse and which gives the character of discourse to the discourse that entwines two people. This is a correlation, established by the physical part of the discourse, between physical and psychological beings who are mutually implicated, a correlation experienced by people playing their respective roles in communication. "Sprechen und Hören, Kundgabe psychischer Erlebnisse im Sprechen und Kundnahme derselben im Hören, sind einander zugeordnet."

Certainly the type of communication to which Husserl refers is very different from that which concerns us, literary communication, but for the further developments of our vision of realism it is important to leave intact this phenomenological idea of intersubjective intentionality or co-intentionality. For example, S. Y. Kuroda (1976: p. 126) assumes, as does Husserl, that "the essence of linguistic performance consists in meaning-assigning acts (*Bedeutungsverleihende Akt*) and meaning-fulfilling acts (*Bedeutungerfüllende Akt*), and that in phenomenological terms meaning is 'an intentional object.'" (Kuroda, 1976: p. 130) Nevertheless, "the meaning-realizing act . . . takes place in both the author's and the reader's consciousness" (1976: 137); but since the only contact possible between them is one through the text, and since it cannot exist identity between the respective contexts of author and reader (which is most often the case), this co-intentionality we are referring to will never be identified with the genetic fallacy which Wimsatt and Beardsley denounce.

With regard to conventions, it is certain that in order to intersubjectively communicate an intention of realism, for example, it is necessary to revert to them. Quite apart from reality itself's, including its artistic or literary representation, being largely the result of social conventions, the fictional *epoché* constitutes a true conventional complex. From this comes the sharp assertion of Warning (1979: p. 331) to the effect that fictionality is essentially contractual.

From a review of the pragmatic theories of "speech-acts" applied to fiction texts, one would deduce that result not from the presence of certain semantic or syntactic properties but from intentional modifications effected by the agents — emitter and receptor — of the communicative action. In the case where both realize the same modification, the fictionality will be complete through co-intentionality; but it is enough if the authentic owner of the text — phenomenologically speaking, the reader — practices intentionality in order that the work be fictionalized in its totality.

Formal logic and philosophic semantics have recently developed a theory of "possible worlds" which has application to the literary problem which occupies us. In effect, the "imaginative worlds, narrative worlds, or the worlds of the work of art" in general are a variation on these possible universes where true assertions are those which adjust to the proposed terms, and false assertions are those which do not.

However, we must take note in such an approach of the limitations of the autonomous and immanent character of the possible or artistic world created by the fictional discourse. We are situated in an *intensional* semantic, one which attempts play with the forms of expression, is more concerned with connotation than with denotation or reference, which easily causes us to incline toward "formal or immanent realism."

In this respect, we judge to be of maximum utility the distinction that Gottlob Frege (1892: p. 85—98) made in his theory of meaning in natural languages. The distinction between *Bedeutung*, or the referent object of a sign, and *Sinn*, or the meaning of the same (the manner in which the expression designates that object, the information given about it so that we can identify it).

One tends to relate both of these notions to Carnap's counterpoising of the *extensional* and the *intensional* and to the theories of "reference" and "meaning" proposed by Quine (and we could extend these parallels still further to include the *signification/signifiance* distinction of Michael Riffaterre). (See R. Barthes *et al.*, 1982: p. 93—94) But there is not a full correspondence here between these systems and the phenomenological linguistic theory found in the first of the *Logische Untersuchungen* (§§12—16), which was studied in his day by Jacques Derrida (1967). There, Edmund Husserl also differentiates between the object — *gegenstand* — or non-verbal phenomenon denoted by a word, and the way in which the object is presented, or the *meaning* (for which he employs the term *Bedeutung* in usage completely contradictory to that of Frege).

Frege's counterpoising was the catalyst, as is well known, for the expansion of the semiotic scheme of Morris, which consists in making the distinction between Semantics, properly speaking, that which is concerned with the relations between signs and their mental representations, and Sigmatics, which is devoted to the relations of signs to the objects to which they refer.

In this we get closer to the comprehension of realism not from the side of the author or the isolated text, but primordially from that of the reader, in an analysis which draws necessary support from Phenomenology which does not conceive a work of literary art as having ontologic plenitude if it is not actualized, and from a Pragmatics that does not consider meanings only in relation to the merely enunciated, but from within the dialectic between the enunciation, the reception, and the referent. Realism in action, where the activity of the ultimate addressee is decisive, represents one of the most conspicuous manifestations of the "principle of cooperation" formulated by H. P. Grice (1975: pp. 41—58) — on which Teun van Dijk has read as an unpublished work (1976: p. 44), and which the late Jon K. Adams applied to the specific field of literary fiction.

The basic idea that animates this analysis is that of linguistic conduct as a type of intentional social interaction, driven by a cooperative willingness, which, according to Grice, J. Lyons exemplifies, with the sentence "John is a tiger" and the metaphoric key which corresponds to it.

In any case, we have come close to a clear formulation: literary realism is a fundamentally pragmatic phenomenon that results from the projection, onto an intentional world that the text suggests, of a vision of the external world that the reader — every reader — possesses.

Because of this, Paul Ricoeur in his interpretation of mimesis, to which we have already made reference (1981), underlines its identity — strengthened by the very suffix of each — with poiesis and with praxis, which yields a meaning that makes expressive, not reduplicative, reference to the world.

The pact of fiction, the voluntary suspension of disbelief, renewed by literary *epoché*, certainly corresponds to Grice's "principle of cooperation." It likewise accords with the projection of the external field of reference (EFR) onto the internal field of reference (IFR), and carries it from the reading of all that which the schematic structure of the work claims in order to reach its full ontological abundance.

By this same impulse of cooperation the reader tends to bring the intentional world of the text near to his own, to the extensional referent. Grice adds that in standard linguistic communication all the agents expect rational, serious, collaborative conduct from each other. We propose, correspondingly, that the reader who makes, among the variety of ways of apprehending the work of literature enumerated by Roman Ingarden, an optimal aesthetic actualization of the work, spontaneously and naturally assumes the "seriousness" of what is written, which, although fictitious in its origin, lends itself to a realistic decodification.

This would consist in the hermeneutic task of giving real meaning to the text, illuminating its immanent IFR from the EFR, which amounts to the vision and interpretation of this multiple reality by every one of the readers.

To this third conception of the mimetic literary phenomenon (which we found in the aesthetics of the Polish disciple of Husserl and in Pragmatics), we give the label *intentional realism*. The question lies not so much in the imbrication of the text with reality as much as in how the readers make use of the text to articulate their own reality. Effectively, intentional realism amounts to the giving of realistic meaning to a text on which one has performed a hermeneutics "of integration," not "of reconstruction" (Gadamer) from the referential horizon proportionate to the experience of the world which every reader possesses.

Karlheinz Stierle, in his theory of the "quasipragmatics" of fictional texts, departs from the perspective which we have made our own wherein the fictitious (and the IFR) and the real (and the EFR) interrelate in such a way that one acts as the horizon of the other (1979: 313), without prejudice as to which of the essential characteristics of the literary text may be among a series of unverifiable assertions. But, we add that the reader does not desire such verification, nor could it be reached, as neither can it in a great number of the communicative situations in which verification occurs by virtue of the cooperation illustrated for us by H. P. Grice (1975). "Le texte de fiction s'efface," continues Stierle (1979: 300), au profit d'un au-delà textuel, d'une illusion que le récepteur — sous l'impulsion du texte — produit lui même. L'illusion (comme résultat de la réception quasi pragmatique de la fiction) est un hors-texte comparable à celui de la réception pragmatique."

As we pointed out earlier, Stierle's thesis seems extremely accurate. He places his valuative attribution toward the side of ingenuous forms of reception. For us this quasi-pragmatic reading is that of intentional realism. Everything begins with the *epoché* of the pact of fiction, with the "voluntary suspension of disbelief." Later, there comes a process of increasingly intense involvement with the represented world that interests us. We identify with the characters — if it is a narrative text (novel or play) — or with the lyric enunciator and his internal affections, at the same moment that we are no longer mindful of the discourse's being illusory and even experience that thanks to him (if the discourse is not "eminent" as Gadamer would say, then everything fails). Finally let us not regress to the anterior epistemological attitude, to our voluntary *epoché*. The virtuality of the text and our intentional living of the same brings us to qualitatively elevate the standing of its internal world of reference until we integrate it, without any reservation, into our own external, experiential world. This is Realism, in a word.

There stands the truth of literature, which is, as Pablo Picasso said of art in general, "a lie which makes us fall into the reckoning of truth."

(Quoted in Harry Levin, 1963: p. 39) This is a truth which, if it is to be accepted, does not require us to deny the artistic text's appearance of being an autonomous construct, apart from all previous reality and the intentions of the author. It is in us, the readers, as we appropriate the text to ourselves as an actualized aesthetic object, that realism does its work in that non-reintegrated *epoché* that Husserl's phenomenology can easily justify.

University of Santiago de Compostela

REFERENCES

Abrams, M. H., 1953, *The Mirror and the Lamp. Romantic Theory and the Critical Tradition.* New York: Norton & Co.

Adams, Jon-K., 1985. *Pragmatics and Fiction.* Amsterdam & Philadelphia: John Benjamin Publishing Co.

Barthes, Roland, *et al.*, 1968, "Le Vraisemblable." *Communications*, 11.

Boyd, John D., 1968, *The Function of Mimesis and its Decline.* Cambridge: Harvard University Press. Second edition, New York: Fordham University Press, 1980.

Bruck, J., 1982, "From Aristotelian Mimesis to 'Bourgeois' Realism.'" *Poetics* **II**, pp. 189—202.

Derrida, Jacques, 1967, *La Voix et le phénomène. Introduction au problème du signe dans la phénomènologie de Husserl.* Paris: P. U. F.

Derrida, Jacques, 1978, *La vérité en peinture.* Paris: Flammarion.

Derrida, Jacques, *et al.*, 1975, *Mimesis des articulations.* Paris: Aubier-Flammarion.

Dijk, Teun A. Van (compiler), 1976, *Pragmatics of Language and Literature.* Amsterdam: North-Holland.

Durand, Gilbert, 1960, *Les structures anthropologiques de l'imaginaire.* Paris: Bordas. Second edition, 1969.

Durand, Gilbert, 1961, *Le décor mythique de 'La Chartreuse de Parme'. Les structures figuratives du roman stendhalien.* Paris: Jose Corti. Third edition, 1983.

Durand, Gilbert, 1964, *L'imagination symbolique.* Paris: P. U. F. Fourth edition, 1984.

Frege, Gottlob, 1892, *Kleine Schriften.* Edited by I. Angelelli. Hildersheim: Georg Olms, edition of 1967.

Gadamer, Hans-Georg, 1965, *Warheit und Methode.* Tübingen: J. C. B. Mohr.

Grice, H. P., 1969, "Meaning." *Problems in the Philosophy of Language.* Edited by T. M. Olshewsky. New York: Holt, Rinehart & Winston. Pp. 251—259.

Grice, H. P., 1975, "Logic and Conversation." *Syntax and Semantics, 3: Speech Acts.* Edited by P. Cole and J. L. Morgan. New York and London: Academic Press. Pp. 41—58.

Husserl, Edmund, 1958, *Die Idee der Phänomenologie.* The Hague: Martinus Nijhoff.

Husserl, Edmund, 1976, *Ideen zu einer reinen Phänomenologie und Phänomenologischen Philosophie.* The Hague: Martinus Nijhoff. First edition, 1913.

Husserl, Edmund, 1984, *Logische Untersuchungen*. The Hague: Martinus Nijhoff. First edition, 1929.

Ingarden, Roman, 1931, *Das literarische Kunstwerk*. Tübingen: Max Niemeyer Verlag. Third edition, 1965.

Ingarden, Roman, 1962, "A Marginal Commentary on Aristotle's *Poetics*." *Journal of Aesthetics and Art Criticism* **20**. Pp. 163—173 and 273—285.

Ingarden, Roman, 1968, *Vom Erkennen des literarischen Kunstwerks*. Darmstadt: Wissenschaftliche Buchgesellschaft. English translation by Ruth Ann Crowley and Kenneth R. Olson, *The Cognition of The Literary Work of Art*. Evanston: Northwestern University Press, 1973.

Iser, Wolfgang, 1972, *Der Implizite Leser: Kommunikationsformen des Romans von Bunyan bis Beckett*. Munich: Wilhelm Fink. English translation, *The Implied Reader. Patterns of Communication in Prose Fiction from Bunyan to Beckett*. Baltimore and London: The Johns Hopkins University Press, 1974.

Iser, Wolfgang, 1975, "Die Wirklichkeit der Fiktion." *Rezeptionsästhetik. Theorie und Praxis*. Edited by Rainer Warning. Munich: UTB. Pp. 277—324.

Iser, Wolfgang, 1975b, "The Indeterminacy of the Text. A Critical Reply." *Comparative Criticism. A Year-book* **2**. Edited by Elinor Shaffer. Cambridge: Cambridge University Press. Pp. 27—47.

Iser, Wolfgang, 1976, *Der Akt des Lesens: Theorie ästhetischer Wirkung*. Munich: Fink. English translation, *The Act of Reading: A Theory of Aesthetic Response*. Baltimore: The Johns Hopkins University Press, 1978.

Iser, Wolfgang, 1983, "Akte des Fingierens, oder: Was ist des Fiktive im Fiktionalen Text?" *Funktionen des Fiktiven*. Edited by D. Henrich and W. Iser. Munich: Fink.

Kuroda, S. Y., 1976, "Reflections on the Foundations of Narrative Theory from a Linguistic Point of View." *Pragmatics of Language and Literature*. Edited by Teun A. Van Dijk. Amsterdam: North-Holland.

Lattre, Alain de, 1975, *Le réalisme selon Zola. Archéologie d'une intelligence*. Paris: P. U. F.

Levin, Harry, 1951, "What is Realism?" *Comparative Literature*, III, Vol. **3**, 193—199.

Levin, Harry, 1963, *The Gates of Horn* (*A Study of Five French Realists*). New York: Oxford University Press.

Levin, Harry, 1972, "On the Dissemination of Realism." *Grounds for Comparison*. Cambridge: Harvard University Press. Pp. 244—261.

Lewis, Thomas E., 1979, "Notes Towards a Theory of the Referent." *PMLA* **94**, 3, Pp. 459—475.

Lyons, J., 1981, *Language, Meaning and Context*. London: William Collins & Sons.

McKeon, Richard, 1936, "Literary Criticism and the Concept of Imitation in Antiquity." *Modern Philology* **XXXIV**, 1—35. We quote from *Critics and Criticism: Ancient and Modern*. Edited by R. S. Crane. Chicago: The University of Chicago Press, 1952, pp. 147—175.

Peckham, Morse, 1970, "Is the Problem of Literary Realism a Pseudo-Problem?" *Critique. Studies in Modern Fiction*, Vol. **XII**, 95—112.

Prendergast, Cristopher, 1986, *The Order of Mimesis. Balzac, Stendhal, Nerval, Flaubert*. Cambridge: Cambridge University Press.

Reisz de Rivorola, Susana, 1979, "Ficcionalidad, referencia, tipos de ficción literaria." *Lexis* **3**, 2, pp. 99—170.

Ricoeur, Paul, 1981, "Mimesis and Representation." *Annals of Scholarship*, II, Vol. **3**, pp. 15—32.

Ricoeur, Paul, 1983, *Temps et récit. Tome I.* Paris: Du Seuil.

Sartre, Jean Paul, 1940 *L'Imaginaire. Psychologie phénoménologique de l'imagination.* Paris: Gallimard. New edition, 1982.

Souche-Dagues, D., 1972, *Le développment de l'intentionalité dans la phénoménologie husserlienne.* The Hague: Martinus Nijhoff.

Spariosu, Mihai, 1982, *Literature, Mimesis and Play: Essays in Literary Theory.* Tübingen: Gunter Narr Verlag.

Stierle, Karlheinz, 1979, "Réception et fiction." *Poétique.* Vol. **39**, pp. 299—320.

Verdenius, W. J., 1949, *Mimesis: Plato's Doctrine of Artistic Imitation and Its Meaning to Us.* Leiden: E. J. Brill.

Warning, Rainer, 1979, "Pour une pragmatique du discours fictionnel." *Poétique.* Vol. **39**, pp. 321—7.

Watt, Ian, 1957, *The Rise of the Novel. Studies in Defoe, Richardson and Fielding.* London: Chatto & Windus.

Wittgenstein, Ludwig, 1921, *Tractatus Logico-Philosophicus.* London: Routledge & Kegan Paul. Edition of 1962.

Wittgenstein, Ludwig, 1945—9, *Philosophische Untersuchungen* (*Philosophical Investigations*). Oxford: Blackwell. Edition of 1958.

MARÍA RUBIO MARTÍN

THE READER AND THE REALITY OF THE LITERARY TEXT: TOWARDS THE CONSTRUCTION OF AESTHETIC MEANING

The deeper and more exhaustive knowledge of the literary work, which arose mostly from the formalist studies carried out in the twentieth century, has created its own theoretical framework, and this framework — leaving aside any critical approach it — necessarily goes through its integration in the general semiotic space in the way it was initially conceived by C. Morris.[1] For his definition of the above-mentioned space, Morris considers three types of relationships which are established in the communicative framework between the sign, the participants, the referent and the context. Thus, he talks about *syntax* where relationships are established between signs and within signs, about *semantics* where there are relationships between a sign and the referent expressed by it, and about *pragmatics* where the relationships are drawn between participants — the producer and the recipient — signs, and the communicative context.

Taking into account this tripartition but, above all, the relationships which are generated between those elements, and considering the literary work as a sign — a fact which has given way to talk about *Literary Semiotics*[2] instead of *Linguistic Semiotics* — it is logical to limit the study of the artistic literary phenomenon to the analysis of one of its components — the text — although the former continues to be the central one.

Researching the ultimate values of the poetic, discovering the most hidden and sublime meanings of the verbal work of art through its external manifestations, these are the goals which the Theory of Literature has aimed at from its origins, and their realization is impossible nowadays without all the factors that surround the literary text, and which are fundamental in the construction of its meaning.

This theoretical-critical reading of other components of the communicative framework has been made through the studies of Extensional Semantics — among other trends in linguistics — and they have given rise to the elaboration of a theory of possible literary worlds[3] and to the creation of a new poetic theory of imaginative creation, which is

A-T. Tymieniecka (ed.), Analecta Husserliana, Vol. XXXVII, 237–245.

inherent in the values of poetry and which makes a literary work an artistic object full of anthropological and transcendental resonances.[4] Therefore, this reading brings with it the necessity to study other sciences, such as Psychology, Anthropology, Sociology, Philosophy, etc.

Within this framework of collaboration and interaction between these different disciplines, the literary work will be studied in itself and in relation to all those other elements which take part in the literary communicative process, their activity and presence being decisive in the construction of the overall meaning of the literary work.

On this occasion, at the center of my interest is the study of the reader as a recipient of the literary work, since the reader is one of the elements whose influence is being valued — and in some cases over-valued — by much of contemporary literary criticism.

From the very earliest theoretical-literary documents to the most recent critical proposals, the presence of the reader has been contemplated with more or less intensity. This can be proved by such concepts as that of *catharsis*, as proposed by Aristotle in his *Poetics*,[5] and which already has a vast tradition in theoretical-literary studies, or such as the doctrine of *scaz*, proposed by the Russian formalists.[6]

As a communicative category, the reader — as recipient — was the subject of study in the first works carried out in modern Text Linguistics. All authors have recognized his importance and they have situated him appropriately in the communicative process. He is also fundamental in Pragmatic Theory, where he has been proposed as the most important element in the definition of the text. Therefore, it is not surprising that, out of the seven criteria that Dressler and Beaugrande establish for textuality,[7] four of them refer to him, either directly or indirectly. Thus, whereas cohesion and coherence belong to the text, and intentionality to the producer, acceptability, informativeness, situationality, and intertextuality are, in some way or other, linked to the recipient.

Apart from Textual Linguistics, the recipient has also been present in most of the reflections that are made nowadays about the artistic literary phenomenon, although it is obvious that nowhere has he attained the role of protagonist more than in the role that has been awarded him by what is now known as the *Aesthetics of Reception*. This critical and methodological orientation — which has the nature of a school — has its roots in the theory of Phenomenology, in Husserl's general principle and in Heidegger's phenomenological foundations,

Ingarden and Gadamer being among the first who postulated it. It was Gadamer who, in his work *Wahrheit und Methode*,[8] applied Heidegger's situational view of Literary Theory. The literary work stands before the reader as something unfinished whose sense will depend on the reader's historical situation. This conception has developed in one of the two trends which can be considered most important within the *Aesthetics of Reception*.[9] I am referring to the line of thinking proposed by H. R. Jauss which is characterized by the belief that reception is part of a historical-normative phenomenon, its importance being even greater within Literary Historiography.[10]

However, at the present moment, I am more interested in Ingarden's hermeneutic line of thought,[11] which was proposed by Iser. For the latter, reception is a central component in the constitution of textual meaning, and this textual meaning is considered the result of the interaction between textual signals and the reader's acts of competence.[12] That proposal has the advantage — against other proposals which also emphasize the reader's presence — of distancing itself from those dangerous subjectivities and relativities in which it is so easy to fall. Iser remarks, in an explicit way, the differences between himself, who speaks of a *Theory of the Effect*, and Jauss, who speaks of a *Theory of Reception*.[13] The latter has more to do with the decisions of readers, who are historically consituted, whereas the former has a greater textual orientation, in the sense that it takes the text as its starting point, although the text is considered as a reformulation of a reality which has already been formulated. Therefore, a theory of the effect, as Iser formulates it, must face the problem of how this analysis should be accomplished — and also understood — something which has not yet been done.

We must, then, take the literary text as a reality in order to understand those proposals better. Iser places the construction of meaning in a process of reception, in the reader's activity. What is already a reality, following Iser, is subjected to a process of reformulation in the act of reading, whose function is to actualize the potentialities of the text. Thus, the problem is to delimit those potentialities and see their origins. To do this, I move away from Iser's stance and center my attention on the complexity of a literary text.

One of the possible means of studying literary work is consideration of it as representation. This representation, however, is a double one. On the one hand, it is the representation — as I shall explain later — of the author's inner world and, on the other hand — and perhaps this is what has been best studied in literary theory – it is the representation of the author's outer world, which is intensionalized in the text.[14]

If we accept that reality is what exists in itself, what is objective, the true being, the real and objective existence of a thing, the literary text – taken as something that is there now, as a realization – appears to us a reality, in the sense that it has a real and effective existence. But when the attitude of the observer becomes analytic and the text is considered as an actualization, then all that, which seemed to be evident, becomes enigmatic.

Accepting, as I have just said, reality as it is, there appears clearly its opposition to what is merely apparent on the one side, and to what is possible on the other. Thus, reality is more or less equivalent to what is actual and realization to actualization.

The first possibility (reality against illusion) has no interest for me, since it implies admitting that the literary text is something that we find in the external and physical world. But I am interested in the second possibility (reality taken as what is actual, as opposed to possibility) because reality becomes a synonym for existence. If we take what is concrete in the term 'reality,' the literary text will be something denotative of a particular being – a reality; it will indicate the real being as a state of being. It would be Kant's empirical reality, as opposed to transcendental ideality – the literary text as an empirical reality, a reality, and the literary text as an actualization, a state of being.

Ferrater[15] claims that the modes of expression of reality are among the most important of all the problems that reality throws up. This problem tends to be labeled that of reality and language. It deals with the way of knowing how it is possible to talk about what is real and which are the most important linguistic qualities that allow us to do that.

Following Bunge,[16] we can study the relationships of designation and reference both in the term 'reality' and in the term 'literary text' in order to discover the concepts they designate, and the facts, things, and actions to which they refer. This would allow us to establish the connotations and designations of the concepts, and would lead us to a more adequate intension and extension of what we intend to know. In a similar sense, Husserl studies the word in its double reality of sign-indication and sign-symbol.[17] In this way, the linguistic sign has two ways of signifying: through what has been said and through what has not been said but has been made evident.

I mentioned earlier that what is evident becomes enigmatic. Leaving aside the first way I pointed out and getting into the enigma, I remember the Lacanian assertion which says that what is real is not the world and that which is real is impossible in reality, which leads us to the hidden potentialities in the literary text.

Freud — in *The Interpretation of Dreams*[18] — claims that three main imprints are made on the psychological apparatus: a memory of a mental tension which is related to necessity, the image of the object which can satisfy necessity, and, finally, the experience of a satisfaction which is what makes the subject go back to the previous experiences in which the three elements associated with the three imprints were found. All this serves the principle of homeostasis, which describes how the subject finds his repose. It sends us back to the literary text, whose creator is a subject who is an author and acts in the moment, or a succession of moments, in which he makes — brings to reality — the literary text from itself in a process of actualization.

There is a tension in the author which leads him to the formalized activity of writing, writing becomes necessary as an actual need. Therefore, writing becomes actualized and what is written is an actualization. However, when writing, the author experiences the memories of previous satisfactions and a present satisfaction as well. The object which can satisfy his need is the literary text, realized not as a physical reality but as that in which the satisfaction of a need has been realized, a satisfaction which makes the state of tension arising from stimulation disappear.

Many people have said that writing is no more than a form of representation or symbolizing (as Bühler has said of language). The literary text would be, then, a 'res extensa' which has been actualized in a 'res cogitans.' The question is, which of the two is what it is, or better, is reality as a quality of what is real?

The writer is a being who tends to repose, to balance, in whom numerous internal and external articulations disturb repose — if he has it — and generate a displeasing tension from which he escapes by means of his literary activity. However, that satisfaction which calms his tension prompts those same tensions and, among them, the mnemonic traces of past experiences can be found. The literary text is, therefore, a materialization with a double satisfaction: that of an actual need, and that had in the attempt to repeat the primary satisfaction already experienced. In this sense, the text is understood as a representation of a state experienced in a moment of double need, but a state which actualizes the process of a real being, the writer. What is actualized, the realization, is not the reality from which it has emerged. The objective reality of a literary text is not what is real in the subjectivity of the acting author. The text would be, rather, the configurational space of the psycho-anthropological, spatial and temporal reality-realization — of what is real and transcending in the author.

The tension provoked by a stimulus — to continue on with Freud's theory — becomes a movement which goes on until an act relieves that need, being movement and tension until it finds relief in some realization; thus, if the realization does not provide complete satisfaction, tension and movement will continue, prompting new activity, realizations, and realities. The interest of this theory is that it brings another element to the probing of the enigma: the relative knowledge of previous experiences, which really implies conditions of reality. Reality is made up of knowledge, but what is the knowledge that constitutes that reality? If knowledge is the introduction of the world into the knower, knowledge is also representation; that is, the apprehension which results from the imposition of what is imaginary and what is symbolic. Thus, it can be said that if reality is knowledge, knowledge is representation, and representation is the imagination and symbolization of what is real, reality is representation, and what is represented in reality is what is real. The literary text is an actualization of the author's imagination and, at the same time, a set of symbols for the part of the external reality which has been incorporated in the text. The literary text as reality is representation. What is in the text is not what is represented by its author but, rather, the representation the author makes, which is what is real. What is real is, in fact, impossible, and reality, successive realization, is a continuous search for what is real.

All these reflections are made with the intention of contributing an overview of the literary text at its creation to the theories of reception, knowing that it is always going to be a materialization in words of a reality which is greater than the literary text itself, and which is beyond the limitations of the written word. In this context, part of the nature of the 'points of indeterminacy' — following Ingarden — or the 'gaps' — following Iser — can be found.

If I have explained all these considerations it has been with the object of placing an inherent meaning in the text itself. I have only pointed out what is corresponding and relevant in the theories of

Ingarden and Iser. In Iser's *El acto de leer*, and in Ingarden's theory, we find the following corroboration of what I have explained here:

La obra es más que el texto, puesto que sólo cobra vida en la concreción y, por su parte, ésta no se halla totalmente libre de las aptitudes que le introduce el lector, aun cuando tales aptitudes sean activadas según los condicionantes del texto. Allí, pues, donde el texto y el lector covergen, ése es el lugar de la obra literaria y tiene un carácter virtual que no puede ser reducido ni a la realidad del texto ni a las aptitudes definitorias del lector. De esta virtualidad de la obra artística brota su dinámica, que a su vez establace la condición del efecto producido por la obra. Consecuentemente, sólo mediante la acción constitutiva de una conciencia que lo recibe el texto llega a su realidad, de manera que la obra artística es el proceso de constituirse el texto en la conciencia del lector.[19]

It is also necessary to point out, as has been done in the most accurate and modern editions of the *Poetics*,[20] the necessary radicalization of these theories, together with their opportunism, which has led us, as Garcia Berrio and T. Hernández say, to forget the evidences at the starting point of the authorial codification in the essential recording and imprinting of messages in the memory,[21] something that is evident, if we add to this the radicalized attitudes of some post-structuralists, such as Eco and Barthes, who have maintained the Principle of the Plurality of Readings of a work.

Where can we find, then, the aesthetic meaning of a literary work? From what I have said in this paper, it is obvious that it is not to be found solely in the text written by the author. It is, rather, a concretion made by the reader, which is — following Ingarden — a way of filling the points of indeterminacy in the work through the incorporation not only on the reader's subjectivity but of the relationships between the reader's experiences and the objectivities constituting the textual structure of a literary work. In spite of this, the potential effects of the literary work continue to be unlimited, renewing themselves in every act of reading. We have to go back to them if we still wonder about the nature of the literary work, about its summoning power, and about its strange and, many times incomprehensible, communicability, especially when it is not very clear for us which is the way to follow the summons.

The poetic mystery becomes more and more irreducible to the theories of the different schools, and even to the topics and interests of a particular theory. It is obvious that a General Poetic Theory and a Theory of Rhetoric are needed, and especially an integral ideal of literary criticism, as has been pointed out by well-known critics such as

García Berrio.[22] The mystery of the poetic text still towers over us. The explanation of its material structure does not seem to be sufficient nowadays. And the theory of the reader does not account for all those imaginary expectations we have of it. Perhaps, the deficiencies and limitations of a particular critical method are only, and precisely, a reflection of the greatness of the verbal work of art. Who would be able to subject to radical thought something whose nature already possesses the irrational and inexplicable in an experience which is not to be captured by conventional formulae or physical laws?

A further and final comment must be made. Poetry is, beyond its material substance, its whimsical reading. As García Berrio puts it: "Poesía, por tanto, como eco azaroso y rumor sentimental imprevisible, pero no por eso poesía como efecto infundado, caprichoso, integramente convencional."[23]

University of Valladolid

NOTES

[1] C. Morris, *Writings On the General Theory of Signs* (The Hague, Paris: Mouton, 1971).

[2] T. Albaladejo Mayordomo, *Teoria de los mundos posibles y macroestructura narrativa* (Alicante: Universidad de Alicante, 1986), pp. 15—21.

[3] *Ibid.*

[4] The primary bibliography for the Poetics of the imaginary can be found in most of Garcia Berrio's latest works, from *La construcciòn imaginaria en 'Càntico' de Jorge Guillén* (Limoges: Université de Limoges, 1985), to his most recent work, written in collaboration with T. Hernández, *La Poética: tradicion y modernidad* (Madrid: Sintesis, 1988).

[5] Aristotle, *Poetics*, trans. V. Garcia Yebra (Madrid: Gredos, 1974).

[6] A. Garcia Berrio, *Significado actual del Formalismo Ruso* (Barcelona: Planeta, 1973), pp. 253—254.

[7] R. de Beaugrande and W. Dressler, *Einführung in die Text — Linguistik* (Tubingen: Max Niemeyer Verlag, 1972).

[8] H. G. Gadamer, *Wahrheit und Methode. Gründruge einer philosophischen Hermeneutik* (Tubingen: 1960).

[9] A very good comprehensive review of the proposals of the Aesthetics of Reception can be found in Pozuelo Yvancos, *Teoría del lenguaje literario* (Madrid: Càtedra, 1988), pp. 105—127.

[10] H. R. Jauss, *Aesthetische Erfahrung und Literarische Hermeneutik* (Munich: Wilhelm Fink Verlag, 1977).

[11] For this subject see R. Ingarden, *Das literarische Kunstwerk* (Tubingen: Max Niemeyer Verlag, 1972).

[12] Although the first of Iser's works about the Aesthetics of Reception was *Der Implizite Leser* (Munich: Wilhelm Fink Verlag, 1972), I have followed those of his principles found in *Der Akt des Lesens. Theorie ästhetischer Wirkung* (Munich: Wilhelm Fink Verlag, 1976), since it is a later work. The quotations I have used in this paper are from the Spanish edition, *El acto de leer* (Madrid: Taurus, 1987).
[13] Iser, *El acto de leer, op. cit.*, p. 12.
[14] On the concept of intensionalization referred to a literary work, see T. Albaladejo Mayordomo, *op. cit.*, pp. 39—87.
[15] I use one of the definitions of reality which can be found in his *Diccionario de Filosofia* (Madrid: Alianza Editorial, 1979).
[16] M. Bunge, *La investigación científica* (Barcelona: Ariel, 1975).
[17] E. Husserl, *Logische Untersuchungen*, Husserliana XVIII, XIX/1, and XIX/2 (The Hague: Martinus Nijhoff, 1984). For this topic, it is also interesting to read the comment that F. Martínez-Bonati makes in *La estructura de la obra literaria* (Barcelona: Ariel, 1983), pp. 87—88 (English version — *Fictive Discourse and Structures of Literature. A Phenomenological Approach* [Ithaca-London: Cornell University Press, 1981]).
[18] S. Freud, *La interpretaciòn de los sueños*, in *Obras Completas* (Madrid: Biblioteca Nueva I, 1948).
[19] Iser, *op. cit.* p. 44.
[20] I refer to the works of J. M. Pozuelo Yvancos: *Teoria del lenguaie literario, op. cit.* and of Garcia Berrio & T. Hernàndez: *La poética: tradiciòn y modernidad, op. cit.*
[21] García Berrio and Hernàndez, *op. cit.*, p. 85.
[22] A. García Berrio, "Màs allà de los 'ismos': Sobre la imprèscindible globalidad critica," in *Introducción a la critica literaria actual*, ed. T. Aullón de Haro (Madrid: Playor, 1984).
[23] A. Garcia Berrio, "Qué es lo que la poesía es," *Lingüística Española Actual*, **IX**, 2, (1987), pp. 177—178, p. 187.

MIODRAG CEKIĆ

ART AS COMMUNICATION

I

The essential difference between philosophy on the one hand and science, art, morality, religion and ideology on the other is that while the latter spiritual realms develop relatively independently from one another and from philosophy, philosophy depends for its survival on these other activities of the spirit. Philosophy, in fact, lives on the imperfection of science, art, morality, religion, and ideology. If science were perfect, then it would not permit the independent existence of another spiritual discipline which pretends to explain it. If art were perfect, then also there would be nothing left for the spiritual activity of philosophy. The same also goes for the state of things in the realms of morality, religion, and ideology. Science, art morality, religion, and ideology hold their place in the spiritual totality of humanity thanks to philosophy. Only philosophy is capable of giving them the dignity of belonging to a whole, a system, meaning and value. But, in doing that, philosophy is not able to establish her own realm in which it reigns exclusively. It has become, rather, the spirit and soul of all the other fields, or rather, it is capable of being that.

This position of philosophy in the spiritual cosmos, however, does not allow anybody to use it for just any scientific, artistic, moral, religious, or ideological end. Whoever has tried this has necessarily seen for himself that philosophy is — sooner or later — merely an obstacle for achieving such an end. A scientist is required to treat his subjects in accordance with the immanent demands of a given field, and not to offer generalized philosophical reflections as a substitute for a lack of facts or for the search for causal links between facts. An artist must make an artistic ideal sensuously incarnate, and not explicate intellectually that which should have been formed artistically. The moral actor would be abusing philosophy if he used it as a sophistic means of justifying selfishness and cheating, or to frame a theory of a moral fact where firmness in moral action was required. The religious man ceases to be religious the moment that which is divine to him is

A-T. Tymieniecka (ed.), Analecta Husserliana, Vol. XXXVII, 247–259.

caught in the net of a philosophical idea. Finally, the ideologist would be compromising the value of his ideals and mottos if he explained philosophically their real origin. Philosophy can exist only in the minds of those who approach it without any particular interest. Engaging in it has always been and will remain a spiritual satisfaction.

II

A spiritual pauper knows only one kind of life — the one in which he finds himself. It is therefore necessary to him. He sees only the past, he is a total empiricist, a real a posteriori detective who is satisfied only with what is offered to him. Thus he is an apologist of necessity. He swears to necessity because he is free to follow only the course of events already known.

But the Promethean spirit is oriented to the future. He creates, therefore he is free. Freedom is the truth of his lifelong existence, so it is quite natural for him to confess to it theoretically. He must fight for the new, for him it is a necessity to be free.

Freedom and necessity are ways of acting.

III

The path from a sense datum to a concept is always exclusively a path of knowledge. External reality knows no difference between the sensory and the conceptual. External reality is the flow of phenomena. Every possible substratum of phenomena, every "absolute reality" is only a subjective, anthropomorphic hypostasis. What we are involved in is phenomenal-existing-reality (*Dasein*), while the real being of things (*Sein*) is only a rational substance, a human positing of something beneath all.

There is only what is accessible to the senses. All the rest — a general concept, a law, a principle, a universal idea — is a shadow; all the rest is being, that which *we think* it is. The being of things is only created by thought.

Things exist, but their being does not exist. What things *are* is not a question of the things themselves, it is *our* question. Man seeks the being of things because he does not possess it; if he possessed it he would not seek it. And since he does not find it in his search, man produces it within himself, for his own spiritual integration. Every being is a subjective projection. Every ontology dissolves into anthropology.

Some (ontologists) turn being into essence; others, vice versa, turn essence into being. But in both cases the method used is that of reduction. There is a difference only in the direction of the reduction. In the former it is a process of introverting, and in the latter, a process of exteriorizing. Practitioners of the former draw the world into themselves and experience it as their own burden: they feel the pains of the world as their own. Practitioners of the latter exclude the world from themselves and thus turn all their truths and errors into immanent laws of reality itself.

IV

Philosophers say that man is much more a living being than the animals, not due to his intelligence (in fact, due to this, but this is not what is essential), but due to the fact that man singles out objects from the conglomeration of their natural amorphousness, due to the fact that he distinguishes being from essence, that he forms objects and observes them, and knows them ideally in his mind. Man, then, is self-aware, and therefore creates civilization and culture, and animals have none of this.

But is not such a concept of man only a self-complacent illusion? In my opinion, man is a continuing extension of the animal. He does the same things as animals, in fact, only he does them by roundabout ways. Thinking has its source in animal instincts and interests and serves them. Therefore (thinking) man does the same things as the animals, but gives all of that a logical explanation, a cultured form, a humane mask. Thus, while the animal, catching its prey, makes of it what he wants straight away, man does not use his prey (another man) straight away, but first subjects him to the legal comedy of a trial, of accusation and defence, only to destroy him afterwards. Self-awareness, culture, and civilization are only roundabout ways of fulfilling animal needs.

V

There has never been a moment in the history of mankind when man — whether an emperor or a slave — has been able to tell the whole truth, that is, to say all that he thinks and feels. Man's position in society has made him state his opinions of the state of affairs in such a way that he does not violate the sacrosanct nature of the status quo. Man has always taken his profoundest truth to the grave.

Nor have later generations been able to know the truth, i.e., what the dead thought about themselves and their time, for the truth about past people and events is always revised according to current needs.

Writers have striven in various ways to communicate their truth, even allowing their heroes to tell many forbidden truths, proclaiming those heroes mad. But criticism always senses the device: if the hero of story is mad, the writer who has created that madman is not mad. This is usually seen from the fact that not even a mad hero is completely free: although he is mad, he nevertheless knows how to make compromises.

But nevertheless, this pessimism about the question of the truthfulness of human communication is not justified. As we know, man is a social being, but what is much less known, he is a social being in that he does not dare to say all that he thinks and feels. He could, in fact, say all that he wishes if he were isolated from the social context, i.e., if he were to become a complete Robinson Crusoe. But in that case, the complete truth would lose all its *raison d'être*. Here too we see the confirmation of what I have stated elsewhere, that only that which is partial has truthfulness and value.

Many philosophers who follow Kant's thinking have shown us that man cannot think and know the world in the divine way because he is restricted to his human sensory organs and categoric structure of thinking. However, we today know that man cannot know the world from the human point of view either.

True, man must get used to the fact that the distorted truth is our human truth. If we have called what we see or think, but cannot speak, the truth, then this is a case of an unsolved terminological problem. We see that the emperor is naked (according to Andersen's story), but we nevertheless claim that the emperor is clad in the most beautiful clothes. Which is the truth — that which we see or that which we claim? If we wish to use terms adequately, then we should say that only the latter is true, for truth relates only to statements.

VI

No philosopher describes the world in general, nor the human world, i.e., the world of all or the majority of people, but only his own personal world. I shall cite two examples. Hegel defined man as a self-aware, spiritual and intellectual being. In doing so, he took himself as the model man (he could not have done otherwise), for in his life he changed everything into thought, into the modification of thinking, into the predicate of thinking. Marx saw man as an active being, and a being who acts above all in a revolutionary way. Here again the writer himself was the living model and illustration of that idea, for Marx was an active revolutionary. So it is with all philosophers. Philosophy is the exposition of the mental structure of its creators.

VII

Many people live in error, thinking themselves to be what they rather think, that they are determined by what they think. But for the majority of people, and perhaps for all, only very little of what they think is realized in their life. And man is only what is realized. Man is defined by what he has done, man is what he does. Otherwise man is full of all kinds of good and evil instincts, inclinations, passions, intentions, desires, feelings, and thoughts, but only a very little of all this is realized. And man is only what he has realized of all these psychic deposits, he is only that spot which he has left behind himself.

VIII

Man, like an electron, is indeterminate. Just as the parameters of the movement of an electron can never be assigned coordinates, and the spatio-temporal position of an electron can never be completely determined, so man can never be completely determined by all the possible anthropological, psychological, sociological, and other co-efficients, however much the attempt is made to determine him. For man is an undetermined potential of active energy, a being in antithesis to all else, a being of possibility, the being who alone surprises himself with his creativity and self-surpassing.

IX

Man is a reservoir of activity, a core of activity, a center of creative forces, an atom of power who only needs to be freed and allowed to act, and who is then capable of working wonders.

X

In the social science there are no pure claims, there is no truth per se. The pure science of man, of society, and of the human world is only *contradictio in adiecto*. The question is always posed: who is that which you claim truth for, who does it work for? Or in other words: that which you claim must be connected in some way with your interests (naturally not only your economic but also your emotional, intellectual, and other interests), or otherwise you would not claim it. Man is constantly engaged humanly, for he exists and works, and in that engagement of his it is not the most insignificant matter what he thinks and claims about himself, about others, and about things which affect all of us. And, therefore, it is a deception (or self-deception) when someone calls his trickery, fabricated out of personal or vested interests, a (pure) scientific theory.

XI

The literary form of expression has the ability to say in a more explicit, more striking way that which is ordinary, prosaic, that which we all feel, experience, perceive, think, and know. Literature is an original, peculiar, and unrepeatable transformation of an effectively realized sensation, perception, feeling, experience, or knowing. The writer gives power to the ordinary man. The way in which something ordinary and everyday is communicated constitutes the essence of literature. A banal expression is beyond the framework of literature. It is only the ecstatic potential of literary formation which reveals to us life in its essence.

XII

Literature is in one sense truer than life, because it is the condensed expression of life itself, its quintessence. But in another sense literature is only a pale copy of life, because it is always something arranged, stylized, but life itself is full of inexhaustible possibilities which can be realized.

On one occasion the artist's thinking and imagination are broader and more powerful than life, because they link that which is rarely linked in life. On another occasion the artist's initiative is merely a poor imitation of the real and inexhaustible imagination of life itself.

XIII

The only thing which is interesting about art is that life is transposed just as it is, that is, just as it is experienced. Life, life — that magic label of all that is human! Only that creation in which the real (ecstatic) breath of life is felt — whatever that life is like, whether one of wandering or of conformity — can be qualified as artistic. And the same goes for the true philosophy of life.

XIV

It is sometimes said that literature (for example current, contemporary literature) ought to reflect life or reality, as if it did not do that, or as if it could fail to do that. Literature is a function of life, literature is the transposition of reality — this is a tautological truth. Literature is quite certainly the mirror of life, and if somebody does not like the image in the mirror, if he feels that life is disfigured by it, then it is clear that the person is afraid of his own baseness. Even here literature is guilty of nothing, it is only a symptom and possibly a means of change for the better. If literature wanders, if it is disturbed and paranoid, then that is a sure sign that life itself, which is within its subtext, is wading through a swamp of meaninglessness. And this does not negate, but rather proves, the thesis that literature always reflects reality.

XV

Art is one form of communication. Art links the isolated, the concrete, the individual and the temporal with the universal, the abstract, the general and the external. Experiencing a work of art, we link ourselves with all people and all words, we become capable of living the life of mankind, the universe, the gods.

XVI

A poet loves to keep a secret, a philosopher loves to discover a secret.

A poet exists only when there is at least one secret. A philosopher has unveiled at least one secret. A poet turns reality into myth, a philosopher turns myth into reality.

The poet and the philosopher take opposite roles: the former is the Sphynx, the latter Oedipus.

XVII

It is sometimes said of a writer or philosopher that he "escaped into the past," in that he deals with something which once happened. Such a statement often sounds like a reproach. The first thing that may be said about that, however, is that that escape into the past is necessary, because the given writer could not express himself in a contemporary theme. And then it should be added immediately that this treatment of the past is one way, and in no sense the worst way, of discussing the present itself. "The escape into the past" only shows that the social and spiritual climate of the present is too cold, and that the past is resorted to as a way of speaking about the present. In fact, every discussion of the past — even the very recording of history — informs us much more about the conceptions and value judgements of the speaker and writer than about the views of the world and life of the people of those past times spoken and written about.

This "escape into the past" can be compared with the predicament of a stalk which has been tied up in a certain way by a gardener thus turning it from its normal pattern of growth so that the stalk stretches out in another direction imposed upon it. But just because of this, the new direction of its growth expresses its now, existing, "present" predicament.

XVIII

Art, all in fine nuances, corresponds to the indeterminateness and elusiveness of the adjective. The predicate of science is the exact recording of processes, and therefore it is a verb, occurrence, technique. Philosophy, that luxurious contemplation, is a lucid system of nouns.

XIX

Buffon said: style is the man, and I say, method is the man. The fact that someone used this or that method or a combination of such and such methods does not mean that he made that choice on the basis of certain rational motives, but because he is just that sort of man. A man creates (his) method, and then others using scientific logic or literary theory come along later and always reduce that method to its logical or poetic components. Everyone applies *his* method, which is either the one he himself has evolved or the one he has chosen from the arsenal of already existing methods, and, of course, he always does so in accord with his personal idiosyncrasies and not on the basis of so-called scientific methodology or literary poetics.

XX

Some modern logicians or philosophers of science are fond of interpreting certain scientific discoveries or theories (for example, Einstein's theory of relativity) and finding in them an internal logical structure by which they establish a necessary logical continuity between the state of science or theory prior to that discovery and after it. It would be better if they would establish that continuity themselves rather than merely interpret it with hindsight. ("When the cart breaks down, all roads are open.")

Similarly, some modern literature theoreticians take great delight in studying the internal poetic structure of a literary work by a great writer and discovering in it the poetics of that writer, as if the writer had first created his literary theory and then constructed his work around it. If that were really the case, we should have to ask this theoretician of literature why he himself does not evolve a literary theory and then apply it in the writing of a great work!

In fact, we do not find a strict logic or methodology of scientific discovery or theory preceding scientific creation itself, and in the same way as we do not find literary theory preceding actual literary creation. In both science and literature what we have first is a great creative mind bowing to its inner vocation and producing something of considerable scientific or literary worth. Afterwards this can happily be interpreted in the light of a later established scientific or literary theory.

XXI

It is sometimes said that literature, art, science, and philosophy are creations which make up "another world," that is, a world which is different from our first, real, ordinary, everyday, banal world. In my view, the world of art, science, and philosophy ought not to be another world; it must be our same everyday world and life, only slightly transformed, better understood and better interpreted. True, in this way it is already becoming another world. Therefore, I have to admit that the world of art, science, and philosophy is another world, i.e., a world different from our everyday world. However, the creation of this other world is our sin, though at the same time it is also the justification of our human existence, our *Dasein*. Man exists as man only because of that "sin," for that "sin," thanks to that "sin," and in that "sin."

XXII

Man is a creator because what is amorphous and undefined (what is given, though the term "given" is inaccurate since nothing is given, rather something is "given" the moment it begins to be *processed*) he transforms into something with a shape and a definition. In other words, he is a creator because he takes the amorphous and produces something created (by him and according to him). This is the process of man's living, or, as it is called with great pathos, of his creation. And this process is always and everywhere the same: in real life, in economics, in art, in science, and in philosophy. Naturally, man will never find it easy to process what is "given." There will always remain something unprocessed and incomplete; what he does manage to process will forever be paler and more artificial than the original – life. Therefore, there is no point in admiring the original, "life," which is incomplete, and underestimating its creation , just as there is equally no point in admiring "art" while underestimating "life." The first is done by the "ontologists" of art, while the "gnoseologists" of art are guilty of the second.

XXIII

The variable (in mathematics) gravitates towards its own limit, but that limit, too, is variable. A limit, therefore, is never emphatically a *limit*, for it not only moves away from us and appears unattainable, but it also ceases to be something final and complete (that is, it stops being a limit in the sense of an obstacle). And the limit itself becomes something undefined, variable, unlimited, something which helps to constitute the unlimited. A limit is turned into a regulating idea, a bait which leads our intellectual exploration in a certain direction. Like some sort of vessel, a limit frames our mental searching and exists for us, not in itself. A limit or a limiting value is not something we gradually approach; it is that approaching itself.

XXIV

If a man today summons up the courage to make value judgments, he is beset on all sides by the question: but how can you avoid subjectivity? I always answer: subjectivity cannot be avoided nor shoud it be. If we avoided subjectivity in making a judgment, that judgment would not mean anything to anybody, seeing that the subject always speaks to another subject. An "objective" judgment would not relate in any way at all to the subject. As a subject (i.e., the possessor of my own subjectivity), I can think, speak, and act in society, and hence communicate with another subject, who is also subjective, i.e., who also is the possessor of his own subjectivity. Communication and agreement (like quarrelling and conflict) are only possible between subjects and, vice versa, only subjects can agree and quarrel.

XXV

Is every man free? For the true philosopher this question is not sufficiently justified. Every man is free. The greatest dogmatist or conformist is as free as the greatest sceptic or non-conformist. Let us imagine the greatest conformist, and then go one step further and imagine a conformist who is an even greater conformist than the greatest conformist. He, too, is free. There he is, this poor soul, forever trying to avoid doing anything that would cost him something. It is his constant effort, to not hurt himself, while not hurting others, that is his freedom, his use of freedom.

XXVI

Throughout their history all the peoples of the world have had their happier moments, their victories and celebrations, but also their defeats, their disasters, and their suffering. The Serbs are no exception. Nevertheless, while many other nations have borne their sufferings more or less stoically, the Serbs have frequently lamented and wailed. This wailing echoes not only throughout our history, but all through our poetry, both popular and literary. Our poets cry out and lament together with the people (and how could it be otherwise!) and in this way consecrate this national mourning and elevate it to the status of a poetic cult. Not only that, the theoreticians of our poetry raise our wailing, which has been "artistically" processed in our literature, to the level of theory and so even add to it a metaphysical meaning. However, I should prefer to see less of this national weeping, wailing, and lamenting. I should like it better if poets stopped treating it in their own poetic way and if our theoreticians ousted it from our national metaphysics as a primitive and uncultured way of experiencing the sufferings of our people.

XXVII

Is it victory or defeat which brings man self-awareness and leads him to a discovery of himself? One thing is certain: victory leads man away from himself? Victory prevents man from seeing his own weaknesses; he becomes the slave of his victory. Victory enables man to rule men and things, therefore he is necessarily transformed into something other than what he was before victory. The victor must project himself out of himself and as a consequence, lose himself. However, even defeat, simply as defeat, does not help man to find himself. Defeat only has a beneficial effect when it prepares man for further struggle, i.e., when it helps man to remain what he is and when it allows him to recognize his own strength. Hence man only discovers himself when, in the face of defeat, he remains the victor.

XXVIII

Systematic thought is something final and immutable. It is the sadness of quietude. If a man submits to systematic thinking, he is no longer stirred to reflect. Systematic thought is infertile. It does not belong to history or a museum, but to the museum stockroom. Yet, when thought is not systematic, this makes man feel uncomfortable and arouses in him the urge to arrange his disordered thoughts, i.e., to create a whole. And during man's attempt to put his thoughts in order, new sparks inevitably fly, new thoughts appear which are fruitful even when they are erroneous. That is why I place a higher value on unsystematic thought.

University of Belgrade

FERNANDO CABO ASEGUINOLAZA

PHENOMENOLOGY AND THE RECEPTION OF LITERARY TEXTS: THE IMPLIED READER AS AN ELEMENT OF A GENRE

In a short article designed to refute Anthony Close's Hirschian approach to *Don Quijote*,[1] i.e., the search for an understanding based on Cervantes' possible intentions in the novel as opposed to subsequent Romantic interpretations, Inés Azar attacked the theoretical basis of Close's work in the following words:

> A written text is the permanent mark of an exclusion: the absence of the reader in the act of writing, the absence of the writer in the act of reading, the absence of the reader and writer in the text itself.[2]

According to this tenet, a written text is an exclusive presence constructed upon absences, the result of the banishment of those involved in the act of communication in which it originated, and it is therefore vain to try to imbue it with covert intentions in later interpretative reconstruction.

The radical impossibility of reconstructionist interpretation, i.e., of recovering the author's original intention, may also be thought of as being fundamentally due to the fact that, even conceding the existence of an intention in the text, this intention must itself be recognized, in Terry Eagleton's words,[3] as itself a complex text that cannot be unequivocally and lastingly determined; or, perhaps better still, as the literary text itself on its own terms. But this implies that the notion of a text cannot be based on the idea of the absolute exclusion of author and reader; as the core of a process of communication, the text — in this case the literary text — bears the marks of this process, the stamp of actions past or future. The literary communication process is indeed characterized by absences, by the non-copresence of author and reader, but not by their *exclusion*. On the contrary, both are very evidently implicit in the literary text; they are necessarily involved in its coming into being.

If phenomenology has had any influence on the study of literature other than to have prompted penetrating remarks on the existential status of literary works,[4] that influence has consisted precisely in

A-T. Tymieniecka (ed.), Analecta Husserliana, Vol. XXXVII, 261–273.

helping us to appreciate this important fact. This is true for each of two phenomenological approaches to literature. One, followed by Ralph Freedman,[5] for example, has applied the phenomenological notion of intentionality to the relationship between the author and his work, subject and object, as it appears in the text. Freedman claims that the poem — and by extension, perhaps, literary works in general — can be understood as "that mental context expressed in language which contains both the intended [?] object (of perception) and the intending consciousness,"[6] without there being any obstacle to any other kind of phenomenon or consciousness being understood as its object.[7]

The other avenue by which phenomenology has influenced literary criticism, and that with which I shall be primarily concerned in this paper, has approached the literary text as an intentional object determining its own perception, in other words, as the arena of the relationship between the intentional object — the text itself — and the intentional consciousness of the reader. In Martinez Bonati's words:

> la obra artística no sólo presenta un objeto, sino que, además, 'objetiva' o fija una perspectiva, un mirar, una forma de recibir o aprehender al objeto . . . En la obra de arte, los actos intencionales que dan constitución al objeto están, ellos mismos, semi-objetivados. Se los ha sacado de su habitual transparencia.[8]

It is this line of thought that has given rise to the well-known concept of the "implied reader," which was first succinctly formulated by Wayne Booth[9] and later developed extensively by Wolfgang Iser, and which Darío Villanueva has accurately described as being halfway between — or, if one prefers, the synthesis of — the opposed concepts of the *immanent* and the *extrinsic* reception of the text.[10] The implied reader identifies himself as a reading strategy that is implicit in the structure of the text; this is "the feasible reader's rôle in the text."[11] He can be thought of as adopting the attitude that the text both assumes in its reader and — in ways that differ from text to text — tends to lead him into. According to Iser,

> the scheme described as the reader's rôle is a structure of the text. As such, however, it embodies an intention that is only fulfilled by acts induced in the recipient. . . . The structure of the text and the structure of the act jointly behave as intention and fulfillment. Both are merged in the concept of the implied reader.[12]

Thus the implied reader is, to use the words with which Didier Coste defines his *lecteur virtuel*, a "potential prestructuring of interpretation,"[13] and as such is written into the text itself. He is not, therefore, as Genette would have, just a possible reader, nor, as Chatman claims, "the audience presupposed by the narrative,"[14] nor the author's desired reader. Neither is the reading strategy that he represents a specific interpretation of the text — though it naturally influences any attempt at interpretation that takes the structure of the text itself as its starting point, if for no other reason because, as Hans Robert Jauss sustains:

Le rôle du "lecteur implicite" est donc considerée comme le point de reference qui permet de comprendre la réception effective par les lecteurs historiques comme des "divergences".[15]

The hypothesis that I wish to put forward in this paper has to do with the possibility that the reception that is present in intentional form in any literary text may be relevant to its classification in one or another genre; in other words, that texts related by their genre — I shall not dwell here on what this relationship may consist in — may exhibit implied reader structures characteristic of that genre, i.e., may share a structural element, one of whose peculiarities is its close connection with extrinsic reception and all that it involves. Specifically, my hypothesis concerns the Spanish picaresque literature of the *Siglo de Oro*, which with its wealth of apostrophes, represented readers, internal recipients, anthypophoras, and other intratextual signs of readers or recipients in general is perhaps unrivalled in extent of influence of reception on text.[16] Certainly, the evident obsession of Spanish picaresque narrations with the way they are to be read makes it of immediate interest to investigate the characteristics of their implied readers. My hypothesis is that in these works the place of the implied reader is, as it were, outside the act of communication whose poles are the narrator and the narratee of the picaresque tale, i.e., that there is no direct relationship between narrator and implied reader of the kind that critics have often thought to perceive.[17] Quite the contrary, the implied reader of the picaresque tale is the witness of an act of communication that does not affect him directly, that he takes no part in, and that is basically foreign to him.

To see this, we must first of all reject the ingenuous (or perhaps simply vague) interpretation of what are usually called picaresque novels as the autobiographies of *pícaros*. Rather, they should be defined as the fictitious enunciation of those auto-biographies, which means not only that the autobiography, as history, is fictitious, but that the entire autobiographic act, with its narrator, its narratee, and the situation that surrounds the two, is also part of a prior enunciation, and hence of a prior enouncing. As an example, Guzmán, as any reader of picaresque novels knows, is the author of a book in which he narrates the story of his life and reflects at length upon it; but this book is not the book written by Mateo Alemán and entitled *Guzmán de Alfarache* (even though the two are to a large extent identical), just as a certain anonymous writer is the author of a book that is not the same as the letter Lázaro addresses to "Vuestra Merced." A tangible sign of this enunciative duality in *Guzmán* is the "Declaración para el entendimiento de este libro," in which the author accepts responsibility for the "poética historia" and at the same time affirms that Guzmán "escribió su vida desde las galeras." It is thus absolutely necessary to distinguish in all picaresque works (I am tempted to say "in all literary works") a twofold enunciation: on the one hand there is an *effective* enunciation whose enouncer is the writer; and on the other *enounced* enunciation, or to use José María Nadal's terminology, the narrative.[18]

The effective enunciation contains within it an enounced enunciation — the picaresque tale — whose enounced enouncer is the picaresque narrator and whose recipient or enounced enouncee is the narratee of the picaresque tale or narrative, who appears in very diverse guises in different works.[19] The two enunciations, the enunciation *sensu stricto* and the narrative, are, as is evident from what I have said above, separated by a difference that cannot be ignored by literary theory and criticism, and which gives rise to a dual series of figures (narrator/author, narratee/reader) corresponding to the effective enunciation alone and to the narrative that it contains. The components of these pairs are evidently not identical, nor are they even intended to coincide, not, at least, as a basic trait of the genre, as is clear from the two examples just mentioned. Nobody would think to identify the picaresque narrator strictly with the author of the novel, even in those cases in which they may plausibly be related by the presumed historical truth of the deeds narrated. On the contrary, the two are separated by what Benveniste and Greimas call a *débrayage*.[20] And exactly the same may be asserted of the pair located at the receiving pole. It must moreover be borne in mind that from the moment we accept the existence of an enounced enunciation, it is likewise necessary to accept that the referential context of the narrative, though often implicit, is also part of this enunciation, and hence cannot without question be identified with the context of the effective enunciation. Thus the narrator, in the picaresque novel, is the *pícaro* or *ex-pícaro* who in a given situation and with more or less specific aims sets down his autobiography, while the narratees are the readers or recipients at whom the autobiography is aimed, who are written into the same context as the *picaro* himself. The narratee is therefore to be identified, albeit implicitly, with these readers of the enounced situation, and in no direct way with the real reader, nor, of course, with the implied reader.

It is true that the situation of the narrator and narratee cannot be said to be always merely implicit. One might even speak of an ordering of picaresque works according to the amount of evidence provided by the text in favor of this status. This is an issue which undoubtedly invites much more detailed analysis than is possible in this communication; here I must perforce abstain from any but a very succinct exposition. In the first place there are works in which the framed status of the picaresque tale is absolutely explicit, which feature an outer narrative frame that includes the *pícaro* and his future addressees prior to their adopting the roles of narrator and narratee. Examples of this format include *El diálogo del capón, La desordenada codicia de los bienes ajenos* and *La hija de Celestina.* In a second kind of work this outer frame is suggested but not explicitly presented. The clearest instance of this group is undoubtedly the original *Lazarillo,* in which the prior relationship between the *pícaro* and whoever it is he addresses as "Vuestra Merced" is an unknown whose interpretation is fundamental for the sense to be given to what Lázaro relates. Finally, a third group comprises those tales in which the pre-narrative relationship between those who will become narrator and narratee is almost — thought never quite — totally unstated; we may recall Alemán's *Guzmán* and *El guitón Onofre,* for example. This is not to say that there are not picaresque works — such as *Marcos de Obregón,* Juan de Luna's *Lazarillo* and *Estebanillo González* — whose place in the above ordering is difficult to establish, and which deserve closer attention. What is of primary interest here, however, is that what is overt in the first two groups — that all the narratee's attributes, whether *qua* narratee or of any other kind, are factitious, i.e., enounced — must hold also for the third. If on the grounds that Lázaro addresses a specific narratee the reader does not identify himself as the person addressed, there do not appear to be sound theoretical arguments for his not adopting the same attitude towards Guzmán's pleas. From the point of view of enunciation, the only difference between the two cases is perhaps no more than that in one it is a letter that is written and in the other a book.

In view of the foregoing, I do not regard as tenable the opinion of Gérard Genette and Shlomith Rimmon-Kenan[21] concerning the total identity of implied reader and narratee (the only exception they allow is when the narratee is intradiegetic rather than extradiegetic). According to Genette,

> le narrataire extradiégétique n'est pas, comme l'intradiégétique, un "relais" entre le narrateur et le lecteur virtuel: il se confond absolument avec ce lecteur virtuel — relais, lui, avec le lecteur réel, qui peut ou non s'identifier (en ce sens) au narrataire intradiégétique, qui est après tout un personnage comme les autres.[22]

To begin with, I believe, like Santiago Renard, that the narratee, *qua* narratee, is a "creation" of the narrator's, [23] so that it is perhaps not very meaningful to distinguish between intra- and extradiegetic narratee. A more convincing distinction is that which Darío Villaneuva draws between narratee and para-narratee.[24] But this apart, we have seen that once we acknowledge the existence, in the picaresque literature of the *Siglo de Oro,* of so evident an enounced enunciation, or narrative, then for sheer theoretical consistency we are obliged to deny the identity of implied reader and narratee. In the *Lazarillo,* "Vuestra Merced" is part of the first narrative; he is, in Genette's terminology, an extradiegetic narratee. But what grounds have we for identifying him with the implied reader? In fact, Iser himself states that "the role of the reader written into the text cannot coincide with that of the fictitious reader, where the author injects an imaginary reader into the world of the text; in so doing he creates an additional viewpoint, enhancing the perspectivist richness of the text."[25]

Far from being identifiable with the implied reader, the picaresque narratee is one of the elements of the text that configure him. Previously, I drew attention to the mixture of intrinsic and extrinsic aspects in the implied reader; we can now reformulate this borderline nature by asserting that he straddles the divide between enunciation and narrative. His enunciative character lies in his defining an interpretation of the global communicative act involved in the picaresque novel, rather than in his representing solely the receiving pole of this act; his narrative side lies in his being, after all is said and done, an element of the text constructed chiefly with the resources proper to a text.

It is the way in which the implied reader is configured in the text that shows his radical separation from either of the poles of the picaresque narrative, and how instead he stands as a spectator who contemplates the narrative as a whole comprising both the emitting and the receiving poles. This configuration is achieved in the picaresque novel by means of a great variety of artifices. Though their detailed discussion lies outside the scope of this communication, it will be helpful here to mention one or two of the most important. As we have already seen, one evidently derives from the characteristics of the picaresque narratee, who in spite of the differences among different novels is an element whose invariable presence is continually borne home by frequent allusions, the only real exception to this rule being Juan de Luna's *Lazarillo*. The explicit allusions to the narratee are furthermore very often accompanied by a whole series of different kinds of represented readers that at times even results in internal contradictions.[26] The immediate effect of this recipient-centered rhetoric is to set the implied reader at a distance from this pole of the picaresque narrative, as if kept away by the throng of other instances of intratextual reception. As Didier Coste noted, "en thématisant une gamme limitée des possibilités constituant le lecteur virtuel, ce processus implique une sélection et suppose une distantiation du lecteur virtuel par rapport au texte dans son ensemble."[27] This is just what happens in the picaresque novel; the stronger the presence of the narratee, the less room is left for the implied reader to stand opposite the narrator as the direct recipient of his narrative.

Chatman goes so far as to acknowledge the theoretical possibility of there being an "unreliable" narratee.[28] Rimmon-Kenan considers that such an narratee must necessarily be intradiegetic — Montúfar in *La hija de Celestina* would be a clear case in point.[29] But my previous remarks imply that in the case of unreliable narratees too we must question this restriction; for if the extradiegetic narratee and the implied reader are not identical, if the very existence of non-intradiegetic narratees is doubtful, and if the narratee is to be considered to be the narrator's creation (and hence his subordinate), then we cannot but admit the possibility of other kinds of untrustworthy narratees. One consequence of this is that a narratee created by an unreliable narrator is as distant from the implied reader as the unreliable narrator is from the implied author — which is what we find in *Guzmán, El Buscón, La pícara Justina, El guitón Onofre*, and numerous other picaresque works.

When we examine instances of the "represented implied reader,"[30] who chiefly appears in what Genette called paratexts (prefaces, introductions, etc.), we discover further characteristics of the peculiar kind of implied reader that seems to recur throughout picaresque literature. The represented implied reader is the recipient addressed directly by the author-figure present in the text, and the author-figure's attitude to him is of prime importance in the construction of he implied reader. Not that the represented implied reader may be taken "as is" as the representation of the implied reader, i.e., that everything predicated of the former applies to the latter; but the configuration of the represented implied reader does contribute decisively, rather like that of the narratee, to constructing the attitude of the implied reader. An example is seen in the irony often aimed at the represented implied reader, especially in prologues, which helps to dissociate the implied reader from the narrative act that he is to witness, as is evident in the prologue of the *Lazarillo de Manzanares*:

Christiano lector, o lo qu'eres, ¿quién me mete a mí en enfadarte con un Prólogo que me tenga más costa que el mismo libro, disculpándome en unas cosas y dándote a entender otras, que si tú las quieres condenar, no importa gaste yo todo el papel de Génova en defenderlas? No me pase por el pensamiento! Si te parecieire bien, págale y llévale, y sino [*sic*], debalde te puedes yr sin él. Vale.[31]

In other cases, such as *Buscón*, such wry apellations both remind the reader, as above, of his status as purchaser of the book and also directly introduce what is to follow as burlesque. Another good example is to be found in *Estebanillo González*. These attitudes towards the represented implied reader have in common the effect of stressing the text as enunciation rather than as narrative. This dissociation between the two, together with the irony with which the enunciation itself is treated (think, for example, of Machado de Silva, at the beginning of whose novel there figures, as its author, one *Félix Márques, catedrático de picardía*, who dedicates the book to the street porters of Madrid), has as its chief result the construction of an implied reader lying off to the side of the narrative, and hence of both narrator and narratee, who, as Barry Ife holds in his recent book, prevent the flesh-and-blood reader from identifying with the narrative.[32]

Ironic treatment of the represented implied reader is not universal in picaresque works, though the presence of a whole series of artifices bringing about a comparable effect can be guaranteed. Such features are fundamentally of secondary importance, because what primarily makes the implied reader a spectator of the picaresque narrative as a whole, i.e., of the relationship established between narrator and narratee by the autobiographic act, is precisely the bond relating the two parties of the enounced enunciation. The implied reader is a third party, and at the same time serves as a link between narrative and enunciation. Claudio Guillén, referring basically to theater, remarked not long ago that "el diálogo abarca no dos sino tres partícipes; y el monólogo no uno sino dos,"[33] and more recently extended the scope of this observation to epistolary literature;[34] while Mikhail Bakhtine wrote in "El problema de texto" that "comprendre c'est, nécessairement, devenir le *troisième* dans un dialogue": as well as the recipient presupposed by any enunciation, the existence of an author also entails the existence of a *sur-destinataire* endowed with a capacity for global comprehension of the enunciation.[35] In the same way, in picaresque literature, the implied reader is a third party with respect to the relationship established by the *pícaro* with his narratee, though a third party with certain peculiar traits that should not be overlooked.

As enounced enunciation, the picaresque narrative is not effective enunciation, so that the implied reader of the picaresque novel cannot be Bakhtine's super-recipient (which would make him, the implied reader, the construct of an enounced enouncer, the *pícaro*); and for the same reason he cannot be identified with Claudio Guillén's epistolary third party either (in spite of the similarities occurring in *Lazarillo*, for example). Rather, we must consider the role of the picaresque implied reader as being similar to that attributed by Guillén to the spectator of the theatrical dialogue, for both the picaresque implied reader and the theatrical spectator witness, from a privileged vantage point, the unfolding of an enounced enunciation (dramatic dialogue in one case, the picaresque narrative in the other), and at the same time constitute the nexus that, however subtle, always exists between the effective enunciation and the enounced enunciation. The enunciative mechanisms of the picaresque novel present the narrative as a spectacle that, though at times ambiguous and disturbing, does not call for the reader's direct involvement (despite the multitude of written in readers) because it is fundamentally foreign to him.

In view of the above considerations it seems to be a plausible hypothesis that the manner in which the implied reader is constructed exhibits certain regularities that are related to genre. Certainly, in the picaresque literature of the *Siglo de Oro* a general trait of the implied reader is his dissociation from the picaresque narrative, its narrator and its narratee, which are set against him as distinct, foreign elements. The reader of the picaresque novel is not encouraged to identify his own situation with that which is set before him, but to contemplate the latter in the same way, perhaps, as that in which the theatergoer may contemplate a drama. In turn, the existence of this model in picaresque literature confirms that the implied reader is an instance of reception that is quite distinct from other intrinsic recipients — in particular, and in spite of the confusion that may be caused by other kinds of text, he is distinct from the narratee. The implied reader is thus not a mere intermediary, but rather the intentional inclusion of a reading style, defining a specific angle of approach to the text; he is, in fact, only metaphorically a recipient.

The general scheme outlined above does not exhaust the insight into picaresque literature that can be gained by examination of the implied reader, and neither is it limited to the picaresque novel of the *Siglo de Oro*. What we have highlighted is a certain kind of genre-related artifice involving what, more generally, might be termed the zero level of literary reception. Again, a quotation from Martínez Bonati is relevant here: "¿Cabe contemplación pura de la situación comunicativa, conocimiento de ella sin participación en ella? Tal conocimiento es posible, teniendo por objeto situaciones comunicativas imaginarias. Es la perspectiva del lector de literatura."[36] What we have stressed here is that the perspective described in this quotation is not independent of the text, but is written into it as the receptional consequences of the relationship between effective and enounced enunciation, or to use Martínez Bonati's own terms, between the effective and the imaginary situation. As an intentional object, the implied reader comprises not only what is contemplated, but also, perhaps as a distinct phenomenological stratum, the act of contemplation.[37]

To conclude, I should like to point out that the value of the implied reader for characterizing genres becomes especially evident when we examine the ideological implications of his construction at the level of effective enunciation. It is illuminating, for example, to compare our findings regarding Spanish picaresque literature of the *Siglo de Oro* with what G. Molinié has recently published concerning the reception of French baroque narrative in general and *Le Toledan* in particular.[38] Briefly, in both genres there is the need to distinguish between different levels of enunciation and treat the great complexity of the relationships between these various levels. In the French case, however, the enounced enunciation serves, not, as in the Spanish case, to make the reader the spectator of something that is independent of him, but instead to subject the enunciation to the taste of the reader, who with a much stronger desire for identification seeks his own satisfaction in the text. The two literary situations, though created upon a common literary foundation, exist within quite different institutional contexts, and this is reflected in the difference between their implied reader structures. I hope to be able to expand upon this theme in the future.

University of Santiago de Compostela

Translated by Ian Coleman.

NOTES

[1] Anthony Close, *The Romantic Approach to 'Don Quixote'* (Cambridge: Cambridge University Press, 1978).

[2] Inés Azar, "Meaning, Intention and the Written Text: Anthony Close's Approach to *Don Quixote* and its Critics," *Modern Language Notes*, Vol. **96** (1981), pp. 440—444 (p. 444).

[3] Arguing against Hirsch, Eagleton states that "an author's intention is itself a complex 'text', which can be debated, translated and variously interpreted just like any other." (Terry Eagleton, *Literary Theory. An Introduction* [Oxford: Basil Blackwell, 1983], p. 69).

[4] Examples are Roman Ingarden's well-known works *Das literarische Kunstwerk* (Halle: Max Niemeyer, 1931) and *Vom Erkennen des literarischen Kunstwerks* (Tübingen: Max Niemeyer, 1968); and in the Spanish world, Félix Martínez Bonati's *La estructura de la obra literaria* (Barcelona: Ariel, 1983).

[5] Ralph Freedman, "Intentionality and the Literary Object," in Murray Krieger and L. S. Dembo (eds.), *Directions for Criticism. Structuralism and its Alternatives* (Madison: The University of Wisconsin Press, 1977), pp. 137—159.

[6] *Ibid.*, p. 151.

[7] A possible example is discussed in the third volume of *Temps et récit* (Paul Ricoeur, *Temps et récit III. Le temps raconté* [Paris: Seuil, 1985]).

[8] Félix Martínez Bonati, "Fenomenología y crítica (notas para una discusión)," *Dispositio* Vol. **IX** (1984), pp. 91—107 (p. 103).

[9] Wayne Booth, *La retórica de la ficción* (Barcelona: Bosch, 1978), p. 129.

[10] "Ciertamente, las teorías de Wolfgang Iser se sitúan en una zona de transición entre las dos grandes orientaciones críticas de la Recepción que antes apuntábamos: la intrínsica, estructural o formal, y la fenomenológica, porque su *lector implícito es* tanto una instancia de recepción inscrita en el texto como la actualización fenoménica de la misma a través del acto de lectura." — Darío Villanueva, "Narratario y lectores

implícitos en la evolución formal de la novela picaresca" in Luis González del Valle and Darío Villanueva (eds.), *Estudios en honor a Ricardo Gullón* (Lincoln (Nebraska): 1985), pp. 343—67 (p. 344).

[11] Wolfgang Iser, *El acto de leer* (Madrid: Taurus, 1987), p. 69.

[12] *Ibid.*, pp. 66 and 67.

[13] Didier Coste, "Trois conceptions du lecteur et leur contribution à une théorie du texte littéraire," *Poétique*, Vol. **43** (1980), pp. 354—71 (p. 360).

[14] Gérard Genette, *Nouveau discours du récit* (Paris: Editions du Seuil, 1983), p. 103. Seymour Chatman, *Story and Discourse. Narrative Structure in Fiction and Film* (Ithaca: Cornell University Press, 1978), p. 150. It should be pointed out that in discussing the implied reader, whom he prefers to call the "*lecteur virtuel*," Genette makes no reference to Iser, but directly develops the ideas of Booth and Rimmon-Kenan.

[15] This, at any rate, is the point of view of Joseph Kurt in "L'esthétique de la réception. Une nouvelle approache de la littérature?," *Lettres Romanes*, Vol. **37** (1983), pp. 199—220 (p. 215).

[16] Apart from the above-mentioned work by Darío Villanueva, other books on this aspect of picaresque narrative include Helen Reed's *The Reader in the Picaresque Novel* (London: Tamesis, 1984) and Barry Ife's *Reading and Fiction in Golden-Age Spain. A Platonist Critique and Some Picaresque Replies* (Cambridge: Cambridge University Press, 1985).

[17] This confusion pervades a large proportion of the critical literature on the picaresque novel, including the above-mentioned book by Helen Reed and classics of criticism such as L. J. Woodward's "The Author-Reader Relationship in the *Lazarillo de Tormes*," *Forum for Modern Language Studies*, Vol. **1** (1965), pp. 43—53; Joan Arias' *Guzmán de Alfarache: The Unrepentant Narrator* (London: Tamesis, 1977); and Harry Sieber's "Apostrophes in the *Buscón*: an approach to Quevedo's narrative technique," *Modern Language Notes*, Vol. **83** (1968), pp. 178—211.

[18] José María Nadal, "La enunciación narrativa," *Investigaciones semióticas*, I, Madrid, CSIC, 1986, pp. 367—90.

[19] I do not believe that it is fruitful to carry the *mise en abîme* to the point of distinguishing, within the enounced enunciation, a second, lower level of enounced enunciation. Whether the narrator (the enounced enouncer) is supposed to coincide with the real enouncer or, as in the picaresque novel, to differ radically appears to be a rather unimportant issue. As we shall see, this does not prevent the narrative's referring to its own situation in a way that is often fundamental for appreciation of the text.

[20] Greimas and Courtés define *débrayage* as "la operación por la cual la instancia de la enunciación — en el momento del acto de lenguaje y con miras a la manifestación — disjunte y proyecte fuera de ella ciertos términos vinculados a su estructura de base, a fin de constituir así los elementos fundadores del enunciado-discurso." (A. J. Greimas and J. Courtés, *Semiótica. Diccionario razonado de la teoría del lenguaje*, Spanish translation by E. Ballón Aguirre and H. Campodónico Carrión [Madrid: Gredos, 1982].)

[21] According to Rimmon-Kenan, the extradiegetic narratee is "parallel to or identical with the implied reader" (Slomith Rimmon-Kenan, *Narrative Fiction: Contemporary Poetics* [London: Methuen, 1983], p. 140).

[22] Genette, *op. cit.*, p. 91. Genette thus confirms the thesis he defended in *Figures III* (Paris: Seuil, 1972), p. 266.

[23] According to Renard, the characteristics that define the narratee's relationship with the narrator are its symmetry and the hierarchic subordination of narratee to narrator that is involved in the former's dependence on the latter's voice (Santiago Renard, "Sobre el narratario: problemas teóricos y metodológicos," *Cuadernos de Filologia*, Vol. **1** (1985), pp. 273–289).

[24] Villanueva, *op. cit.*, p. 307.

[25] Franc Schuerewegen too criticizes Genette's identification of *lecteur virtuel* and narratee: "On peut se demander en effet si, plutôt que de privilégier certains 'cas de syncrèse' somme toute assez rares . . . il ne vaut pas mieux durcir la barrière entre texte et lecteur et radicaliser *l'alterité* du destinataire inscrit" (Franc Schuerewegen, "Réflexions sur le narrataire," *Poétique*, Vol. **70** [1987], pp. 247–254 [p. 248]).

[26] Examples include *Buscón* and *Marcos de Obregón*, the former of which has on this account often been singled out by critics as a shining instance.

[27] Coste, *op. cit.*, p. 360.

[28] In this case, Chatman explicitly acknowledges the possibility of dissociation between narratee and implied reader, since in one of his forms the existence of an unreliable narratee implies that "the implied reader invoked by the implied author [is] at odds with the narratee invoked by the narrator" (Chatman, *op. cit.*, p. 261). This qualifies an earlier assertion that approaches what appears to be a commonplace in narratology (as we have seen in the cases of Genette and Rimmon-Kenan): "The narratee-character is only one device by which the *implied author* informs the real reader how to perform as implied reader" (*ibid.*, p. 151); especially if, as appears from the context, we are to suppose that this information is an invitation for one to identify with the other.

[29] Rimmon-Kenan, *op. cit.*, p. 104.

[30] Regarding this concept, see the article by Dario Villanueva cited above.

[31] Juan Cortés de Tolosa, *Lazarillo de Manzanares*, ed. G. E. Sansone (Madrid: Espasa Calpe, 1974), p. 9.

[32] According to Ife, the development of print culture and of the habit of individual reading and the attacks on narrative fiction by moralists and intellectuals led together contemporary authors of fiction, picaresque fiction included, to exaggerate the structural artificiality of their works so as to avoid encouraging in the reader too great a proneness to identify with the fictitious world.

[33] Claudio Guillén, *Entre lo uno y lo diverso* (Barcelona: Crítica, 1985), p. 203.

[34] Claudio Guillén, "Notes toward the Study of the Renaissance Letter" in Barbara Lewalski (ed.), *Renaissance Genres. Essays on Theory, History and Interpretation* (Cambridge [Mass.]: Harvard University Press, 1986), pp. 70–101 (p. 100).

[35] M. Bakhtine, *Esthétique de la création verbale* (Paris: Gallimard, 1984), p. 336.

[36] Martínez Bonati, *La estructura . . .*, *op. cit.*, p. 118.

[37] The implied reader is perhaps the answer to the following question that Martínez Bonati asks: "¿Dónde ubicar aquella ironía que (como en obras de Thomas Mann, por ejemplo en el *Doktor Faustus*) es parte esencial de la obra y *no* es tono irónico del narrador, sino óptica total, que distancia e ironiza al narrador mismo junto con *toda* la situación comunicativa imaginaria?" (*Ibid.*, p. 188)

[38] G. Molinié, "Sémiotique du narrataire dans les romans baroques," *Cahiers de Littérature du XVIIe siècle*, Vol. **10** (1988), pp. 17–25.

CONSTANTIN CRISAN

L'OEUVRE LITTERAIRE, LA CONSTRUCTION INTERIEURE ET LA RECONSTRUCTION

Faute d'une explication préalable, le sous-titre* pourrait apparaître une contradiction *in adjectu* sinon comme une présupposition irrationnelle. Ainsi, devrions-nous élucider tout d'abord les termes conceptuels`a tour de rôle et avec une *patience* que nous aimerions être aussi celle du lecteur lui-même, avant qu'il bute sur la matière de notre pensée ou qu'il bondisse sur la *forêt* des choses, sur cette *Hyle* (*ὕλη*) de la réalité primordiale indistincte.

Par la construction *intérieure* de l'oeuvre nous voudrions avancer dans une vision, en principe nouvelle, une autre sémantique du concept de *feeling* au-dela même des dix sens accrédités par ce terme anglais épuisé tour`a tour par les recherches esthétiques post-husserliennes. En ce qui nous concerne, ni le *sentiment*, ni le *sentir*, ni la *mentalité*, ni la *sensibilité, la conviction* ou même la *conscience* ne couvrent pas deux autres sens que nous tenons pour fondamentaux. Il s'agit précisément de *tempérament* ou du *comportement in actu* en tant qu'agents déterminatifs du langage personnel de l'artiste-créateur. Un artiste peut exprimer une mentalité, une conviction ou une sensibilité, déterminées ou provoquées par des déterminations pre-données plus ou moins proches en réflectant d'une manière plus ou moins trahissante sa pensée sensible. Mais son code psychique et affectif ou son comportement résulte nécessairement de son *chiffre* caché — qui s'acheve ou subit une détérioration dès qu'il se décrit ou s'exprime — et ce chiffre (dans l'acceptation d'Anna-Teresa Tymieniecka) est censé être le miroir de son tempérament inconfondable et par conséquent impossible à être gouverné par des agents extérieurs entièrement dominateurs. Deux exemples sauraient satisfaire à une conviction quasi-incontestable: même dans sa "période néronéene", la pensée de Sénéque n'est pas le fruit entier du néronisme; la pensée du philosophe relève en partie de l'idéologie impériale, mais son langage personnel, son style et/ou tempérament intérieur(s) sont en propre ceux de Sénéque l'homme-penseur soumis avant tout à son moi. Il en va de même si l'on pense à Céline: *Voyage au bout de la nuit* ou *Mort à crédit*, ces romans de grande valeur, reflètent-ils les mentalités ou les idées souvent aber-

A-T. Tymieniecka (ed.), Analecta Husserliana, Vol. XXXVII, 275–281.

rantes et reprobables, que l'homme Céline a publiquement exhibées dans certains moments de sa vie? Pareille opinion nous semble pratiquement insoutenable.

Voici deux exemples qui renforcent notre volonté de dissocier le comportement existentiel du code de l'oeuvre, de *son langage*, dont la construction se dévoile dans son *anatomie autonomique* malgré toute fusion ou agencement du monde et de la vie personnelle de l'artiste dans le tissu intime de ses écrits.

L'autre terme, la *reconstruction* n'exclut pas en principe l'idée de réception, mais pour une fois oublions le sens strictement sociologique du mot réception qui abuse en lui-même d'un concrétisme "insensé"; selon nous, reconstruction signifie quelque chose de plus important que *le plaisir du texte* — dont parle, entre autres, Roland Barthes, un plaisir tel, que le lecteur se sent en fait appréhendé par l'oeuvre, extérieurement. Seule la *faculté de compréhension*, la *volonté* du/de *comprendre* par le fait même de saisir réellement les moyens et les outils employés à élever l'édifice aboutissent à la figure d'un vrai lecteur. D'un lecteur qui serait pour nouns une somme et un sommet (rêvons-les, pourquoi pas?), pouvant en l'occurrence refaire la démarche logique et affective de cet autre "être-en-situation" qu'est le createur lui-même. Plus exactement, il s'agit là d'un lecteur qui démonte les mécanismes, les engrenages, qui ne crains pas à démolir les murs de la . . . maison, afin de deviner le miracle de la construction et *revoir* (au moins imaginairement), à l'aide du miroir de son propre langage, *le langage même*, le langage *ex originae* de l'oeuvre. Indéniablement, c'est à ce lecteur plus ou moins utopique que rêve souvent tout créateur, qu'il soit écrivain, peintre, sculpteur ou compositeur . . . .

Il y a un texte husserlien qui ne se réfère pas *stricto sensu* à l'oeuvre d'art mais que nous avons lu avec le présentiment d'une proposition théorique sur la genèse de la *perception par la réflexion* comme première instance, et tout de suite après, comme seconde instance, la transformation de cet acte en jugement ou, comme dit Husserl lui-même, en "accomplissement des jugements". C'est cet "espace — continue le philosophe — qui est à l'origine des concepts relatifs à l'état des choses et à leur étant (dans le sens de copule); ce n'est pas dans ces actes considérés en tant qu'objets, mais dans les objets de ces actes nous identifierons le fondement de l'abstraction pour la réalisation (*Realisierung*) de ces concepts."[1]

L'écrivain perçoit le monde, pareil à tout autre être humain et dans cet état il ne fait que représenter les choses, l'univers en géneral, par le plus simple des processus de *l'attention*: la saisie ou l'identification de leur propre existence. C'est par une opération autre qu'il diffère en essence de l'individu ordinaire, lorsqu'il transforme ces choses en "complexions d'idées" — pour donner un autre sens à cette définition de Berkeley, citée par Husserl[2] — en arrivant à ce que Husserl entend par *Vorstellung* et surtout par *Stelvertreter* c'est-à-dire par représentation représentative qui inclut selon l'auteur des *Ideen* un *Anschauung* (représentation intuitive).[3] Faut-il ajouter que ce genre de représentation — qui appartient en essence à l'artiste — est une représentation visée, consciemment voulue ou dûment laborieuse? On dirait qu'elle ressemble en fait à un fruit qui serait surveillé sans cesse jusqu'à ce qu'il mûrisse.

Nous voici donc d'un coup devant une perception qui englobe la réception imaginaire par le truchement de la création vouée à être réceptionnée et qui l'est d'ailleurs à partir du premier lecteur imaginaire sinon assez souvent affectif: l'auteur lui-même. Cette réception est-elle immanente au créateur et par cela même en aucune façon ne diffère-t-elle pas de celle appartenant à l'homme quotidien? Ou bien elle est en clivage dès le début étant donné son but autre, son *telos* qui vise la création est qui est présuposée dans l'acte, dans le fait artistique simultanément. Il va de soi, l'acte de percevoir le monde extérieur est jusqu'à un certain point commun mais à la différence de son semblable "de la rue" — qui se laisse irrigué par des sensoris externes — l'artiste engage un dialogue avec ceux-ci, envoyant à son tour vers ceux-ci ses sensoris à lui et souvent même à côté d'une série de sensoris, une bonne foule ou tout une "pluie" de stimulis; c'est grâce à ces stimulis que la simple induction du particulier devient une déduction, se transforme en *abstractisation généralisée* capable à refaire les mailles d'un événement ou les "vertèbres" d'un phénomène dont il n'a vu que partiellement. Toujours est-il qu'à ce degré, nous sommes à peine dans l'antichambre de sa démarche. L'artiste évalue par analogie et confrontation, opère des sélections, se propose des options, mais il ne *réalise* pas encore ou de toute manière pas dans le sens de l'acte créatif. Dans ce stade ou sur cette marche il ne fait rien d'autre qu'agresser la réalité, en essayant de la refaire comme dans un puzzle pour lequel il ne dispose pas pourtant que des objets-témoins que "l'homme commun" a frôlés à son tour mais que l'artiste à déjà saisis en soupesant leur valeur, avec sa propre balance. Il pèse, mais il n'évalue pas pour le moment. A cet istant-là il semble être un individu aussi *hylétique* que le lecteur, ou l'homme quotidien. Mais lorsqu'il dépasse ce stade — de la réalité qui travaille pour lui dans une réalité qui est à même travaillée par lui-même — il cumule *l'entrer* et le *sortir* en une sorte de *feeling* que nous serions tenté de définir par *mentalité projective*. Si pour l'oeil commun, un accident routier qui provoque la mort d'un homme est un événement (com)passionnel ou constitue un motif d'une probable narration émergeant souvent dans le passionnel, pour l'artiste le même accident (con)tient (d') une structure extra-réelle.

Je dirais même qu'il est le signe de passer outre — ou si l'on veut, de transcender — l'itinéraire sentimental dans le champs d'une sémantique mentale qui recouvre une toute autre vision des choses: la victime devient un héros, avec elle, avec ce héros un homme est tué, avec cet homme disparaît un univers entier ce qui ressemble en essence à la stratégie dramatique de la pièce "shakespearienne" d'Eugène Ionesco, *Le Roi se meurt*.

Une démarche similaire, mais fondée en ensemble sur une autre grammaire fictionnelle, entreprend le romancier qui fait de cet événement un sujet de roman psychologique, policier etc. Il codifie le hasard et son feeling avec, ce qui fait que le "détail" initial devient par à coups quasiment invisible, gagnant des proportions autres que nul lecteur en tant que feu témoin oculaire ne pourrait reconnaître. Le hylétique commun engendre une phénoménalité constructive et cet édifice contient sa propre phénoménologie représentant la représentation d'une représentation que nous pourrions appeler *interprétation* "irrationnelle" d'une réalité rationnelle pour commencer. En ce cas-là le triple topos — *information-connaissance-information* — ne se fonde pas à l'encontre de ce qu'on pourrait prendre pour un événement *ex original* mais le degré de perception du particulier par son investissement avec l'énérgie de la généralité, par la revalorisation des phénomènens, des sentiments et de leur *corrélat*: *la fiction*. Mais encore faut-il préciser que c'est à la fiction — en tant que réduction endogénétique de l'acte artistique proprement dit — que nons voudrions rattacher notre idée concernant la représentation d'une représentation représentée, en "surplombant" ainsi à côté de Husserl l'erreur de Berkeley. Et c'est juste dans la *Critique de la théorie de la représentation représentative de Berkeley* — que Husserl propose dans ses *Recherches Logiques* — que nous puisons sans entrave aucune la définition de la conscience artistique accomplie. Nous disons accomplie puisque rendue publique par le fait d'être, d'exister en tant que généralisation qui singularise, sans pour autant satisfaire à telle ou telle quiddité particulariste se suffisant à elle-même, autrement dit morte-née. Mais Husserl nous offre lui-même la *donnée claire* de ce que nous serions tenté(s) d'appeler le pragma achevé individuellement — par l'auteur créateur —, ce pragma ayant droit à l'exégèse extérieure, sans pour autant être détériorée dans son *nisus creativus*: "La conscience *tous les A* se réalise dans l'unité d'un mouvement, dans un acte homogène et *sui generis*, un acte qui n'a de composante, d'aucune espèce, qui se rapporteraient à tous les *A* singuliers, et qui l'on ne pourrait produire ni remplacer par aucune somme ou combinaison d'actes singuliers ou de suggestions singulières."[4]

Il suffit de mettre cette phrase à côté de l'accident routier dont nous parlions auparavant et dont les reflets et l'entendement se propagent de manière différente chez le "lecteur" oculaire, par rapport à cet autre lecteur qui l'a vu tout probablement de ses propres yeux ou qui semble avoir vu l'événement mais qu'il en est censé refaire entièrement, et à sa manière inconfondable. Qui plus est, pareil accident occupe à nos yeux la situation d'une définition génétique de l'oeuvre d'art y compris une image appropriée de sa perception *in medium*. Nous aurons beau demander où se trouvent-ils ces mots en propre dans la phrase husserlienne, la réponse ne puisse être qu'une seule et indeniable dans/par sa vérite crue(elle): partout, dans la donation souterraine de ses sens qui définissent *in petto* le simulacre phénoménologique du travail artistique et son résultat.

Qu'il déplaise à d'aucuns, nous allons soutenir que cette sentence husserlienne c'est *l'éon dogmatique* le plus ineffable de ses *Recherches Logiques*. Nous employons ici le concept *d'éon dogmatique* — que nous avons emprunté au phénoménologue et philosophe roumain Lucian Blaga — dans le sens conceptuel de son essai publié en 1931 et récemment en version française. Inutile d'ajouter que nous puisons pour une fois le sens de *dogmatique* dans le champ sémantique plus ou moins ouvert de celui *d'axiome*.

Certainement sur une forme autre que Husserl mais désireux, pareil au philosophe des *Ideen*, d'aboutir à une *entéléchie* esthétique — voir, par ex., le chapitre *Esthétique ekstatique et entéléchie* — Lucian Blaga propose à sa manière une différence entre l'intellect *enstatique* et l'intellect *ekstatique*. Le premier serait selon nous et en déviant très peu la pensée de Blaga celui qui se rapproche le plus de la perception *hylétique* de l'être en tant que lecteur (non-créateur) mais qui s'efforce pourtant d'arriver, par le truchement de l'oeuvre "au bout de soi-même". Tandis que le créateur est censé parcourir les deux étapes, subir les deux examens de la vie: enstatique pour commencer, l'intellect créateur devrait (se) dépasser *dans* et *par* l'oeuvre en *ekstatique*, en ramassant les débris du . . . miroir cassé dans une unité harmonique, digne à jouir lui-même de sa découverte, afin de se réjouir encore une fois à côté du lecteur en deça et au-delà de l'intelligence universelle et des *données* de l'expérience humaine *disjointe* et *cumulative* — puisque acquise et cumulée de par sa conscience. "Deux états fondamentaux de l'intellect sont à distinguer — écrit Blaga — : l'état "enstatique" et l'état "ekstatique". L'intellect est enstatique pour autant qu'il reste dans le cadre de ses fonctions logiques normales. Mais dès que le recours à des concepts l'oblige pour formuler quelque chose, à se situer résolument hors de soi, il entre en absolu désaccord avec ses fonctions logiques et il devient ainsi ekstatique".[5] C'est là une sorte d'équivalent ou de synchrogenèse idéative entre les "actes singuliers ou [des] suggestions singulières" — dont parlait Husserl toute à l'heure — qui ne sauraient atteindre au bonheur/malheur ou discours créateur que sous le sceau consciemment imposé d'un "éon dogmatique", de cette foi dramatique en hypostase d'autolégislation du moi artistique *in actu* ou confisqué par le *feeling* de l'oeuvre.

Que la vérité souterraine de cet axiome se rapporte ou non distinctement à l'art ou qu'elle soit un apophtègme à crédit sûrement universel, peu importe. Husserl — et sur une voie autre, le roumain Blaga, excellent lecteur créateur de ses oeuvres, spécialisé notamment dans la phénoménologie de la culture et la morphogenèse des valeurs — suggère sans ordonner, il signifie et propose sans orgueil précis une interprétation intraculturelle essentiellement précisée pourtant. Mais avec ses mots nous pouvons voir grand: et *l'entrer* et *l'avoir*; et *l'être* et le *donner* de l'auteur — créateur face à face avec son: *sortir* (dans le monde) par l'intermédiaire de ses explorations *intramondanes* et *mondanéisées* qui influent sur l'oeuvre *avant, pendant* et *après* son processus d'achèvement.

Ou, dans les termes de la *Poetica Nova* d'Anna-Teresa Tymieniecka, il s'agit là d'une part d'un *Cipher* comme réalité humaine et existentielle agençant l'oeuvre et d'autre part d'une *Self-Examination* qui précède le *Creative "Ciphering"* dont le lecteur lui-même est censé *re-découvrir* en faisant connaissance du produit mental psycho-affectif, philosophique et social qu'est l'oeuvre somme toute.[6]

Par ailleurs sans être pourtant indiqué tel quel par Husserl, le sens que nous avons découvert nous-même(s) dans la phrase husserlienne se trouve évidemment sous autre forme dans le pénultième paragraphe de la *Poetica Nova: Interpretation: The Creative Coordinates for the Existence-Significant. Exfoliation and "Embodiment" of the Literary Text.*[7]

C'est dans ce champs du *Signifiant de l'Existence* que commence la première "marche" vers la découverte de l'oeuvre par le lecteur. Que cette marche puisse atteindre à une démarche — dans le sens intellectuel du terme — c'est au sociologue de l'art et à la sociologie de la sociologie de l'art de se prononcer. Quant à nous, résumons-nous pour l'instant de saluer cet hôte devant la porte de la maison — c'est déjà un *phainomenon* ; s'il va y entrer ou la reconstruire selon ses propres noèmes on noèses, c'est son affaire à lui "partagée" parfois avec l'auteur.

Vu pourtant sa présence consciente et présupposée non-dangereuse saluons-le: un beau jour — sait-on jamais? — il pourrait remplacer à même le maître, à cette condition seule qu'il sache le Grec, examen obligatoire, selon, devant le Temple des mystères d'Eleusis.

Bucharest

NOTES

* *Entrer* et *sortir* de l'auteur-artiste *dans* et *de* l'oeuvre; disjonction et cumulativité existentielle(s) comme *feeling* intérieur. Le rationalisme *hylétique* de la perception.

[1] Voir *La Phénoménologie des "Recherches Logiques de Husserl"*, par René Schérer (Paris: P.U.F., 1967), tome II, p. 318 (Les mots sont soulignés par Husserl).

[2] Edmund Husserl, *Recherches Logiques*, traduit de l'allemand par Hubert Elie avec la collaboration de Lothar Kelkel et René Schérer, Tome Second. Recherches pour la phénoménologie et la théorie de la connaissance, Première partie: (Recherches I et II) (Paris: P.U.F., 1961), p. 211.

[3] *Ibidem*, p. 209. Voir aussi Remarques sur la traduction de quelques termes, p. 276—277.

[4] *Ibidem, loc. cit.*, p. 214.

[5] Voir, *L'Eon dogmatique* — par Lucian Blaga, Version roumaine: *Eonul dogmatic*, Editura Fundaţiilor Regale, 1931; Version française: *L'Eon dogmatique*, surtout le chap. 6; *L'Intellect enstatique et ekstatique*. En français par Jessie et Raoul Marin, Mariana et Georges Danescu, Préface de Horia Vintilă (Lausanne: Editions L'Age d'Homme, 1988), p. 77.

[6] Anna-Teresa Tymieniecka, *Poetica Nova. The Creative Crucibles of Human Existence and of Art*. Part I. — The Poetics of Literature (Voir surtout chap. I et II), (Dordrecht: D. Reidel Publishing Company, 1982).

[7] *Ibidem*, p. 78—80.

SIDNEY FESHBACH

THE HUNDREDLETTERED NAME: THUNDER IN JAMES JOYCE'S *FINNEGANS WAKE*[1]*

I: LONGITUDE AND LATITUDE OF THE THUNDER

In "Literal/Littoral/Littorananima: The Figure on the Shore in the Works of James Joyce,"[2] introducing a "poetics of the seashore," I considered Joyce (or his characters) walking and imagining along the seashore. In *Finnegans Wake*, Joyce made use of two natural-water cycles. These water cycles include, by natural inference, rainbows and thunder. The two water-cycles retain their natural base and are associated humanly and mythologically in *Finnegans Wake* with the ocean-father, the cloud-daughter, and the river-wife. The first water cycle is the spring of water that turns into a stream or lake and river to flow to the sea. This cycle was a commonplace in nineteenth-century English poetry, where the stream was associated with daily individualized experiences and the ocean offered painless impersonality and oblivion.[3] The ocean was also the reservoir of images of a metaphysical collectivity.[4] In this tradition, Joyce wrote to his brother that "Anyway my opinion is that if I put down a bucket into my own soul's well, sexual department, I draw up Griffith's and Ibsen's and Skeffington's and Bernard Vaughan's and St. Aloysius' and Shelley's and Renan's water along with my own. And I am going to do that in my novel (inter alia) and plank the bucket down before the shades and substances above mentioned to see how they like it [. . .]."[5] The river-wife returns to her ocean-father: "And it's old and old it's sad and old it's sad and weary I go back to you, my cold father, my cold mad father, my cold mad feary father [. . .]." (*F*628) The second water cycle is that of the water of the ocean evaporating to rise and, chilled, become clouds that move inland, cool further and rain to join the lakes, streams, and rivers; the river flowing outward joins the sea and the ocean again. The mother-river says, "Yes, you're changing, sonhusband, and you're turning, I can feel you, for a daughterwife from the hills again [. . .]. For she'll be sweet for you as I was sweet when I came down out of me mother [. . .]. Let her rain now if she likes." (*F*627)[6] Joyce's interest in this cycle is not

A-T. Tymieniecka (ed.), Analecta Husserliana, Vol. XXXVII, 283—297.

unique; after all, it is found in *Job*[7]; but unlike the other cycle, it is difficult to recall poets who used this.

As part of the river-ocean-cloud cycles, or as a coeval event, is another meteorological activity, *rainbows,* which are mentioned significantly in the *Wake.* Rainbows are the "skysign of soft advertisement" (*F*4) that occurs after a battle or a storm, the sign of the renewal of peace; this recalls, of course, the biblical rainbow of Noah after the Flood: "archlight and rory end to the regginbrow was to be seen ringsome on the aquaface." (*F*3) Joyce makes use of the rainbow's several colors throughout.[8] A point at or before the appearance of the rain and the rainbow is the thunder in the *Wake* which I want to introduce into these studies in the literary phenomenology of the elements.

In this strangest and most complex of literary works, *Finnegans Wake,* any comments are, of course, necessarily incomplete, but with a new definition of incompleteness. Not only do we know that every text is capable of as many interpretations as there are occasions for interpreting, but this text humbles absolutely all its interpreters: every statement about it is reductive in the worse sense of that term. With this failure fixed in mind before we begin, like Beckett characters, we continue nevertheless. For a phenomenology that begins with these thunder words, it is necessary to include the presence of profound ambiguity in the text in general and the deeper ambiguity in the thunderwords. What appears here as a brief introduction to the *Wake* is important: it is part of the process of locating the passage to be considered and of bracketing materials not used but understood as being present to the mind nevertheless. All of Joyce's work participates with all contained anywhere in it, either by synecdoche, metaphor, or analogy. Thus, the portion of this essay that appears merely an introduction, preparing for the analysis, is also part of the analysis. Furthermore, it is impractical to consider all ten thunderwords in this essay: only the first is presented itself and considered in any detail. Most of this essay is introductory to the text and the questions.

The first thunderword is: "(bababadalgharaghtakamminarronnkonn bronntonnerronntuonnthunntrovarrhounawnskawntoohoohoordenenthur nuk!)" (*F*3) Joyce used one-hundred letters for nine thunder-claps and the tenth, the last, one-hundred one letters. This precise numbering compels us to consider initially the thunderwords in relation to quantification. In the numerology of the *Wake* ten is completion, eleven renewal; one hundred is also completion and one-hundred one is renewal. The total of all the thunderwords' letters would be one thousand for completion and 1001 for renewal. Among the multitude of literary, rather than numerological, analogies and identities, this 1001 in the *Wake*, a book that appears to be a dream, for which it invents a style of "night-talk," suggests Scheherazade's 1001 stories known as the *Arabian Nights* and the 1001 names of Allah spoken of in Muslim theology.[9] A Scheherazade-Arabian-Muslim motif runs through the book. The number eleven appears in a year's designation which, Clive Hart suggests, may also be the beginning minute of the book, designation, 11:32.[10] If ten is completion, then this eleven is a renewal combined with 32 which is the number in mechanical physics for falling objects, i.e., 32 feet per second per second. Hence, renewal and fall, or rise-and-fall, occurs in a single year, or even more intensely, a single minute of time. That beginning-again and fall-again occur in the same moment suggests the Orphic and Gnostic myths where creation is division or fall. The total number of occasions of the *Wake*'s thunderclaps is ten, the number of completion, and the total of letters in these ten is 1001, the number of renewal; hence, altogether they are completion, renewal, fall, and they turn, by that extra-letter, toward an eleventh thunderword and another beginning and another fall. It is believed that the last word of the *Wake*, "the," leads to the first word, "riverrun": this makes the book a circle, or a cycle of cycles. The last thunderword may lean toward the first thunderword (although the first is not short a letter). Thus, we see in this example of how the numbers Joyce uses stretch into strings of associations that combine both numbers and meanings, or allegorical numerology. And beyond that to profoundest ambiguity.

Before considering the thunderwords more directly, we must consider the context and for this I must construct an absurdly brief summary about the entire *Finnegans Wake*, all and parts, going from the large to the small.

Title: *Finnegans Wake*. Tim Finnegan, builder, hod-carrier, falls down a ladder and dies. This is followed by a wake. But, of course, in this book of completion and renewal, death is also renewal, and a wake is also an awakening, that is, "*fin*-again." It is not necessarily absurd to consider the idea of a plot for this poetic epic but it cannot be reduced to the kind of plot one could do for the typical nineteenth-century novels; yet it has an association with the large family-novels of that period, only in this case the family is analogous, in Vico's designation, to all "gentile nations," that is those nations that are directly described as foreign or non-believers in both the Hebrew or Christian Bibles.

While Finnegan is referred to throughout, overall, the *Wake* is the story of the relationships of a family in a business: a father-husband, Humphrey Chimpden Earwicker, wife-mother, Anna Livia Plurabelle, two sons, Shaun and Shem, a daughter, Isobel, two helpers, and others, numbering from a few more, as in the Phoenix Park, to many more at the wake. The basic dynamic may be reduced, albeit brutally reductive, to a fusion in concrete practice of theories of Giambattista Vico, his *New Science* and *Autobiography*,[11] Sigmund Freud, his *Psychopathology of Everyday Life* and *Jokes and the Unconscious*,[12] and C. G. Jung, his *The Significance of the Father in the Destiny of the Individual*[13] – that is, autobiography, history, world-transformations, and archetypalization, perceived in layers.

Joyce uses a language and style that sustain a basis in the language and syntax of English, using many languages that, in sum, may be called Indo-European resources. Other languages are present, such as Asian and African languages, and, therefore, other language-bases, but these are relatively less significant and always placed on an armature of the English-base. The thunderclaps are distinctive not only for their obvious length, but they stand apart from the environing style. Joyce used mainly three overall styles for his works. In terms taken from traditional rhetoric, he used for *A Portrait of the Artist as a Young Man* the longer periodic sentence of the thoughtful Ciceronian style, for *Ulysses*, the fragmenting associative Senecan style, and for *Finnegans Wake*, the dreamy-visionary Isocratic style.[14] The Isocratic style uses language for its most obvious musical resources in repetitions and rhythms in the sound and syntax. The thunderwords use the sound-values and phonemic-units, but there appears to be no clear phonemic or structural unity for the entire one-hundred letters to make it a practical or useful word. The thunderwords are distinctive, onomatopoetic jokes, funny, and, despite their importance, useless for communication. Within these precise yet vague boundaries, the style takes on a different patterns in association with different characters.

The *Wake* is divided into numbered sections, or books (I—IV). Each of its four books is a different Viconian cultural-period: the Age of Gods, the Age of Heroes, the Age of Man, and *Ricorso*. Each book is divided into chapters associated with a different main character. In a volume of six hundred twenty-five pages, beginning with page 3, the ten major thunderings in the *Wake* (I–X) occur on pages 3, 23, 44, 90, 113, 257, 314, 332, 414, and 424. The number of thunderwords is – by books – 5 in Book I, 3 in Book II, 2 in Book III, and 0 in Book IV. The sequence, to my eye, approaches a pattern, but then dissipates – 5, 3, 2, 0, suggesting that were there to be an extension it would be a variant of the classical Fibonacci series, to be solved with a recursive formula. That there are none in the last book may be explained by its being a period that is not an "age," in Viconian terms. The last thunderclap is one-third from the end of the book: a question to consider is why the thunderwords stop where they do.

Vico was concerned with the "gentile nations." "Gentile" has two definitions: from Latin "gens" or tribal groups and from its use in the Hebrew Bible, to mean those people that are not within the folds of the faith. Vico's map of the gentile nations is Eurocentric, with reference to Egypt, Greece, Rome, Germany, and nations of the Middle East. While Joyce's map is much larger, indeed, in some ways, world-wide, it is still dominated by Eurocentrism. The implication in every case – biblical, Viconian, Joycean – is that the gentile nations are whatever groups that are not explicitly Hebrew or Christian. For Vico, the Christian Bible is the book of God that presents a providential history for believers, but not for the gentile nations. Similarly, nature itself can be read in this light as the book of Nature, understood still within the purview of the Christian God. Instead the gentile nations have left words and texts that through imaginative philological analysis can reveal the lived history of these peoples. While Vico points the way to a book of gentile history, the gentile nations of the world have no book that is comparable to the one book of the faithful. I believe Joyce has written a bible for the gentile nations.

Joyce's motive for writing such a volume is as old as his first fiction and theories, that his life's task is to describe the secular world for which his poetic and aesthetic theories and practices were secular transformations of theology. This motive was deeply personal with Joyce and, at the same time, continuous with the European Renaissance and English Romanticism.[15]

Some of Joyce's models for the *Wake* are Ovid's *Metamorphoses*, which constructs a seemingly continuous sequence of transformations from the "beginning" to his present. Joyce's own epic of metamorphoses begins with a deliberate echo of Ovid's beginning verses. Ovid has a sequence of "not yets" for the period preceding the creation of order: "My mind is bent to tell of bodies changed into new forms [. . .]. Before the sea was [. . .]. No sun as yet [. . .], nor did [. . .]."[16] Joyce changes Ovid's "not yet" to double the time, so that "not yet" is also a repetition, i.e., "not yet again": "Sir Tristram, violer d'amores, fr'over the short sea, had passencore rearrived from North Armorica [. . .]: nor had topsawyer's rocks [. . .]: nor avoice from afire [. . .]: not yet, though venissoon after [. . .]."(*F*3) This is followed immediately with the first of the thunderwords.

Ovid in poetry and Vico in cultural history used myths for different goals, but they both saw them as integrally related to the political state, one for his immediate Caesar and one for political states in general. For both history is poetic transformation, but for Ovid these are not psychological as such, while for Vico, poetry, history, and epistemology are co-functional. The Viconian sense of that unity makes him particularly useful to Joyce who places the "world" in the language of a dreamer, that is, an intentional world. Where Ovid maintains in his transformations clear differentiation of characters, Joyce makes transformation an event at all levels from the references of names to syllables of words. He even rolls over and changes the size of the letter E to look like a W and a 3. Ovid used Hesiod's *schema* for historical sequences – the golden age, the silver age, the bronze age, the heroic age, and the iron age – but then at the end suggests that his present world would be seen as a perfect time with Augustus as leader. Joyce also follows Hesiod's *schema*, but uses it in the synoptic revision of Vico's much more complex, elaborately baroque, *schema*. Another model is the *Mahabharata*, with its complex of stories contained within stories within stories, each layer possibly having a different ontological presence. Another model Joyce uses is the circular reading of the *Torah* in which the Jewish congregation finishes the complete text every year and returns immediately to *Genesis* to start again; seemingly, Joyce leaves the last line of *Finnegans Wake* incomplete and reaches around to the first line, with its initial lower-case letter. The last sentence brings the book to an incomplete ending with a "the," and the first sentence begins the book: "A way a lone a last a loved a long the" (*F*628) "riverrun, past Eve and Adam's, from swerve of shore to bend of bay, brings us by a commodius vicus of recirculation back to Howth Castle and Environs." (*F*3) (I should add that though a cycle is clearly indicated the time/space between the end and the beginning is "ambiguous": should another daytime, e.g., as *Ulysses* is one day, from waking to sleeping, be inserted?) The day or night is March 18—19, 1922, by the account of Nat Halper.[17] The implication is that the readers should read through to the end and continue to the first line. His differs from both the *Metamorphoses* and the *Torah* in that they are linear histories, starting at the beginning and heading toward a conclusive Golden Age, whether under the aegis of Augustus or a Messiah, a descendent in the lineage of King David. That these models are encyclopedic and, for some, sacred is not accidental, for the *Wake* is by intention a secular bible in which the thunderwords signal massive cultural changes among the gentile nations.

II: ON THUNDER

Joyce's own attitude toward thunder may be expressed in *A Portrait* in Stephen Dedalus's comments: "I fear many things: dogs, horses, fire-arms, the sea, thunderstorms, machinery, the country roads at night." (*P* 243)[18] This list is Stephen's, several of these items appear in the first chapter of *Ulysses*; when Joyce wrote this he had already read Vico, and may have had Vico in mind. Vico's "starting point was [. . .] roughly the state of nature of the natural-law theories. In order to reach it from *Genesis*, as his scruples required, he imagined the descendants of Ham and Japheth, and those of Shem except the Hebraic line, as dispersed after the flood, wandering in 'the vast forest of the earth,' forgetting the speech and customs of their ancestors, and descending to the level of beasts.

During the two centuries Vico allowed for this process of bestialization, the earth slowly dried, until the first thunderclaps caused by its exhalations startled these brutish men here and there in the act of shameless canine copulation with captured women, and terrified them into the shelter and secrecy of caves for their intercourse. Out of these retreats of fear and shame the first families arose, with a settled life apart in the caves of the earth, sanctioned by 'the frightful religions' started in their minds by the thunderbolts of the sky god. With settled habitations came clearing and tilling of the soil, ownership, property, morality, and burial of the dead. Marriage, childbirth, burial, the sowing and reaping of crops, were surrounded with religious ceremonial. The family-father was its king, priest and prophet, sacrificing to the gods to win their favor, and taking auspices to declare their will. He was the arbiter of right and wrong, rewarding the good and punishing the wrongdoer. This 'state of the families,' 'monastic, Cyclopean, monarchic,' and not the antecedent chaos, with its 'bestial communism with women' and 'confusion of human seeds,' was the true state of nature from which the civil state emerged; and the three pre-political institutions of religion, marriage, and burial of the dead were the first principles of the new science. This primitive stage of social evolution Vico called 'the age of the gods.'"[19]

Not yet had the cycle begun again, when, in Book I, a thunderclap occurs and the Age of Gods begins. Thus, before commenting on the first thunderword, it is useful, perhaps necessary, to note the last, the tenth which has one-hundred one letters, that is, tilts toward the first. A brief discussion of this thunderword can also help prepare for a more detailed phenomenological approach to the first thunderword.

Thunder X

Two thunderwords (IX and X) occur in Book III: at the stroke of twelve, as the father Earwicker falls asleep, the dreamer imagines he hears the laughter and celebration of Earwicker's son Shaun, who is the main interest of first three chapters of this book. The two occur in the same chapter (*F*414, *F*424): Thunderword IX is Shaun's cough at the beginning of his telling of the Ondt and the Gracehoper story and X is part of a conversation after the end of that story. Perhaps the two should be considered as a pair of thunders, but there is not enough space here to discuss more than X, and that only briefly.

Shaun's audience asks him fourteen questions, with one result being a revelation of attitudes toward his brother Shem. Question 8: asks him to sing; he clears his throat, with a cough, which is a thunderword IX (*F*414) Question 12: "Since Shaun ended his last answer by excommunicating Shem, they ask why. Shaun answers that he hates Shem's 'root language' (*F*424). Shaun's reason for hating Shem seems peculiar, even mysterious, until we probe deeply into its implication. The 'root language' of Shem is filled with thunder echoes of the divine judgment. Shem's words are the hammer of Thor which could destroy the civilization of which Shaun is the representative. Joyce is here following Vico's notion that all language has its origins in man's effort to formulate the meaning of the primal thunderclap. Shem's language threatens to make that meaning clear, and is thus fraught with judgement on Shaunian society. Shaun's fear of Shem's language shows that he, Shaun, very well knows the secret and power of his brother."[20] "The thunder noise is here ascribed directly to the thunder-god Thor. This variant of the word is full of syllables suggesting the monsters that are to go against the gods at the time of Ragnarok: Midgard serpent, Fenris wolf, Sur. And we recognize Molnir, Thor's hammer, and Loki, father of the monsters."[21] As he answers, he crosses himself out of fear of the artist's power over words. The result is another thunderword, the tenth and final occurrence of the thunder in the circular *Finnegans Wake.* (*F*424) Question 12 begins " — But for what, thrice truthful teller, Shaun of grace?" (*F*424) And it ends with the audience's response: " — The hundredlettered name again, last word of perfect language": "Ullhodturdenweirmudgaardgringnirurdrmolnirfenrirlukkilokkibaugimandodrrerinsurtkrinmgernrackinarockar!" (*F*424)

J. S. Atherton's analysis of this thunderword, which differs in some details with the previous interpretation, should be quoted in full so as to indicate the kind of verbal activity Joyce uses. "Ull' — Thor's stepson. 'Hod' — the blind god who killed Balder. [. . .] 'Tordenvier,' stormy weather. The god Tyr's name may also be included, and weird is Fate. According to one version Tyr warned the gods of the coming of Ragnarok, and the danger from Fenrir the wolf. The next — 'mudgaard' — is obviously based on Midgard, but here may refer to the Midgardsormr, or Midgardr, the 'World-serpent.' If 'grignir' were Grimmir, it would give us a name for Odin, but I cannot see how such a change could happen. 'Urdr' is the old Norse word for Destiny, which became the name of the eldest of the Norns; 'molnir' is Thor's hammer Mjolnir, The Destroyer; 'fenrir,' the wolf; 'lukkilokki' Loki. 'Baug' was the giant who helped Odin to steal the cauldron Odreri;' the syllable 'mand' adds the word Bogeyman to Baug-Odreri. 'Surt,' the fire-giant, fought in the final battle agains the gods. [. . .] Finally, the Norse 'Ragna rokkr,' means 'the twilight of the gods,' not Ragna rok, 'the end of the world.'"[22] This thunder predicts rises after falls. The terminal Viconian historical period in which it is contained is the Age of the Human.

Thunderword I

Let us turn to the first thunderword but now placed within its immediate context for not only do all the thunderwords appear in a context, but, for the phenomenology of the thunderwords in Joyce's *Finnegans*

Wake context is necessary. Here are the first three paragraphs of the *Wake*.

"riverrun, past Eve and Adam's, from swerve of shore to bend of bay, brings us by a commodius vicus of recirculation back to Howth Castle and Environs.

Sir Tristram, violer d'amores, fr'over the short sea, had passencore rearrived from North Armorica on this side of the scraggy isthmus of Europe Minor to wielderfight his penisolate war: nor had topsawyer's rocks by the stream Oconee exaggerated themselse to Laurens County's gorgios while they went doublin their mumper all the time: nor avoice from afire bellowsed mishe mishe to tauftauf thuartpeatrick: not yet, though venissoon after, had a kidscad buttended a bland old isaac: not yet, though all's fair in vanessy, were sosie sesthers wroth with twone nathandjoe. Rot a peck of pa's malt had Jhem or Shen brewed by arclight and rory end to the regginbrow was to be seen ringsome on the aquaface.

The fall (bababadalgharaghtakamminarronnkonnbronntonnerronntu-onnthunntrovarrhounawnskawntoohoohoordenenthurnuk!) of a once wallstrait oldparr is retaled early in bed and later on life down through all christian minstrelsy. The great fall of the offwall entailed at such short notice the pftjschute of Finnegan, erse solid man, that the humptyhillhead of humself prumptly sends an unquiring one well to the west in quest of his tumptytumtoes: and their upturnpikepointandplace is at the knock out in the park where oranges have been laid to rust upon the green since devlinsfirst loved livvy." (*F*3)

The thunderword is not a serious onamatopoetic imitation of the sound of thunder, as a flute may be scored to follow mimetically the sound line of particular bird. Furthermore, even though these are called the thunderwords, its first sound reference is not to that of thunder, but, for example, in Thunderword I, to that of someone, Finnegan the hodcarrier, falling down a ladder. That the first of Viconian "thundering" in the *Wake* is that of someone falling down a ladder recalls that Vico, in his *Autobiography*, reports of himself that "He was a boy of high spirits and impatient of rest; but at the age of seven he fell head first from the top of a ladder to the floor below, and remained a good five hours without motion or consciousness."[23] Similarly, the hodcarrier is stretched out on the ground. This thunderword is a collection of syllables, perhaps small words from different languages, perhaps a semi-controlled gibberish that resides between the dictionaries of languages in use and that lexicon of an Indo-European construct inferred from the world languages by a few linguists. In many words in the *Wake*, Joyce directs the readers' attention to the letters of the alphabet; in this case, though we are aware of the reading process, we are reading small words, syllables, or phrases. At my stage of reading the *Wake*, I feel the need for a multilingual, including Indo-European, syllabary rather than a dictionary of each word: this suggests that much of the reading takes place at a level that of the syllabic-unit as well as word-unit, just as the longer units are paragraphs that form a web of words rather than sentences sometimes running in endless strings where subject, verb, tense, and the rest of the predicate have all parted company.

The first level of Thunderword I is made up of words for thunder in Japanese (*kaminari*), Hindustani (*karak*), Greek (*Brontao*), French (*tonnerre*), Portuguese (*trovao*) Old Rumanian (*tun*), and Danish (*tordenen*).[24] Altogether, they signify Finnegan's falling down the ladder. When the thundering sound of the word for the falling person comprises words whose denotative definition is to thunder, there occur several unstable effects: a unity of sound and sense that falling off a ladder and thundering are related, a joke that falling off a ladder would sound like thunder, a split between reading words for thunder and sounding the word that is for that fall (a contrast to these unstable fusions of sound and sense would be if a word for the fall could have been constructed of words for a human's falling off a ladder — assuming such a family of terms possible. Unions and splits of eye and ear throughout form one of the pleasures in reading the *Wake*, but I wonder if this recapitulates a "fall" of the *sensoria*). The lack of stability in perceiving the word may be mimetically proper for this moment when an entire cultural change occurs. But, then again, all of the *Wake* is manifestly unstable, appropriate to change in all moments of personal, familial, and cultural history. Furthermore, the speaker of this portion of the chapter appears to be a travel-guide bringing the traveler-readers in from Dublin Bay to the River Liffey: is it an aspect of the guide that so many languages are known or that the fall off the ladder is presented in parenthesis?

Thunderword I seems to begin with stuttering, almost baby talk, perhaps "Babeling": "bababada." The intent of the story is to indicate that we feel (perhaps when up a ladder), then we fall (hard to the hard ground), that, in Viconian terms, fallen civilizations spring out of fear and discontent, that guilt is a constant in a fallen world, but so is getting up again, if only to fall again. Some of this is expressed in the stuttering of HCE when confronting some event or events that cause him to feel guilt.

Joyce's thunderwords, more so than the other words in the *Wake* that have similar linguistic ambivalences, point in the two directions just mentioned, the dictionaries of languages in use and an Indo-European lexicon. Joyce has collected in one thundering word, the fall of Finnegan, Vico, thundering verbalized, and, as I show in a moment, the people of the world. One reason this is done is so that an incredibly sophisticated language-oriented author, Joyce, can transcend his own ultimate sophistication and his readers' by returning to the primal matters of language as much as he understood them. That is to say, the thunderwords, especially this first one, which is "merely" falling down a ladder, is preceded by allusions to the flood and the rainbow of the story of Noah in *Genesis*, and to other figures in the same and other books of the Hebrew bible. Two paragraphs later Joyce epitomizes the titles of the Hebrew bible in "Bygmester Finnegan, of the Stuttering Hand, freemen's maurer, lived in the broadest way immarginable in his rushlit toofarback for messuages before joshuan judges had given us numbers or Helviticus committed deuteronomy (one yeastyday he sternely struxk his tete in a tub for to watsch the future of his fates but ere he swiftly stook it out again, by the might of moses, the very water was eviparated and all the guenneses had met their exodus so that ought to show you what a pentschanjeuchy chap he was!) [. . .]." (*F*4)

The thunderword is, in terms of world-history, a trivial event — *N.B.* that Joyce places it between parentheses with its own exclamation mark; it is also a cause, expression, and product of significant cultural change. Joyce's doubling the trivial and the significant by incorporating one person's fall with allusions to the biblical is found in the beginning of *A Portrait of the Artist* where the telling of a story, "Once upon a time and a very good time it was," to a little boy alludes to the Christian formula in the Gospel of John, "In the beginning was the Word" and the implication of the birth of an axial event, of personal and world-historical importance. It was Joyce's mission to make of the insignificant breath of a baby the birth of a new world, as in his poem "Ecce Puer," with its allusions to the death of his father and the birth of his grandson: "Young life is breathed/ On the glass;/ The world that was not/ comes to pass./ A child is sleeping: An old man gone."[25] Yet the readers who read the *Wake* aloud find the thunderword too long to be read within a single normal breath. Too long for reading aloud, too long for comprehension as a single word, the thunderword violates Aristotle's recommendation in the *Poetics* that the magnitude of a work should be such as to be able to be encompassed by the audience. This view of this thunderword shows it be analogous to the *Wake* itself, which appears also beyond any encompassment by individual readers. Joyce's thunder that is beyond the ability of readers' reading silently and aloud is appropriate to Vico's thesis that thunder was felt by the men to hold such awful power that their lives were changed totally: they were fallen. The thunderword is one sound of obliteration and is the loss of some world. And it points to the beginning of another. It was stated before that Thunderword I occurs with an Ovidian and biblical context. It is Joyce's, or Vico's, analogue to God's catechism of Job, when "the Lord answered Job out of the whirlwind, and said: Who is this that darkeneth counsel/ By words without knowledge?" "Who hath cleft a channel for the waterflood,/ Or a way for the lightning of the thunder [. . .]?" Is Joyce's thunder his answer to the question, "And canst thou thunder with a voice like Him?"[26] Yes, if only within Joyce's bible for the gentile nations.

In tentative conclusion I suggest that Joyce presents problems for a phenomenology of thunder in literature that when stated, if not solved, can lead the way for future work in this category. One of Joyce's characters says of the tenth thunderword that it was "— The hundredlettered name again, last word of perfect language." (*F*424) And is the first thunderword, the first word of perfect language or is every thunderword the last? In *Finnegans Wake*, Joyce parodies T. S. Eliot's *Waste Land*, which was derived in part from his *Ulysses*. Eliot's masterwork ends in a fifth section, "V. What the Thunder Said" and that section ends with "Shantih shantih shantih" (Peace peace peace).[27] So too, in Joyce's more complex work, did Anna Livia Plurabelle, the river-mother speak of the "peace and silence" before the thunder and the fall. "My great blue bedroom, the air so quiet, scarce a cloud. In peace and silence. I could have stayed up there for always only. It's something fails us. First we feel. Then we fall." (*F*627) And then came the thunderword. "The hundredlettered name again, last word of perfect language" names an experience of fear, falling, fragmentation in a world perfect in its fallenness.

NOTES

* For Alice Fredman.

[1] James Joyce, *Finnegans Wake*, Viking Press, New York, 1958; references are given in text in parenthesis.

[2] In *Analecta Husserliana*, Vol. XIX, pp. 325—42), ed. A-T. Tymieniecka (Dordrecht: D. Reidel Publishing Co., 1985).

[3] Examples typical of the mid-century are found the poetry of Matthew Arnold.

[4] This metaphysical basin is found in Ralph Waldo Emerson's concept of the "Over-soul," tranformed into the "collective unconscious" of C. G. Jung, and discussed as "the ocean feeling" of mysticism by S. Freud.

[5] James Joyce, *Letters*, 3 vols., ed. Richard Ellmann (New York: Viking Press, in Vol. II, 1966), p. 191.

[6] A second male figure, HCE, is associated mythologically with the earth and mountains. This figure has twin sons who are mythologically the two banks, "rivals," of the river, "*rive*."

[7] *Job*, tr. Victor E. Reichert (London: Soncino Press, 1946), ch. 38.

[8] Roland McHugh in "Rainbows," *A Wake Newslitter*, New Series **15**/5 (October, 1978), 76—7, suggests that the *Wake* has twenty-eight complete rainbows.

[9] J. S. Atherton, *Books at the Wake: A Study of Literary Allusions in James Joyce's Finnegans Wake* (New York: Viking Press, 1960), see Ch. 12, "The Koran." M. Makir Alwan in "Another Interpretation of the Thunderwords in *Finnegans Wake*," *A Wake Newslitter*, New Series **X**/2 (April, 1973), 20—1, suggests a very elaborate connection with the Islamic rosary. Richard M. Kain in "Why is the Thunder a Hundred-Letter Word?" *A Wake Newslitter*, New Series **VIII**/1 (February, 1971), 15, suggests that reason is that the temple of Jupiter Tonans, on the Capitoline Hill of Rome, has one-hundred steps.

[10] Clive Hart, *Structure and Motif in Finnegans Wake* (Evanston: Northwestern University Press, 1962), p. 71.

[11] Giambattista Vico, *The New Science*, trs. Thomas Goddard Bergin and Max Harold Fisch (Ithaca: Cornell University Press, 1968), and *The Autobiography*, trs. Max Harold Fisch and Thomas Goddard Bergin, Cornell Paperback (Ithaca: Cornell University Press, 1975).

[12] Sigmund Freud, *Psychopathology of Everyday Life. A Childhood Memory of Leonardo da Vinci*, and *Jokes and the Unconscious*.

[13] C. G. Jung, *The Significance of the Father in the Destiny of the Individual.*

[14] Sidney Feshbach, "'Writ Our Bit as Intermidgets': Classical Rhetoric in the Early Writings of James Joyce," *James Joyce Quarterly*, Vol. **19/2** (Winter, 1982), 208—13.

[15] Among many studies of this cultural history, the best for English literary studies is M. H. Abrams, *Natural Supernaturalism*, Norton, Norton Library (N.Y.: W. W. Norton, 1973).

[16] Ovid, *Metamorphoses*, Bk. I. 1—20, tr. Frank Justus Miller, Loeb Library (Cambridge: Harvard University Press, 1956).

[17] Nathan Halper, "The Date of Earwicker's Dream," *Studies in Joyce* (Ann Arbor: UMI Research Press, 1983), pp. 29—45.

[18] *A Portrait of the Artist as Young Man*, Viking Press, New York, 1968. Joyce's fear

of thunder, mentioned in memoirs, is used by J. Mitchell Morse in *A Conceptual Guide to Finnegans Wake*, eds. Michael H. Begnal and Fritz Senn (University Park: Pennsylvania State University Press, 1974), p. 3. However, so far evidence of Joyce's fear of thunder is contradictory, with one friend saying he was fearful and another not fearful at all. There is similar contradiction regarding Joyce's fear of dogs. The rule is that Joyce and Stephen are not easy identities.

[19] Vico, *New Science*, 49—50.

[20] J. Campbell and H. M. Robinson, *Skeleton Key of Finnegans Wake*, 268n25

[21] Campbell-Robinson. 268n26

[22] J. S. Atherton, in Begnal and Senn, pp. 168—9. I was forced to make some silent changes in the quotation; I refer readers interested to the original.

[23] Vico, *Autobiography*, 111.

[24] Roland McHugh, *Annotations to Finnegans Wake* (Baltimore: Johns Hopkins Press, 1980), p. 3.

[25] James Joyce, *The Portable Joyce*, intro. Harry Levin (N.Y.: Viking, 1949), p. 663.

[26] *Job* 38. 1—2, .25, and 40.9.

[27] T. S. Eliot, *The Complete Poems and Plays* (New York: Harcourt, Brace and Co., 1952), p. 50.

BRUCE ROSS

REFIGURING NATURE: TROPES OF ESTRANGEMENT IN CONTEMPORARY AMERICAN POETRY

At a recent poetry reading held at Southern Illinois University, Czleslaw Milosz suggested that all true poetry is essentially daemonic and confessed that he was "forced into a corner" to respond poetically as a witness to the brutal social realities depicted in his well-known poems on the Nazi period. For Milosz true poetry is an inspiriting of the soul rather than a committed response to a historical occasion. In primitive and archaic cultures the daemonic and the historical converged in myth and ritual. The thrust of the Romantic Period was to reorient man to that convergence by taking classical Greek culture as its model. The word "daemonic" is derived from the Greek and may be defined as a god, the spirit of a place, or a man's attendant deity. The Greek lyric poet was thus inspired by the Muse Erato, goddess of lyric poetry. He also responded to the spirits inhabiting nature as does Sappho in her poem on daybreak, which she perceives as the sandalled Eos, goddess of dawn. William Blake tried to create his own mythopoeic world by inventing gods that presided over individuals, cities, social revolutions, and nature with an overlay of Christian humanism. But the Industrial Revolution, which inaugurates the modern period of dislocated sensibility, undermines the poetic attempts begun during the Renaissance to reclaim the classical daemonic world. Yeats, the so-called last romantic, would rely on Celtic and world myth, spiritual lore, and private symbolism to manifest such a world, but the realities of technological innovation, cultural revolution, and world war, the "inanimate fantasy" of "Meditation in Time of War," would test the authenticity of such a world.

The contrary poetic claims felt by both Yeats and Milosz in response to world war may be characterized, to borrow Matthew Arnold's terms, as Hellenism and Hebraism. In the former the poet stands in a god-inhabited world and expresses that world in a mythopoeic consciousness. For this poet nature is sacred in itself and to be valued literally. In the latter the poet stands in a world from which the One God has removed Himself and in which the poet expresses himself through a

A-T. Tymieniecka (ed.), Analecta Husserliana, Vol. XXXVII, 299–311.

moral consciousness. For this poet nature is a manifestation of the One God's will and might be valued as a metaphor of that will. One poet produces lyric poetry. The other produces moral-ironic poetry. One sees the gods in nature. The other applies the moral law of the Unseen God to man and nature. After the Romantic Period and responding to the complexities of a modern industrialized age, few poets could confidently express themselves in the mythopoeic idiom or place themselves comfortably within nature. The Modernist Movement retained the ironic stance of Hebraism but replaced the One God and Nature with art as such in a godless consciousness. Post-modernism substitutes an expressive consciousness for the idea of art. No longer standing beyond nature like the modernists, the post-modern poet places himself contemptuously above nature.

In a demythologized world, how does the contemporary American nature poet account for the uncanny and daemonic that were once the province of romanticism? Such a poet is heir to the tradition of American poetry that reflects the slow failure of Christian Hebraism and the periodic attempt to reclaim a romanticized Hellenism. Our earliest poets interpreted nature through Christian allegory, as does Edward Taylor in his "Upon a Spider Catching a Fly" in which the spider becomes a symbol of the devil. During the Romantic Movement, our poets tried to reinterpret nature through myth rather than simple Christian allegory. Relying upon Greek, Hindu, Egyptian, Islamic, Christian, and other religions for his metaphors, Emerson evolved an Americanized version of Eastern transcendental philosophy that is reflected in the poem "Brahma" with its repudiation of even the great world gods and the Greek goddesses of art:

> The strong gods pine for my abode,
>   And pine in vain the sacred Seven;
> But thou, meek lover of the good!
>   Find me, and turn thy back on heaven.

Thus freed from theological near-sightedness Emerson can write a poem like "The Snowstorm," a Keats-like mythologizing of winter. No longer a collection of religious symbols, nature is endowed with its own spirit-like life. Similarly, in Section 41 of *Song of Myself* Whitman invests himself and his fellow men with the truths of the ancient religions and apotheosizes ordinary existence, equating firemen with gods and farmers with angels. Whitman offers a mysticism of the present moment in nature and society that even superimposes itself over the cruelty of war, as in the poem "Reconciliation" in which the speaker of the poem kisses his dead enemy. In fact *The Leaves of Grass* and the later poems in response to the Civil War represent the first major attempt in American poetry to fuse the daemonic and history. Whitman's characteristic pantheism which is tinged like Wordsworth's with Christian morality remains the central model for an American mythopoesis. Emily Dickinson offers a homey orientation to nature while maintaining resonances of the old Christian allegory and Poe explores the realm of the uncanny that reflects the darker side of romanticism. Dickinson's calm presentations of the natural life of New England and of imagined genteel occasions of death contrast with Poe's depictions of dreams, haunted palaces, and exotic regions. Frost and Jeffers are among the few poets writing into the modern period who consistently produce nature poems, Frost's tone and prosody extending from classical pastoral idioms and Jeffers's poetics depending on Greek myth.

T. S. Eliot's *The Wasteland* demarcates the failure of the romantic strain of American poetry under pressure from the modern world. Instead of incorporating world myth into a synthetic world view like Whitman, Eliot rummages through such myth in a desperate attempt to shore up his faith. Yeats's passionate despair has become in the early and middle Eliot an ironic presentation of the collapse of civilization, and this mode in the stream of failed Hebraism leads to the moral-ironic tone of academic poetry. Wallace Stevens's modernist aestheticism and William Carlos Williams's revelatory realism offer two alternatives to Hebraism. Stevens attempts to create myths within poetry itself so that poetry as a whole becomes an alternate universe valued precisely for its abstractness in what Stevens terms a supreme fiction. Williams, contrarywise, explores the basic concreteness of existence, the that upon which "so much depends" of his poem "The Red Wheelbarrow." In *Patterson* he, like Whitman in *The Leaves of Grass*, attempts to mythologize a particular place and culture. In a direct reaction to *The Wasteland*, Hart Crane, perhaps the last American romantic, mythologizes America in *The Bridge* within the context of historical and modern culture. The younger Robert Duncan attempts in turn to recreate the exact feeling of classical Greek daemonic poetry. Most of the middle generation of nature poets begin in the Hebraic strain by exploring traditional religious themes but gradually evolve as

Hellenistic poets of the particular in nature. Robert Bly and W. S. Merwin are examples. The slightly younger Roethke and James Dickey, however, maintain an exquisite sense of the daemonic in nature from their earliest poems, Roethke mythologizing his childhood and nature and Dickey the southern countryside. Social currents during the sixties and a reaction to the academic poetry of the fifties resurrected the Whitmanesque strain of poetry. The major figures here are Alan Ginsberg, who uses oriental modes and a romantic sensibility to create Blake-like poems on nature, love, and social criticism, and Gary Snyder, who applies oriental aesthetics to produce poems that evoke the beauty and wildness of nature. With the seventies, however, nature recedes as an important subject. The legacy of modernist aestheticism leads to a post-modern poet like John Ashberry whose poems reflect an intricate play, often ironic, of consciousness with an undertone of nostalgia.

The visionary, romantic strain of the daemonic does not often appear in the lyric poetry of contemporary American poets — those who were born in the forties and fifties and who have produced at least one significant book of poems.[1] Their poems are anecdotal, narrative, personal, conversational, cool in tone, and basically non-figurative. Their poems use simple irony for a concluding turn or for an overall modulation. These poets use simple metaphor and simile and, occasionally, extended metaphor, although a few of them work up their subjects and metaphors through metaphors-within-metaphors. There is little use of the techniques of surrealism, expressionism, or symbolism. Overall, a direct account of a subject is presented through casual narrative, favored subjects being love, death, innocence, and manhood. Unlike the fifties and sixties poetry, this poetry is rarely visionary, mythic, projectivist, confessional, or moral. And most of it stands outside of nature as such, with a general focus upon human emotions and activities, all more or less recognizably familiar.

However, a great number of younger American poets have written poems that observe in part the otherness of nature. Nature becomes a metaphor for the contemporary poet's estrangement from the province of visionary poetry and poesis — the conundrum of *natura naturata* and *natura naturans*, to borrow Coleridge's distinction. Aside from matters of prosody, such as the campaign to establish free verse as an acceptable form, the major direction of modern American poetics has been to orient the poet's consciousness to objective reality or nature.

Once such an orientation is established, a poet may confidently write poems about objective reality's metaphysical character as well as its daemon-like ongoing being. In 1913, Ezra Pound's manifesto on Imagism argued that poetry must address its subjects forthrightly, whether subjective or objective, in order to distance itself from nineteenth-century conventions of sentimentality and clichéd figurative expression. His definition of poetry as "an intellectual and emotional complex in an instant of time"[2] redefined poetry as the mental union of imagination, emotion, and perceived external reality. Williams's Objectivism, an extension of Imagism, assumes that consciousness is of the same nature as the external world and that therefore the poet may faithfully perceive, even enter, objective reality. Williams views human imagination as a mechanism that is activated by its encounter with reality itself. As he declares in *Patterson,* "Say it, no ideas but in things." For Williams reality presides over poetry and the imagination. Opposed to Williams's revelatory realism, Stevens's poetics of the supreme fiction holds that the imagination is an aesthetic order imposed upon things. His poems are about reality but not of it, thus establishing the modernist's turn from the romantic engagement with nature itself. For Stevens and the modernists, the imagination, working through symbols, presides over poetry and reality. In response to the conservative formal values of the modernist and academic poetry of the thirties and forties, Charles Olson extends Williams's poetics in his theory of Projectivist Verse. Like Blake, Olson views the reasoning ego as undermining poetic vision through a formal appropriation. For Olson, true poetry evokes a kinetic flow of concrete perceptions recorded in a non-formulaic prosody. And like Williams, Olson feels that the imagination is of the same nature as objective reality. In reaction to the Pound-Williams-Olson poetics with its insistence on "objective" presentation (and in Olson on the supremecy of the line-by-line sound values of a poem), Robert Bly and others developed the idea of the "deep image" in which the unconscious mind is considered to be the real source of poetic imagery. Derived from surrealism, the deep image emphasizes the subjective nature of the self and opposes itself to a poetics that confidently rests upon an "objective" ego and an "objective" external reality. For Bly the subjective image presides over poetry and reality. Thus Bly was able, like Blake, to produce committed poetry in which mythic, hallucinatory, and surrealist images master authoritatively seemingly overwhelming social and political subjects like the Vietnam

War, as in a poem like Bly's "The Teeth Mother Naked at Last." Unlike Yeats' despairing over World War I or Milosz's reluctant moral witnessing of World War II, Bly fuses the daemonic and the historical, Hellenism (here the deep image) and Hebraism (here Protestant mysticism), with impassioned confidence.

The younger American poets display implicitly the Pound-Williams-Olson insistence upon "objective" reality. At the surface they seem to be at home in their culture and in their imaginative orientation. They seem not interested in recreating America in myth as did Whitman in *Leaves of Grass*, Williams in *Patterson*, Crane in *The Bridge*, and Olson in *Maximus*. They do not theorize their poetics. Nor are they strongly interested in surrealist deep images in the visionary mode of Bly, the confessional mode of Lowell, Plath, and Berryman, or the artistic mode of the New York Poets like O'Hara. Their emotional tone has neither the Hebraic ironic-moral rhetoricism of the New Criticism academic poets like Brooks or Tate nor the Hellenistic mythopoeicism of the Whitman-Williams-Crane-Olson line. They are not interested in either a conservative classicist remarking of poetry and culture or a radical romanticist remarking of poetry and culture. They do not comprehend nature and human existence in the tradition of ironic despair or ecstatic inclusiveness. Their emotions and modes are homespun, ordinary, and middle-of-the-road, perhaps reflecting the cultural trends of the seventies and eighties.

Yet men and poets remain grounded in consciousness and nature though they would insulate themselves with the comforts of a decidedly affluent culture. Surprisingly, the cult of narcissicism has not produced strong, idiosyncratic poetry. Rather, in general, the poetry of younger American poets may by characterized as tepid narrative that would be better served as short fiction. However, nature as a subject and, particularly, as a metaphor for poetic estrangement enters much of this poetry, even the poetry of those few strong younger nature poets consciously standing within nature, such as David Bottoms and William Pitt Root. In Bottoms's "The Copperhead" the serpent's menacing strangeness, observed on a fishing trip, becomes a metaphor for the otherness of nature. The first of the poem's two sections describes the serpent with a naturalist's precision. In the second section, despite his skill, the fisherman doesn't catch any fish. He is in nature because of his commonplace sport. But the serpent enters the narrator's foreground in an eerie transitional period in which the otherness of nature might be tentatively breached. This transition is demarcated by the occasion of shadows which make the serpent merge with the trees, nature becoming an unbroken presence, and prompts the fisherman to want to himself merge with nature:

> sat for a long time watching the shadows
> make him a part of the tree,
> and wanted more than once to drift into the shaded water,
> pull myself down a fallen branch toward the trunk
> where he lay quiet and dangerous and unafraid
> all spine and nerve.

The real danger to the speaker is not the serpent but that he might be overwhelmed by the otherness of nature because there is no orientation between his consciousness and the reality of nature. Without mythic, pastoral, imaginative, or moral idioms, a poet's consciousness might be overrun by this formidable otherness of reality. So in Bottoms's "Under the Boathouse" the ordinary act of diving into a lake leads to the narrator's catching a fish hook in his palm. The clichéd figurative language ("kicked off the bottom like a frog") heats up in an extraordinary extended metaphor in which the narrator becomes a human lure and his wife's image on the water above him becomes the shadow of an angel. Like the serpent, the water becomes a metaphor of a strange and unmastered reality. The narrator thus remarks that he "knew the shock of all/things caught by the unknown." In such a state reality is defamiliarized, transformed, and visionary: a man becomes living bait and a woman becomes an angel. The otherness of nature also intrudes in an equally terrifying manner in Root's "Nightswim." In this poem, the play of the couples in a swimming pool at night, expressed in clichéd comparisons to otters and seals, is juxtaposed to the night as a threatening unknown presence. The familiar domesticity of suburban life here encounters an alien reality: "they disturbed a darkness/in the pruned sullen trees." The darkness is embodied in a fiercesome bat that attaches itself to one of the swimmers. Dangerous like Bottoms's serpent and lake, the bat has teeth and claws "like pins" and hisses after being tossed to the pet cat. This horror, which is mistaken for a leaf by the swimmer, provokes in that swimmer an awareness of the unknown as a threatening presence in the poem's last section:

For the rest of the night
that man kept glancing
at the dark, waiting
for another scrap of it
to seek him out.

The swimmer's estrangement may reflect, however, an unconscious first turning to the daemonic wellspring of poesis, to inspiration in the Greek lyric sense, so that the night is inspirited by a bat as the dawn once was by a sandalled goddess.

Root in fact is capable of writing poems in which nature prompts an ecstatic and mythic vision. His "Wheel Turning on the Hub of the Sun" is an epiphany of nature in which the narrator experiences an ecstatic union with a flock of white birds. Observed in the Everglades, these birds allow the narrator to experience an inner essence of the reality of nature. The birds' movements are seen figuratively as the ritual dancing of spirits. Their dance incorporates all of nature: the sun creates mirror images of the birds in the water through which they move and the trees' shadows touch that water. The narrator is also part of this dance, if only as an empathic observer, and regrets the loss of his heightened awareness when the birds fly away. At the poem's conclusion he is left "standing like one/in a darkness after lightening." Although the narrator remains in literally bright sunlight, the dancing birds represent a figuratively brighter light, an illumination of nature' spirit-like essence. In a modern pastoral, "From the Other Shore," Root develops a mythic perception of man and nature without borrowing mythic elements from world culture in the Emerson-Whitman-Pound-Eliot manner. Root inspirits reality through an observation of and sympathy towards nature. He is not creating a unifying myth of America or of the modern world but writing individual poems mythopoeically. Also, like his contemporaries, he is not consciously developing symbols to replace myth. The poem depicts what the narrator sees along a river which flows past the countryside restaurant that he is dining in. The river, trees, the sun, clouds, and sheep are introduced in turn with the descriptive values of landscape painting in the poem's first two sections. The last two sections focus on some washerwomen on the river's opposite shore. The washerwomen are presented as mythic embodiments of nature, being mothers and wives to the shepherds and having an elemental contact with nature itself through the river. Their divergence from the demythologized modern world is at first only suggested in the opposition of their primitive mode of washing, the powerful slappings of which are barely heard by the narrator, and the civilized clearing of the tables in the restaurant. The poem's concluding section, however, explicitly juxtaposes the mythic washerwomen to the demythologized consciousness of our materialistic, consumer oriented world:

> slap these flat stones
> have known for centuries
> or more, before learning
> the roar of cars like ours,
> casually laden
> with items worth
> more than we had known
> before we saw the
> stones of the river rising up
> in the form of women.

Like myths, the washerwomen's acts are perennial. At first the restaurant's charming setting entices from the narrator an almost sentimental pastoral idiom. But that idiom, like nature in Bottoms and Root, offers a transition to the otherness of nature, here precipitating a kind of creation myth, and a protective orientation from which modernity might be criticized. It offers a passage to the other shore of the poem's title, that mythopoeic consciousness in which rocks birth women.

Contemporary American poets have not tried to methodically introduce modern social and economic realities into their poems like Pound or Olson. And none of them has even remotely produced a poetics like Futurism. Instead, the best ones have tried to orient themselves in human nature and external reality. Brad Leithauser writes out of an oriental attention to the essences of natural subjects through a mode in American poetics that starts with Emerson and reaches a pinnacle with Gary Snyder. In "Between Leaps," an evident hommage to Basho's famous haiku, Leithauser captures a moment in a frog's life by relying slightly upon Marianne Moore's prosody and poetics. The imagination that sees a frog as a living residue of a fallen tree trunk or the dragonfly in "Angel" as a crystallization of the air recovers the daemonic. Michael Blumenthal tries to mask his discomfort with the otherness of nature, the beautiful but poisonous mushrooms of "Mushroom Hunting in Late August, Peterborough, N. H." and the grotesque but delicious squid of

"Squid," through botanical appropriation and sexual metaphors and, in both cases, actual ingestion. Unlike Leithauser, Blumenthal cannot enter the inner life of nature, viewing nature sentimentally, as in the example of holy mushrooms, and reducing it to its materiality. In "The Orchard" Gretel Ehrlich offers a Frost-like landscape in which man is integrated into both a domesticated nature, the orchard that is the "only city around," and a wild one, represented by the owl that drops mice bones from one of the orchard's trees. Implicit sexual metaphor judges the fecundity of nature as being of a kind with human nature, the sap in the trees resembling the lover's heartbeats. A Burns-like accord with nature is evoked by Simon Ortiz's "My Father's Song," a narrative of the rescuing of a mouse nest from a planting field within the context of the poet's nostalgia for his father's sensitivity towards nature. Nature is used comfortably to supply traditional analogies in Pattiann Rogers's "Concepts and Their Bodies (The Boy in the Field Alone)" in which the forms and behaviors of animals and plants become a primer for a young boy, thus a dandelion seed becomes an emblem of "departure." The poem concludes with a section that unites the boy with the nature that he observes and offers a prayer-like affirmation of the persistence of the basic truths of nature: "And it will be so hereafter." Ehrlich, Ortiz, and Rogers each attempt to bridge the separation between social man and the world of nature, whether through shared physical natures, respect for the integrity of animal life, or a mutual participation in existence as such.

In most contemporary American poetry figurative language heats up when it depicts nature and often betrays therein the poet's uneasiness with and attraction towards the otherness of nature. In "The Small Lizard" Linda Gregg uses a lizard as a symbol of the developed soul because of the seeming stability of an animal's existence. The poem contrasts the lizard with the visibly changing moon and the narrator's own changing emotions. Jon Anderson can expressionistically depict the overpowering irrational otherness of nature through oxymoron in "American Landscape with Clouds & a Zoo": "Everything/Is too brief, eternal, stable, unpredictable." But the narrator wittily acquiesces to this otherness at the poem's conclusion. In Alfred Corn's "Grass" nature becomes an encroaching jungle in which the terrifying implications of the metaphoric grass-jungle are highlighted by the telescopic view of ants devouring a dragonfly carcass and in the concluding image of the resiliance of the grass despite man's presence. Reginald Gibbon's "The Cedar River" is a narrative of how a fish escapes the narrator's lure.

Presenting a non-figurative description of the fish's awesome strength, the poem, in response to this creature's otherness, concludes with a startling simile that compares the narrator's fishing rod to a nodding "innocent stalk of wild rice," nature here imposing its idiom upon the dazzled fisherman. A more emotionally temperate, but no less dramatic, occasion for being moved to simile occurs in Chana Bloch's "Deer in the Bush," a description of the deer that browse in the shrubbery near the narrator's home. These animals, representatives of nature's wildness, cautiously enter the human world, the "green edges of our lives" of the poem. The poem concludes with a long simile that compares a buck that is munching on berries to an uncle "who comes when he pleases, and keeps/the secrets of the tribe," an expression of the poet's figurative attempt to categorize and familiarize the emotion of awe felt before this being. The indomitable will and beauty of the spawning salmon in Jorie Graham's "Salmon" provide a metaphoric correlative to the poem's narrative of coupling lovers. Sexuality here again provides a recognizable metaphor for nature's otherness, the mystery of sexuality observed by the narrator as a child resembles the mystery of the spawning salmon observed, almost voyeuristically again, on television by the adult narrator. But further, the poem speculates on the nature of animal and human beauty, as the dramatic events of creation and procreation move the poet towards the daemonic. In Richard Tillinghast's "Shooting Ducks in South Louisiana" a feather from a dead mallard becomes a symbolic icon of nature. The poem seems wittily distanced from the observed wildness and fecundity of nature: the cottonmouth moccasins are compared to a partygoer at four in the morning, the lush, sonorous activeness of nature is presented with cartoon-like humor: "gurgling, sneaking, murdering, whooping — /a muskrat breast-stroking through the weeds"; and the carnage of the hunt itself is drolly noted with the matter-of-fact comment that the "whole town of Cutoff ate ducks that Sunday." But the poem's concluding lines betray a sensitivity to the deep mystery of nature:

> a drake I had killed:
> sleek neck hanging limp,
> green head bloodied,
> raucous energy stopped.
> I plucked a purple feather from his dead wing,
> and wore the life of that bird in my hat.

The narrator takes the feather not as a good old boy trophy but as something akin to the primitive's hunting fetish or totemistic icon, reflecting the subtlest comprehension of the otherness of nature as an animated presence in its own right.

In the poems of Ellen Bryant Voight, like "The Bat" and "Jug Brook," the moral-ironic dimension of nature's otherness is explored through meditations on the occurrence of death in nature. In "The Bat" the narrator, like Tillinghast's hunter, evokes ironic regret over killing a creature, here a bat that has invaded the narrator's home. In italics the narrator rehearses Biblical, scientific, and social reasons to justify the killing: "*Dominion over every living thing, /large brain, a choice of weapons.*" The bat is seen as an embodiment of the unconsecrated world of nature and is therefore provided with a metaphorically "secular face." The implied divergence between the sacred essence of man and the profane quality of wild nature is underscored when the dead bat is tossed to the family cat "who shuttles easily between two worlds." The ironic undertone of the poem suggests however that the narrator is disturbed by the killing of the bat and by her justifications for this killing. The poem "Jug Brook" explores the metaphysical justification for death in nature. In the first three sections of the poem the narrator appraises the foraging deer, the rebirth of plants, the swelling of a creek with meltwater, the activeness of animal life as a whole, and, also, the dead creatures decomposing in the woods. The second section attempts to evade the rhetorically presented question, "Why grieve for the lost deer." Death is a fact of nature and, as the narrator declares, "The earth does not grieve." The last section, like the first two, presents a view of nature that perceives that both death and rebirth are part of the essence of nature, represented here by the house cat's preying upon field life and also attending to its kittens. But the poem's concluding prayer suggests that there is really no sufficient explanation for the reality of death except for some basic faith in the general goodness of creation:

> Have I learned nothing? God,
> into whose deep pocket our cries are swept,
> it is you I look for
> in the slate face of the water.

Nature, specifically death in nature, is a mystery of existence. This mystery precipitates the surrealist image of God's pocket and, perhaps, the echoing of Genesis in the poem's last two lines. Like many of her contemporary poets, Voight's encounter with the otherness of nature provokes a heating up of figurative expression and a breaching of the visionary and mythic mode.

Rochester, New York

NOTES

[1] The poems of the contemporary poets discussed here are taken from *The Morrow Anthology of Younger American Poets*, edited by Dave Smith and David Bottoms (New York: Quill, 1985).
[2] Quoted in Karl Malakoff, *Crowell's Handbook of Contemporary American Poetry* (New York: Thomas Y. Crowell Company, 1973), p. 145.

PART FOUR

METAPHYSICAL ISSUES IN AESTHETICS

IWONA LORENC

ANTI-METAPHYSICAL THINKING ON ART

Heidegger and Merleau-Ponty

Sanctioning by his system the apparently unbridgeable abyss between the sphere of the thinking subject and the thought-of object, between thinking and being, Descartes became the patron of a definite philosophical tradition. This tradition, originating with Aristotle, was characterized by the conviction that the essence of the philosophical experience of the world lay in applying the measure of human thinking to the objective dimensions of reality that are external with respect to man. Even if it was a common measure, and this was the case in ancient idealistic rationalism (this measure was the laws of Logos which determined the rational character of both being and thinking), the fact remained that at the root of the truth of philosophical cognition was the subject-object relation. The classical definition of truth presupposes "comparing" knowledge and being. It places them "in ready form," as it were, opposite each other. It does not take into account the fact that knowledge belongs to being and, on the other hand, that being is nothing "ready," but that — if it comes within the scope of our cognition — it is in some measure human being, the object of our cognition.

This way of philosophical thinking, distinctly expressed in the Cartesian system, has been described by many contemporary philosophers as metaphysical thinking. (Naturally, the meaning of the term "metaphysics" here is different from its meaning in Aristotle's writings.) Metaphysical thinking isolates and absolutizes, on the one hand, the subject of cognition and, on the other hand, the object. Thus, it is reflected in philosophical idealism and subjectivism as well as in philosophies reconstructing the factuality and objectivity of reality (from Democritus' materialism to neo-positivism).

The dichotomy of this thinking, made real ever since Descartes, has become a sort of "scheme" of modern philosophy.

An awareness of this scheme led to philosophical efforts that aimed at achieving a definite theoretical distance from each side of the absolutized dichotomy, at finding a plane of philosophical thinking

A-T. Tymieniecka (ed.), Analecta Husserliana, Vol. XXXVII, 315—328.

which would be "above" or "before" the division into subject and object.

Kant saw the task of philosophy in defining the conditions of knowledge, in reaching the a priori of what Descartes had described as the sphere of thinking and of what at the same time constituted the a priori object of this thinking. He therefore attempted to situate philosophy (as critique) on a plane which outdated the traditional ontological division into the cognizing man and the world which is being cognized.

Schelling, too, attempted to overcome the division, in his opinion initially inevitable, into the philosophy of the subject (transcendental philosophy) and the philosophy of the object (the philosophy of nature). According to Schelling, what was above these divisions was the philosophy of absolute identity, which, however, in effect provided simplified and schematic tools for the reconciliation of the real and the ideal.

From the point of view of the later tasks of nineteenth century philosophy, Hegelian dialectics was very important. It shows philosophy as the crowning, and at the same time an internal element, of the development of human socio-cultural praxis (the work of the spirit), perceiving the essence of this development in the incessant overcoming of the continuously emerging forms of "strangeness" between subject (spirit) and object. Thus philosophical thinking here is not something "given," but exists only as a result of a constant turning toward that from which it originates and of which it is the overcoming. In other words, philosophical thinking is an element of the historical world and, at the same time, contains this world. It manifests itself only in a turn toward that which is not thinking, while that which is not thinking achieves its reality as the object of thinking.

As we know, both Marx and post-Husserlian phenomenology, and with them the phenomenological-hermeneutic and existential orientation of contemporary philosophy, draw on the experience of Hegelian dialectics.

The Hegelian inspirations of Marxism have been widely discussed. We only want to emphasize here, citing, for example, Lukács' conception of Marxism, that the categories of labor and socio-historical praxis, fundamental in this philosophical orientation, create the possibility of showing the determinants of thinking in all its forms. This philosophy demonstrates that human thinking is an element of being and, at the same time, reveals the fact that subjective and objective factors, if artificially separated, are ontologically interdependent within the framework of social being.

In turn, phenomenology is a harking back to this same tradition, in the sense that it turns toward the question of the world created by theoretical cognition's being rooted in the sphere of pre-reflective, intentional experience. Especially interesting in this respect are the assumptions of phenomenological hermeneutics and existential phenomenology.

Sharing Husserl's conviction that all awareness is an awareness of a certain thing, hermeneutic reflection turns toward the domain of meaning, which is neither entirely constituted by thinking nor "given" beforehand to thinking as a task to be resolved. In this sense, it annuls the traditional scheme of metaphysics. Hermeneutical thought, inspired by Heidegger, shows that both the process of comprehending the world and the forms of expressing this comprehension (speech) are rooted in the situation of "being in the world." This "being in the world" is open, fundamental, and primary with respect to all forms of human expression. It also constitutes the starting point, and at the same time the point of arrival, of hermeneutic interpretation, which assumes moving around the circle: being in the world – understanding – interpretation – exposition (being in the world). Understanding the symbolic forms of human "being in the world" enables us to reach the sources of meaning, which is a prerequisite for any understanding of the world, and at the same time co-constitutes this meaning (being in the world is always comprehending being).[1]

According to existential phenomenologists, phenomenology is capable of radically overcoming the hiatus between the subjective reality of meaning and the "impenetrability" of reality in itself. They postulate a return to everyday experience, to direct perception of the world. To Merleau-Ponty, that which is given in perception is the totality of the object, while the object is not only a set of stimuli, but also a definite meaning. W. A. Luijpen illustrates this idea thusly:

> every object appears as a definite figure, against a certain background, on a certain horizon of meanings . . . An apple which is not lying on the table, in a dish, in a vendor's basket, which is not hanging on a tree branch or laying in a child's palm, briefly speaking, an apple which appears against no background simply is not a real apple. It is not an object of real perception, but pure illusion, the product of a dream or hallucination.[2]

Philosophical thinking is constantly being enriched by pre-reflective consciousness. Pre-reflective consciousness is a moment of man's being in the world, a part of lived experience; it is consciousness which makes communion with things and people possible, though it is not yet thematic consciousness, it does not yet know that it is consciousness. There is no doubt, Merleau-Ponty writes, that

> philosophical thought starts from it, and even at the beginning is nothing else but the expression of what we can most generally call "life." Life, that is, cognizing, active, affectionate, emotional, etc. being in the world, is conscious being in the world. Since life is conscious, it does not escape me. However, I must "catch" it if I want to philosophize, for the consciousness of my life in the world is merely "lived experience," and life is after all "irréflechi."[3]

The above remarks of a general nature have delimited the area which we shall attempt to examine in more detail, focussing our interest on philosophical reflection on art.

It is no coincidence that, beginning with Kant, philosophy's efforts aimed at overcoming the Cartesian tradition of metaphysical thinking have been accompanied by a specific ennoblement of the philosophical problems of art; for art is the perfect embodiment of the guiding idea of philosophy. The world of works of art, actualizing the idea of beauty, reaches the essence of all modes of man's being in the world, reveals the truth in a manner which does not establish artificial antinomies built on the unity of life, but which demonstrates the ontological interdependence of man and the world, being itself a proof of this unity.

In his fundamental work, *Sein und Zeit*, Heidegger reflected on the achievements of classical European metaphysical thought. His was a critical reflection, and the criticism was directed against what Heidegger believed was the basic deviation of this thought.

Metaphysical philosophy, while asking what being is, sets for itself a task impossible to accomplish. The conceiving of being in its totality, the philosopher asserts, is unfeasible, since we are amidst being in its totality. This totality cannot be wholly "grasped" using the means of scientific or philosophical thinking. Thus, metaphysics can ask the question of being only when it goes beyond this being, when it conceives it in the form of the object being examined (contraposed to the subject), when it extracts it from the natural references of meaning and immobilizes its historicty in the form of the absolutized "now." It avails itself of the measure of truth, which necessitates comparing this "given," limited, objectified being with the "given," ready perception of this being by the subject.

Is a way of thinking possible which would remove this objectivity, a way thinking which would mean the liquidation of metaphysics as the mere presentation of being and would enable philosophy to return to the unity of the presentation and experiencing of being?

According to Heidegger, the total liquidation of metaphysics is impossible by means of thinking alone, for metaphysical thinking is indissolubly linked to the scientific-technical civilization of the West and declines parallel to the decline of this civilization. It is an inalienable feature of European philosophizing. Therefore, its end means the end of philosophy in its traditional sense.

What does Heidegger propose in exchange? He postulates the overcoming of metaphysics. The ways and means of this overcoming suggested in his early works are different from those indicated in his later works. Initially, this transcending of metaphysical philosophy was to occur within the framework of thinking; later, it was to go beyond thinking.

In Heidegger's early stage, this end is served by his postulated reaching for the sources of thinking. This is the fundamental, experiential, historical sphere of the relationship between man and the world, in which consciousness and thing melt into a single, still inseparable whole. In accordance with the phenomenological inspiration of his philosphical intentions, Heidegger postulates a philosophy which reverts to "things themselves," overcomes the naiveté of the conceptions of the philosophy of "being" detached from human experience, and which attempts to discover the "ultimate truth," formulating an ultimate aim of philosophy. As one researcher interprets Heidegger's thought,

> the source of all meaning is the factualness of human existence, factualness that is inevitably historical and finite. Therefore, the task of philosophy is not the definite grounding of the "conception of the world," or the grasping of its ultimate truth (which cannot exist), but the task of accompanying and revealing finite meanings constituted by finite human being in its changing historicity (in the structures of its historically changing existence).[4]

To the "late" Heidegger, the overcoming of metaphysics consists not in counterposing to it some "ready" truth, but in counterposing to this ready truth the constant readiness to search. While transcending metaphysics, philosophy is "constantly on its way, is a continuous effort, an uninterruptedly repeated attempt — because we possess knowledge as a result of the very fact that we exist, and being by its nature is never fully brought to light."[5]

As we know, in the center of Heidegger's philosophy is the category of *Dasein* (human existence). To man, the existence of the human being is not something given, something immediately manifest. The task of philosophy is to reveal its meaning through the effort to comprehend the ways in which it is articulated. Philosophy is to be the hermeneutics (the revealing of the hidden meaning) of *Dasein* /human being, the only being entitled to understand Being.

Thus, philosophy is not a ready-made, closed mental truth. It is a "conversation" with Being, in which the philosopher does not have the last word, since this conversation is an incessant dialogue. The philosophy which overcomes metaphysics is something more than self-consciousness. "In Heidegger's thought," Michalski writes, "... something more is said than that which it knows about itself. It also involves waiting for a subsequent statement in the dialogue, a statement which will being to light its hidden assumptions, its forgotten source. It is just this thought — Heidegger's thought — that attempts to prepare for this possibility."[6]

Philosophy's conversation with Being is effected through speech and thanks to speech. For the understanding of Being is given to man, above all, owing to the fact that he has the ability to name. It is speech that is the horizon of our comprehension of the world. It is, as the philosopher puts it, the home of being; through it, it becomes possible to conceive of bring. "Heidegger attempts to demonstrate," Ożarowski writes, "that without a name, a word, a thing is not only inconceivable within our language horizon, but does not exist at all, since all that exists is that which can be named and expressed in speech." In language, being is revealed, and without it, nothing would exist.[7] Of course, what is involved here is not the idealistic conception of the creative force of the word, but the stating of the fact that speech determines the limits of our *Dasein*, that it is evidence of our meaningful relations with the world, and that it at the same time makes these relations possible.

In the process of speech, we do not create things, but only name and interpret them; but even this naming and interpretation must be consistent with Being itself (it cannot be entirely arbitrary). And yet, speech reveals and gives meaning to things. Therefore, it is precisely in speech, especially in poetry, that the truth of Being expresses itself. Thanks to this quality, speech and poetry are primary with respect to all the kinds of arts.

Poetry enables us to listen to the truth of Being through the barriers of alienation, through the veil of the objectified understanding of the world that is characteristic of the present scientific-technical era. It becomes an refuge for the truth of Being and, at the same time, a "revelation of the essence of the decay" of an era which replaced this truth with false dichotomies.

Above all, however, poetry is the "experience of the unspoken," and as such it constitutes a treasury for philosophy, which adopts as its task the hermeneutics of *Dasein*. To philosophy, the hermeneutical act does not mean a "translation" of the language of poetry into the language of philosophy, but a dialogue between philosophy and poetry as two different ways in which the truth of Being manifests itself. This dialogue itself is a definite manner in which *Dasein* is revealed. Heidegger writes thusly about it:

> The only need would be and is: having soberly considered what has been said in his [Hölderlin's] poetry, one should experience the unspoken. This unspoken is the clearing in the path which the history of being has followed. If we attain it, it will lead our thought into a dialogue with the poetry that has been going on in the history of being. In historical-literary studies, such a dialogue is inevitably regarded as the unscientific violation of what these studies consider to be facts. To philosophy, dialogue is the wilderness in which helplessness [the helplessness of philosophical thinking] escapes into dream. However, the path unconcernedly runs above all this, in accordance with its part.[8]

This dialogue between philosophy and art is by no means an attempt at aestheticizing philosophy, or an attempt at reducing art to the dimensions of philosophical thinking. In its essence, it reveals the structure of *Dasein*, which appears in thinking to be split, while constituting a unity; the latter is not something "given" and is in the process of the revelation of that which is concealed.

To the philosopher, art is a convenient partner insofar as its essential principle corresponds to the principle of truth that underlies Heidegger's phenomenological hermeneutics. It is truth, not in the sense of classical *veritas*, but in the sense of the old Grecian *alethea* – the revelation of the unconcealed. Truth is something not entirely revealed, something that may be only partly unveiled, and if it is unveiled, it is at the expense of losing sight of something else.

One researcher rightly remarks that for Heidegger,

> in a work of art, openness and non-concealment express themselves in the world, while concealment expresses itself in the earth. Thanks to the earth, a work of art ceases to be a product of free imagination, a mental construction, achieves existential independence, autonomy, and immeasurable depth, and cannot be reduced to a stimulator of our experiences. It becomes something infinitely greater — it is an event, a new revelation of being and truth, capable of assembling and transforming the people around itself. Therefore, unlike, for example, in Hegel, truth is not confined to a clearly expressed thought, but is a full manifestation, a full revelation coming from non-revelation and returning to it. It is the fullness of an existential communion whose clear thought is only a narrow segment which does not know its background.[9]

As with Schelling in the period of his philosophy of identity, art and philosophy are for Heidegger two equally valid ways in which the absolute is manifested; these domains are irreducible to each other and as "true" as the manifestations of the truth of Being. But whereas Schelling sought to conceive of the dissimilarity of art by philosophical means (being at the same time fully aware that the essence of art escapes philosophical examination), Heidegger directs the effort of hermeneutic comprehension at the area of the dialogue between philosophical thinking and art. It is the area between the "question" asked by the philosopher when he proceeds to read the meaning of a work of art, and the latter's answer, between the question asked by art itself when it reveals thus far unrevealed areas of meaning and the "word" of the philosopher, who names this meaning and, at the same time, "conceals" the truth, as it were (by means of objectifying categories, the dichotomy of thinking, etc.). This dialogue is an attempt at going beyond the limits of "doomed" thinking to metaphysics. This intention, present in philosophy beginning with Kant, Schelling, and Hegel, was actualized by Heidegger in an inspiring manner — inspiring, amongst others, Merleau-Ponty.

Merleau-Ponty's philosophy is one of the major contemporary attempts at abolishing the traditional form of philosophy as a closed system of concepts, as a purely mental construct. The philosopher attempts to overcome the metaphysical character of philosophy by turning toward the experience which preceded thought.

Meleau-Ponty places in the center of philosophical attention the phenomenon of perception. Perception is what precedes both the order of nature and the order of thinking. Perceptive experience of the world is not something purely natural nor something purely reflective. It is something which precedes the two orders. Thus the philosophy of perception cannot be confined to either of these orders, cannot be the ontology of either "Being-in-itself" (external causality), or the ontology of "Being-for-itself" (the internal world). According to Merleau-Ponty,

What is the most important for us is the cognition of the meaning of the being of the world; we cannot adopt here any assumptions: neither the naive idea of Being-in-itself nor the related idea of Being-representation, Being for consciousness, or Being for man: all these are concepts which we must rethink in connection with our experience of the world, together with the being of the world.[10]

Confining oneself to the scientific cognition of the world means acceptance of the Cartesian picture of the world, a certain type of falsification that removes us from the problem of the meaning of being. Scientific truth is merely a surrogate which needs then philosophical perspective and transformation. For philosophy reaches for the deposits of the experience that precedes science. This plane of reference enables philosophy to maintain a distance from the achievements of science (putting them in brackets), and to reveal its sources and determinants. It is the primary sphere of perceptive experience.

In the light of this experience, the traditional division into consciousness and body, accepted by both the idealistic and the materialistic conceptions, loses its validity.

The first philosophical act would be a return to the experienced world . . . next to the objective world. For it is in this world that we can understand the laws and limits of the objective world, restore to the thing its concrete physiognomy, to organisms their manner of treating the world, to subjectivity its historical roots, and rediscover phenomena, the layer of lived experience through which they are primarily given to us, as well as things; we can discern *in statu nascendi* the "I — the other — things" system, reinvigorate perception, and lay bare the subterfuge, thanks to which the lived experience can be forgotten as a fact and as perception in favor of the object which it provides and the rational tradition which it consolidates.[11]

Turning philosophy toward the order of perception, Merleau-Ponty, in accordance with the phenomenological inspiration of his thought (the return to the thing itself), seeks to reach the sphere that precedes any reflection, to reconstruct its genealogy. Thus perception is a category that precedes intellect, and is not identical with an act of intellect. On the other hand, perception is an inalienable genetic moment, a prerequisite of any reflection.

It determines thinking in the sense that the relationship between the world and the body, which is capable of feeling, and receiving impressions, is effected within it. This feeling ensures to us links with the world-in-itself. At the same time, it is a reference of the world, to ourselves in our corporeality. "Perception is the opening of the subject to that which is different, the opening to reality, to things."[12]

The relation of the subject with the object within the framework of perception materializes within a broader structure of meaning, which this relation both calls forth and co-constitutes. Perceiving the world, we each time re-establish it. While perceiving, we do not so much receive that which comes to us from the world-in-itself, but rather conceive this world within the totality of meaning called forth by this perception of the world and constituted in this perception of the world. It is in the perceptive experience of the world that the order of the body becomes united with that of consciousness, and the factuality of things with that of meaning; this unity "occurs at every moment of the movement of existence."[13]

Unlike Descartes, Merleau-Ponty treats the body as an inalienable condition and means of any communication with the world; at the same time, the body determines the possibilities and limits of reception of the world. It is not the cogito that is primary with respect to all forms of cognition of reality. In order to cognize, this French philosopher says, "the real subject must first possess the world or be in the world, that is, must carry around with itself a system of meanings whose adequacy, interrelations, and dovetailings do not have to be revealed in order to use them."[14] For — let us quote Merleau-Ponty once again —

> any perception, and any action which assumes it, in a word, any use of the body is primary expression, and is thus, not secondary or derivative work which replaces content expressed by signs taken from elsewhere, together with their meanings and rules of usage, but an operation which first constitutes signs as signs, imbues them with content expressed not on the basis of some previous agreement, but through the very signification of their arrangement and configuration, and implants meaning where it was previously lacking, and finally, without limiting itself to the moment of its duration, opens up a new field, initiates a new order, introduces an institution or tradition.[15]

In this sense, each man's being in the world (from the behavior of our bodies to the sublime forms of cultural behavior) is a meaning-creating and expressive being. Human existence does not pertain to some pre-established order of meanings; when we enter into relations with the perceived world, these meanings appear in a definite order, become sensible. "It is enough," the philosopher writes, "if we make some indentations, some crevices among things — and we are doing this all our life — and that which is the most alien to the world, meaning, comes into existence."[16]

Thus, the forms of human existence are forms of expression specific, unique forms, as unique as human existence is. Art also is a form of the expression of our existence, a manner of imparting meaning to the things with which we are in contact. ("The expression of painting takes up and advances the work of forming the world, commenced by perception."[17] Merleau-Ponty — let us stress it once again — seeks the sources of art in the space between the "individual and its situation," where all meaning is born.)

However, is art a distinct form of expression? Does art have special significance here also, as in the previously analyzed conceptions of it?

In fact, art is a select field of human expression, since it demonstrates the genealogy of meaning. For the artist each time "begins from the beginning," does not avail himself of ready, fossilized meanings, but reformulates his world of meaning (and this is why his work is his own, unrepeatable world).

This meaning-creating function of the artist, which harks back to the sources, is convergent with the task which Merleau-Ponty sets for philosophy:

If we really want to understand the genesis of meaning — and without this we shall never comprehend creativity and culture, we shall have to assume the wisdom of the world, in which everything has from the outset had meaning — we must reject any meaning given beforehand and retreat to the initial situation, when the world does not yet mean anything, to the situation of an artist who is only beginning to speak.[18]

A work of art does not offer us ready-made truth (even if we are concerned with mimetic art), but creates its own truth. It is significant that Merleau-Ponty quotes here Sartre's words: "As always in art, one lies in order to be veracious." A work of art will never be able to fully imitate the ties of life. "Therefore, it embodies its own vision of the world, forces us to unreservedly accept a truth which is dissimilar from the thing and which, though deprived of the external model, of tools of expression given beforehand, nevertheless is the truth.[19]

Art is at the same time the model of our relations with the world:

The body impresses its initials on whatever it is doing; despite the diversity of its parts, which makes them very fragile and little resistant, it can concentrate on a single gesture and control their dispersion. Similarly, irrespective of distances in space and time, there exists the unity of the human style which combines the gestures of all painters in a single effort, a single cumulative history, and their products in a single art or single culture.[20]

The fact that art is a materialization of the principle of expression, which underlies human experience, ensures its unity and tradition: "The first drawing on the wall of the cave could lay the foundation for tradition, for it was the continuation of an even older tradition — perception."[21]

Examination of the history of art offers philosophy the possibility of resolving the traditional philosophical problem of the way in which that which is subjective become the objective, of how the internal becomes the external, of the way in which a single act of perception becomes a component of the world of history and culture. Merleau-Ponty expresses this idea outright: "If our concept of history were shaped on the basis of art or speech, something we strongly encourage, it would certainly be much easier for us to find its true meaning."[22]

We already find in Schelling such a way of thinking about art, in which art plays the role of a proof for philosophy. Merleau-Ponty, however, harks back to this tradition through Hegel, who provides him with specific inspiration, specific because the French philosopher takes from Hegelian philosophy only a certain thread: reflection on the question of "when does the internal becomes the external."

Philosophy, like any thought, is conceived by him as a form of expression of the fundamental relation of existence to the world a relation which reveals all meaning. This philosophy also includes art as "a specific deformation thrust upon visible things."[23] From the point of view of the truth about this foundation, art seems to play even a more essential role than philosophy:

What makes a work of art something irreplaceable, not so much a means for providing pleasure, but an organ of spirit, in a way similar to every philosophical or political thought, is the fact that it contains something more than ideas, namely, matrices of ideas; it furnishes us with symbols whose meaning can never be exhausted, for coming along with us to the world to which we do not have the key, it teaches us to see the thing in a manner inaccessible to analytical work, since analysis finds in the object only that which we previously put there.[24]

Therefore, what is left for philosophy as thinking is to either accept — realizing its own limitations — the hermeneutic assumption of progressing around the circle, or to relinquish the traditional, exclusively mental formulae of philosophy and situate it in a position of dialogue with art (as Heidegger's thought shows, these two possibilities do not exclude each other), or to relinquish philosophizing, in accordance with the formula that it is better to say nothing about something which cannot be expressed (for example, in favor of artistic praxis).

Merleau-Ponty proposes the overcoming of the limitations of philosophical thinking by distancing oneself from philosophy and treating it as a form of expression of human being in the world. But does this distance not require the adoption of additional premises? Does it not involve additional theoretical dangers? These are questions which go beyond the scope of the present article.[25]

Our task has been to show, on the basis of two examples, a certain tendency in the contemporary philosophy of art, a tendency which has its source in classical German philosophy. Our remarks concerned the striving to surpass the boundaries of metaphysics in philosophy, in the direction of a solution which (as a model for philosophy) is actualized in the forms of man's artistic communion with the world.

NOTES

1 "Speech is built in such a manner that it is capable of indicating the ground of existence from which it stems and of recognizing itself as the mode of being to which it refers. This circularity between 'I speak' and 'I am' successively gives initiative to the symbolic function and to its existential and impulsive roots. The circle is not a vicious circle; it is a live circle described by expression and expressed by being. "(M. Philibert, *Paul Ricoeur czyli wolność na miare nadziei. Szkic o twórczości. Wybór tekstów* [Paul Ricoeur, or Freedom to the Measure of Hope. An Essay on Creative Work. A Selection of Texts], [Warsaw: 1986], p. 195).

2 W. A. Luijpen, *Fenomenologia egzystencjalna* (Existential Phenomenology), (Warsaw: 1972), p. 120.

3 M. Merleau-Ponty, *Phénoménologie de la Perception*, quoted from: *ibidem*, p. 97.

4 J. Syski, "Heidegger: Fenomenologia jako hermeneutyka Dasein" (Heidegger: Phénomenology as the Hermeneutics of Dasein). *Humanitas*, 1979, Vol. II, p. 212.

5 K. Michalski, *Heidegger i filozofia współczesna* (Heidegger and Present-Day Philosophy), (Warsaw: 1978), p. 221.

6 *Ibidem*, p. 223.

7 J. Ożarowski, "Rola nauki Hegla w myśleniu 'istotnym' M. Heideggera" (The Role of Hegelian Science in M. Heidegger's "Essential" Thinking), *Humanitas*, 1980, Vol. IV, p. 48.

8 M. Heidegger, *Budować, mieszkać myśleć. Eseje wybrane* (To Build, Reside, Think. Selected Essays), (Warsaw: 1977), p. 173.

9 J. Ożarowski, *op. cit.*, p. 43.

[10] M. Merleau-Ponty, *Le visible et l'invisible*, quoted from J. Syski, "M. Merleau-Ponty — droga do filozofii" (M. Merleau-Ponty: The Road to Philosophy), *Humanitas*, 1978, Vol. I, pp. 239—240.

[11] M. Merleau-Ponty, *Phénoménologie de la Perception*, quoted from J. Syski, *op. cit.*, p. 155.

[12] J. Syski, "M. Merleau-Ponty — świadomość i ciało w analizie fenomenologicznej" (M. Merleau-Ponty: Consciousness and Body in Phenomenological Analyasis), *Humanitas*, 1982, Vol. VIII, p. 166.

[13] *Idem*, "Doświadczenie cielesności jako zjawiska transcendentalnego" (The Experience of Corporality as a Transcendental Phenomenon), *Humanitas*, 1980, Vol. V, p. 191.

[14] M. Merleau-Ponty, *Phénoménologie de la Perception*, quoted from: *ibidem*, p. 210.

[15] *Idem*: "Postrzeganie, ekspresja, sztuka" (Perception, Expression, Art). In: *Proza świata* (World Prose) (Warsaw: 1976), p. 212.

[16] *Ibidem*, p. 193.

[17] *Ibidem*, p. 193.

[18] *Ibidem*, p. 190.

[19] *Ibidem*, p. 198.

[20] *Ibidem*, p. 215.

[21] *Ibidem*, p. 217.

[22] *Ibidem*, p. 219.

[23] *Ibidem*, p. 225.

[24] *Ibidem*, p. 225.

[25] H. Lefebvre, amongst others, writes about these questions, "Filozofia dwuznaczności" (The Philosophy of Ambiguity). In: M. Merleau-Ponty: *Proza świata, op. cit.*

CÉSAR MORENO MÁRQUEZ

THE SENSE OF POSSIBILITY

On the Ontologico-Eidetic Relevance of the Character (the Experimental Ego) in Literary Experience

> Pour ce qui est de l'Ame raisonnable ou de l'Esprit, il y a quelque chose de plus, que dans les Monades, ou même dans les simples Ames. Il n'est pas seulement un Miroir de l'univers des Creatures, mais encore une image de la Divinité. L'Esprit n'a pas seulement une perception des ouvrages de Dieu, mais il est même capable de produire quelque chose qui leur ressemble, quoyqu'en petit. Car, pour ne rien dire des merveilles des songes, où nous inventons sans peine (mais aussi sans en avoir la volonté) des choses auxquelles il faudroit penser longtemps pour les trouver quand on veille, notre Ame est Architectonique encore dans les Actions volontaires.
>
> G. W. Leibniz, *Principes de la Nature et de la Grace, fondés en raison*, § 14

In the famous Section four of his work *The Man without Qualities*, Robert Musil defended the thesis that «wenn es . . . Wirklichkeitssinn gibt, und niemand wird bezweifeln, dass er seine Daseinsberechtigung hat, dann muss es auch etwas geben, das man Möglichkeitssinn nennen kann.»[1] Not only did he manage to characterize a certain unrest which permeates literary experience but also, more than that — and in this can be detected the way his ideas link with those of Husserl — a mode of illuminating the *complexity* of the world. According to Musil, the person who possesses the *sense of possibility*

> . . . , sagt beispielsweise nicht: Hier ist dies oder das geschehen, wird geschehen, muss geschehen; sondern er erfindet: Hier könnte, sollte oder müsste geschehen; und wenn man ihm von irgend etwas erklärt, dass es so sei, wie es sei, dann denkt er: Nun, es könnte wahrscheinlich auch anders sein. So liesse sich der Möglichkeitssinn geradezu als die Fähigkeit definieren, alles, was ebensogut sein könnte, zu denken und das, was ist, nicht wichtiger zu nehmen als das, was nicht ist.[2]

The links between the above-mentioned declaration of intentions and those which form the important strands in the methodological program

A-T. Tymieniecka (ed.), Analecta Husserliana, Vol. XXXVII, 329—342.

of Husserlian phenomenology are obvious. We are reminded, for example, of the thesis of Husserl which states «die alte ontologische Lehre, *dass die Erkenntnis der 'Möglichkeiten' der der Wirklichkeiten vorhergehen müsse*, ist m. E., wofern sie recht verstanden und in rechter Weise nutzbar gemacht wird, eine grosse Wahrheit.»[3] Thus, we should not be amazed by the community of interests which is possible between the recourse to fantasy in literary experience and the appeal to the *freie Phantasie* as a way of unfolding phenomenology (as an eidetic science) as Husserl understood it.[4] However, the intentions of Musil and Husserl diverge when it comes to the tone of their reflections. While for Musil the possibilities open to man in his existence seem more "subtle, illusory, phantasmagorical and subjunctive,"[5] for Husserl the passion for possibility was a strand of sober responsibility. However that may be, Husserl would willingly recognize as readily as Musil that the experience of the possible has «etwas sehr Göttliches in sich, ein Feuer, einen Flug, einen Bauwillen und bewussten Utopismus, der die Wirklichkeit nicht scheut, wohl aber als Aufgabe und Erfindung behandlet. Schliesslich ist die Erde gar nicht alt und war scheinbar noch nie so recht in gesegneten Umständen.»[6]

This incorporates a venerable tradition of thought for which a «hundert wirkliche Taler enthalten nicht das mindeste mehr, als hundert mögliche»[7] or in the same vein, if the intentional object exists, nothing changes from the phenomenological point of view.[8] Musil thought that all possibilities which imply, for example, a thousand marks, are included in them, whether they possess them or not.[9] This should not be understood, from this point of view, as a derogatory ignoring of that which exists, but as a subtle indifference before the weight of reality as to its facticity. On the other hand, the appeal to *Möglichkeitssinn* does not imply for Husserl a game of evasion before an always unsatisfactory reality, a game which would reduce the imagined possibilities to a consequence of true reality but, rather a certain intent to find a way to light up our *welterfahrendes Leben.* Literary and ontological experiences can have in common a *passion for possibility* which illuminates the real by means in which the real transcends its usual signification, linked to the meaning of what factually exists, but without losing ontological relevance as a motive, or a point of departure, valid for the illumination of the world.

However, having said that, we are well disposed to explore, however briefly, the intimate link which exists between literary and ontological experiences, as methods of illuminating, radically, and for that matter revolutionarily, the meaning of the world. Our aim is basically to encounter what Walter Biemel has called the *Realitätsträchtigkeit des Irrealen* as it is given through such experiences. For Biemel:

Die Kunst ist wesentlich zum Verständnis der Welt, nicht nur der Welt, wie sie ist, sondern auch der Welt, wie sie sein kann. Jeder Roman erhebt den Anspruch, Realität zu präsentieren. Nur insofern er imstande ist, diesen Anspruch zu erfüllen, ist er wirklich ein Kunstwerk. Realität besagt aber nicht das Universum der Dinge, sondern eine bestimmte Deutung des Seienden, eine Deutung, die wir den Namen Welt einsetzen können . . . . Wir können nie begreifen, was Kunst ist, wenn wir sie von einer bestimmten Realitätsauffassung her als das Fiktive, das Scheinhafte, das Irreale abstempeln.[10]

To sum up, and to be more specific, in this wide and problematic field we have to turn our attention to the importance of Husserlian phenomenology in understanding the transition from author to character as *experimental ego* (Kundera).[11] What it is important for us to emphasize in speaking of transition is the relevance of the notion of *possibility* in the forming of the literary and transcendental figure of the Other. At the same time there is the importance of possibility — as much in the phenomenological-transcendental perspective as in the literary attitude — for the creation and objective validation, as much "a priori" as "a posteriori," of diverse domains of experience through which it would be possible to intersubjectively illuminate the *complexity*[12] of the world and life. That which we are trying to illuminate here is an attempt to elucidate the transcendental effectiveness of the *Fremderfahrung* which can be used against psychologism; to do that we must account for the peculiarity of the literary experience in its freedom with regard to facticity. In its surprising profundity literary experience is able to teach us — from the phenomenological position and with a view to better clarifying the phenomenologico-eidetic project — that the game of literary or fictive *invention* is above all a work of *discovery* and that the passion for possibility becomes a kind of passion for knowledge (Kundera). I believe that in this relationship between invention and discovery the possibility of a transcendental and eidetic phenomenology as a base for ontology is elucidated to a certain extent.

* * *

«*I am Madame Bovary.*» Here we have a well-known declaration of Flaubert (the advocate of methodical impersonality), but a statement no less enigmatic for having his authority. That statement is frankly incomprehensible from the point of view of his immediate untruth if we adopt a psychological or natural perspective, since either it is not true or one has to doubt the gender and good judgement of Flaubert himself. But we will leave to one side such doubts. In order to understand that Emma Bovary «should be» I — and here "I" is not only Flaubert but also each one of us — it is necessary to find a *difference* between mere humanity and the *creative author*, in virtue of which the said Emma Bovary can be understood to be distinct from the *man* Flaubert, but not from the *author* of a work of fiction. If literary creation does not carry with it, thanks to this *difference*, the opening of new ontological space distinct from the space in which "reality" presents itself to us as facticity, creation would be of little ontological significance. Above all, such creation appeals to a *novum* in experience and requires a kind of peculiar and *creative disposition and efficiency*, by means of which the human peculiarity (contingent and factual) of a subject whose I is what it is — and could not be another nor free itself from it — is transcendent. If Flaubert "is" Emma Bovary, this statement does not follow the existential prejudices of the natural attitude. We could say that Emma Bovary is a fictitious character who has emerged from the *human peculiarity* of the author. But to be more precise, we must recognize that this emergence is not mainly psychological; if it were psychological, the implication of the character in the author would prevent the former from *detaching* itself from the individuality of the latter and attaining a peculiar style of real life and "objectivity." The problem is in the way that we should understand this "detachment." At present, and thanks in great measure to the incursions of phenomenology into the field of literary experience,[13] the interpretation of such experiences in the manner of Croce or Vossler — for whom literature resolved itself as a linguistic manifestation of the interiority *hic et nunc* of the author, and every linguistic particularity should correspond to a psychic particularity — has been overcome. In the same way «the image of the work . . . emerges . . . as a spiritual unity of the linguistic form and soul of the author, not as a *product* separate from him.»[14] However, we now know that, as Sábato says, «all the characters represent in some way the author. But all in some way betray him,»[15] and that even «the narration which includes dialogues make clear the inadequacy of the Vosslerian concept. The speech of the characters can in no way be considered as a direct linguistic expression of the author, but must more exactly be seen as that of the characters.»[16]

The decisive thing should be the comprehension of the *correlation* between author and character. As has been well recognized by Gullón it is not strange that in general, «the excesses of those who have read a novel as a manifestation of the author promote a reaction separating the text from the author, a reaction which after having fixed the autonomy of the object, sanctify it.»[17] Between psychological subjectivism and objectivism, or "sanctification" of the autonomy of the text, phenomenology offers the possibility of understanding the link between literary and ontological experience as it shows itself in the creative transition from author to character – which is not simply a transition from a mere human psyche to an irreducible exterior. On the other hand, although the characters have their origin in the author, they are not an immediate representation of him in the same way that the author is able to absorb them[18] or in the same way that each plot or story could be seen as an *autobiography.*[19] We would be mistaken if we took the literary creation and its recreation in reading to be a manifestation of ego-ism and not of us-ism, a point made by Unamuno.[20] Without doubt Emma Bovary is not Flaubert, but rather – and here the Husserlian theory of *Fremderfahrung* is what gives us understanding – there is for Flaubert the essential possibility of *being in another way* in the world. The character has relevance as an *experimental ego* of the author whose independent existence comes out in the "imaginary phrases" of the work, phrases which at the same time reflect in a certain way the author. As Raskolnikov is the essential possibility of Dostoievsky's being-in-another-way, so is Gregor Samsa (in spite of his metamorphosis) for Kafka, or San Manuel Bueno for Miguel de Unamuno – and not only for the author, but also for us, the readers, who become compromised in such imaginary possibilities. For us also the characters are our own "experimental egos," because they express *common shades of intersubjective subjectivity* (see Bonati, note 15).[21] As in the *Fremderfahrung,* the transition from author to character is not a form of destructive psychological transubstantiation of the otherness of the Other, but a transcendental intersubjective transference. The theory of "reception" can be seriously unfolded within this perspective. The character is linked to the author but has to separate itself from him in the literary text, and it is thanks to this detachment that it attains a *communicative or intersubjective efficiency,* being enabled then to open up experimental spaces of agreement and discrepancy, of similarity and *objectivity.* Fantasy makes possible the configuration of fields of unreality (necessarily, "finite fields of meaning," according to Schütz) which can be considered, although not without some shading, as hints of a deeper reality than that to which both the author and the reader are restricted in their immediate and perceptible experience of the world. In spite of the psychological translation (as much on the part of the author as of the reader) from such fields of essential unreality to private fields of the *perceived life-world*,[22] the exercise of fantasy allows a contact with this essential unreality, loaded with reality, to which the astute "sense of possibility" gives us access. Even if immediately motivated, the invention, as an apparent illusion of fantasy, discovers the virgin possibilities of the experiential horizons of the world and of life and collaborates with the passion for knowledge.

If Alonso Quijano is "more alive" than poor Cervantes,[23] it is because he concerns us more than the author. Although characters do not, of course, really exist factually, they are real, and that is why they have to be described with realism (Sábato).[24] The imaginary opens fields of *reality* with a reliability which can be considered "eidetic." It is worthwhile here to quote a reflection of Lotman's:

> The important property of artistic behavior is the fact that the person who practices it, displays two modes of behavior simultaneously; he experiences all the emotions which underlie an analagous practical situation, and at the same time he is clearly conscious that he must not carry out the actions which are related with this situation (for example offering help to the protagonist) . . .
>
> Let us examine a verse from Pushkin:
>
> "*I shall be reduced to tears before fiction.*"
>
> Here is a brilliant characterization of the dual nature of artistic behavior. On the one hand, being conscious of the fact that we are dealing with fiction ought to exclude tears. On the other hand, the fact that we succumb to tears makes us forget that we are dealing with a work of fiction. In fact both types of conduct, although opposites, exist simultaneously — and mutually deepen one another.[25]

The character is real, its "realism" could be even excessive, however it should not be forgotten that we can not expect it to actually exist or to be met "round a corner." In order that we should realize their "reality" it should not be necessary to perceive their imaginary intentional presence. That is why the autonomy of the text can be better understood as the "modification of neutrality" rather than as mere fantasy which is in opposition to the real.[26] As Bonati recognizes

> The author, in order to create, as well as the reader, to be able to understand properly, has to empathize (to introspect) with people and situations, "to feel," to live the life of

those imaginary beings. This is necessary for the full aesthetic contemplation of the reader. From this comes the *expression* of the imaginary discourse which is for the reader, a mental living of that which is said in the text as said by him. One deals here with *contemplative loans*, not with the identification of the author, or reader, with the imaginary story-teller. On the other hand, both the loan and the delivery are possible, as we have said, since one knows that the discourse is imaginary and without practical implications.[27]

I would like to deal with the notion of "contemplative loans." I think it is possible, from the point of view of Husserlian phenomenology, to interpret such "loans" and so understand the acts of both literary creation and reading. In fact we could turn our attention to the important difference between *fantasy* and the *modification of neutrality*, to which we have briefly referred already and which, due to problems of space and time, we are not able to consider here in much depth. Besides this we must deal with the peculiar efficacy of *eidetic variations* (eidetically real despite their *lack of existence* — which makes them seductive and tragic) as modes of access to fields of reality and subjectivity connected with the immediately perceived and experienced, used here as an initial model (*Lebenswelt*) but still independent of it. We intend to make this a little more comprehendible.

One of the most important bases in Husserlian thought with regard to the *Fremderfahrung*, and the important transcendental implication of the intersubjectivity in the egological subjectivity,[28] is in my judgement, and drawing on the writings of Husserl, «the power to be in another way.» This possibility shows the psychological and transcendental plasticity (as an opening into the wide field of possibility) of such a subjectivity and at the same time constitutes the basis for an encounter with the Other. If I were Other — if I were not who I am — signifies for Husserl, given the monadic condition of the reflective subject,[29] if I were-in-another-way:

Ich bin, der ich bin. Aber ich könnte doch anders sein. Es ist ein blosses Faktum, zufällig, dass ich bin der oder so, wie ich bin. Ich habe die Evidenz des Andersseinkönnens, und der Bereich des möglichen Andersseins liegt a priori "in" mir, ich "könnte" alle möglichen Abwandlungen meines Ich intuitiv durchlaufen. *Und dieses Universum der Möglichkeiten meines Andersseins "deckt" sich zugleich mit dem Universum der Möglichkeiten eines Ich überhaupt* . . . Jeder Möglichkeit eines von mir gesonderten, *anderen* Ich überhaupt entspricht eine Möglichkeit *meines* Andersseins . . . .[30]

Therefore, if I were Aladdin[31] my experience of the world would be very different from that, of which I, *here and now*, am capable of having. Equally if I were in Africa or in Ancient Greece[32] I would have access to experiences which at this moment are impossible for me because of my specific facticity. What about my possibility of being Other? I mean Other in the sense of Husserl's idea of being-in-another-way, without being simply reduplicated. The "multiplication" (*Vervielfältigung*) of my subjectivity possible in the "experience of the Other," thanks to the incarnation there of my possibilities of being-in-another-way, is realized in literary experience thanks to the miracle of writing and an exercise in fantasy which overwhelms the immediacy of the fantasy's motives, given by the presence-there of another body/behavior. Turning now to literary experience, we see that the *motivation* of self-fantasizing does not much limit fantasy; the latter is not necessarily compelled to reply to that which is perceived or to account for it. Fantasy *creates and invents.* However invention is at base a *discovery* — of myself in my undiscovered possibilities, of Others, or of the multiple possibilities of world and life experience. In this sense, we are enabled to approach Others real or imaginary through fantasy. Not without reason has Ricoeur indicated the connections between fantasy as it functions in the Husserlian theory of the *Fremderfahrung* and fiction and concluded that «literature could be a more appropriate source for investigating the psychic life of Others than daily relationships with real people.»[33] In conclusion the *passion for possibility is the passion for reality.*

An adequate comprehension of the creative efficacy of literary experience depends on the comprehension of how profoundly reality is able to be exposed "before" the natural-existential prejudice emerges. We can get this adequate comprehension from an *eidetic* fantasy exercise on the *conditionality and subjunctivity* of possibility. In this respect I think that a review of the important differentiation of *alteration* (*Veränderung*) and *variation* (*Variation*) which Husserl proposed to clarify the "contemplation of essences"[34] would be of certain importance in showing the efficacy of phenomenological-eidetic perspective as applied to the *interpretation* of literary creativity.

For Husserl the conditionality and *subjunctivity* of the possibility to be-in-another-way can emerge from the experience known as *Fremderfahrung* in which the transition to the subjectivity of the Other without doubt requires a certain peculiar relationship between perception and fantasy or perhaps can emerge from an *egological-eidetical variation.* In any case it is in this that the Other shows his reality beyond his factiticity (before existing factually). It is in this variation besides that we discover intersubjectivity as being relevant before effective encounter with the Other or with the Others — with all that this implies for the possibilities of self-reflection for monadological intersubjectivity, and, finally, in relation to our theme, for the *disposition*, and more than mere "disposition," of literary *invention and discovery* for communicative *efficacy* or *competence*. Certainly, as much in the *Fremderfahrung* as in eidetic-egological variation, a certain motivation is required (the human peculiarity of the author and the web of interpersonal encounters which have assisted the author as protagonist or as spectator in his life-world). But in variation, access to the Other deviates or departs more deeply from the perceived (I here and now and the Other before me). This is not the moment to deal with what the understanding of eidetic variation is able to contribute to the comprehension of the construction of the web of characters which make up a fictional story. We should consider only that if in the *Fremderfahrung* it is the perception of another body-behavior-there which allows detachment — and therefore the transcendance and otherness — of the Other, in the process of eidetic variation it is a *coexistence* between the facticity of the writer-reader and the possibility of the character which offers a reciprocal autonomy to both of them. Thus, we can say that although the characters have a psychological dependency on their author, their reality has a strictly eidetic dimension.

The Other as character, as an experimental ego, is my eidetic *invention*; it is me, much more than me, in my mode of intersubjective being, and for that reason, the imaginary eidetic-variations which have formed it, can essentially show myself, the Other, or reality from *horizons of possibility* — and in this literature supersedes philosophy — with *liberty* and with attention to the life and world circumstances by which the author dignifies and enriches such variations, predisposing them to alter our emotional dispositions. The character as experimental ego is "real" before being factual; it has a form of reality (an intentional autonomy with meaning) which is not hindered by its non-existence. We can appeal to a relevant exemple, although somewhat tangentially, an example offered by Kundera, in order to show the intrinsic validity of the literary experience as a vehicle for essentially displaying the *possibilities* of being in the world. Referring us to *The Sleepwalkers*, Kundera says that Broch was convinced of the correctness of ascribing the "decadence of values" to a determinate epoch and because of that of the (historical) reality of decadence as well. For Kundera, if this be a mistake, such a mistake should not alter the intrinsic and essential value of the work: the discovery of "a territory unknown to existence," that of the decadence of values as an *undeniable possibility* in the human world. Kundera feels confident to add «whether this possibility becomes reality or not, that is secondary.»[35] It is therefore enough to apply this example to our theme. In the ficticious meeting of me with myself, in my possibilities to be-in-another-way, in the invention of the Other as character, the existence or non-existence of the eidetic variations is not ontologically of any great importance. The variations have an *exemplary eidetic-sense* prior to, and independent of, their *representative* or *mimetic-existential* aspect. It is of no importance for ontological projection, absolutely linked with the sense of possibility, whether they exist or not. In the self-variation it is not only a question of fantasy as modification *of* perception or *of* that which is perceived, but of the product of a series of modifications of neutrality to which the author's possibilities or the experience possibilities are subjected. Although in certain cases the possibilities described could exist, as the fruit of eidetic variation, they do not necessarily have to exist. Ontological experience in its eidetico-phenomenological version and in literary experience has in common a subtle indifference to facticity, without doubt provocative, and a passion for possibility which is, at the same time, a passion for the true reality, passion for knowledge. In a certain sense Mimesis is what governs creative experience (just as the adequation between the intellect and the thing governs the search for "truth"). But that which has to be represented — and this is very important — suffers a deep transformation because it is subjected through fantasy to liberty and to the limits of its *possibilities of being*.

The author lives and expresses himself, he reflects, he invents, he creates, and he discovers; in short, he aspires to a *scientia* of reality, reflecting in a multitude of voices and actions (those of his characters) the essential modes of a being in the world. In saying this we are following Kundera, not only in his thesis regarding the "passion for knowledge" which in his judgement is the "only moral defense of the novel,"[36] but also in his thesis of what constitutes the *raison d'être* of the novel, the "pure" description of the possibilities of being in the world.

We must now conclude. We consider that we have now basically understood not only the profound significance of "reality" as experienced in literary creativity but also the possibility of translating the passion for possibility into a passion for reality. That is why to invent is to discover. (In this sense we could say that "Nature imitates Art.") And thus we have, at the same time, understood some contributions phenomenology has to make to literary theory and also to ontology (always linked to intersubjectivity and to dialectical praxis, but this is another matter).[37] We have besides tried to clarify one of the most brilliant intellectual achievements of phenomenology by going beyond the dryness of theoretical Husserlian concepts. We can go further: if we try to understand the relevance of eidetic praxis to literary creation, perhaps we can open, or at least foresee, a field of investigation. This is what I have tried to do. The theme of the construction of characters which has been central to my argument has really only been a pretext for pursuing this aim.

University of Sevilla

NOTES

[1] R. Musil, *Der Mann ohne Eigenschaften* I (Hamburg: Rowohlt, 1987), p. 16. An interesting work on the similarities between Musil and Husserl is F. Fellmann, *Fenomenología y expresionismo* (Barcelona/Caracas: Alfa, 1984), pp. 48—52.

[2] Musil, *op. cit.*, p. 16.

[3] E. Husserl, *Ideen zu einer reinen Phänomenologie und phänomenologische Philosophie* I, ed. K. Schuhmann (The Hague: Martinus Nijhoff), Husserliana III/1, 1976, p. 178 (§79). For Husserl «Ontology constructs the Logos of possible world in general.» (See *Erste Philosophie* II, ed. R. Boehm [The Hague: Martinus Nijhoff, Husserliana VIII, 1959], p. 213 (Abandlung 3)).

[4] See Husserl, *Ideen* I, *op. cit.*, pp. 145—148 (§70).

[5] Musil, *op. cit.*, p. 16.

[6] *Idem.*

[7] I. Kant, *Kritik der reinen Vernunft*, A 599—B 627.

[8] E. Husserl, *Logische Untersuchungen*, ed. U. Panzer (The Hague: Martinus Nijhoff), Husserliana XIX/1, 1984, p. 387 (V. Untersuchung, §11).

[9] Musil, *op. cit.*, p. 20.

[10] See W. Biemel, "Zur Realitätsträchtigkeit des Irrealen" in *Sozialität und Intersubjektivität. Phänomenologische Perspektiven der Sozialwissenschaften im Umkreis von Aron Gurwitsch und Alfred Schütz*, ed. R. Grathoff and B. Waldenfels (Munich: Wilhelm Fink (Uebergänge, 1), 1983), pp. 269—270. See B. Waldenfels, "Fiktion und Realität" in *In den Netzen der Lebenswelt* (Frankfurt a. M.: Suhrkamp, 1985), p. 231.

[11] See M. Kundera, *El arte de la novela* (Barcelona: Tusquets, 1987), p. 42. Kundera's thesis with regard to the novel has a certain "phenomenological" shading (see p. 43).

[12] For an explanation of the idea of "complexity," see N. Luhmann, "Sinn als Grund-

begriff der Soziologie" in J. Habermas and N. Luhmann, *Theorie der Gesellschaft oder Sozialtechnologie* (Frankfurt a.M.: Suhrkamp, 1971), pp. 32 ff.

[13] With regard to this see F. M. Bonati, *La estructura de la obra literaria* (Barcelona: Ariel, third edition, 1983).

[14] *Ibid.*, p. 142.

[15] See E. Sabato, *El escritor y sus fantasmas* (Barcelona: Seix Barral, 1979), (first edition 1963), p. 130: «The relationship between the author and his characters. All the characters of a novel in some way represent their creator. But all in some way betray him. I am Madame Bovary, what doubt can there be? but I am also Rodolphe in my inability to put up for long with Emma's romantic temperament. I am also M. Homais . . . . Insofar as the characters of a novel emanate from the spirit of their creator they also become independent beings and the writer observes their attitudes, feelings, and ideas with wonder. Attitudes, feelings, and ideas which soon turn out to be the opposite of those which the author has, or normally feels.» According to Bonati: «In the psychological conception of Vossler, the psyche of the author emerges as the substance of the poem. One can ignore the fact that the 'expressed interiority' of the real speaker contains something objective, perceivable for the hearer, and not an innaccessible and reclusive being to whom the signs merely point. That means that the 'expressed interiority' of the real speaker is not the psyche of the speaker, but that it represents it. What is expressed is the common shade of intersubjective subjectivity.» (Bonati, *op. cit.*, p. 163).

[16] The opposite would lead to the intention to derive or deduce the author from his characters. See *ibid.*, p. 149.

[17] See R. Gullón, *Psicologías del autor y lógicas del personaje* (Madrid: Taurus, 1979), Persiles, 109, p. 42. As Gullón himself admits this work came into being following his attendance of a conference on Galdós organized by the Modern Language Association in New York in which the participants seemed to forget that «what they were discussing was a literary text» and got caught up «in a curious debate on characters as living individuals, of fictitious figures as real people and invention as reality.» (*Ibid.*, p. 9).

[18] *Ibid.*, p. 42.

[19] M. de Unamuno, *Cómo se hace una novela*, in *San Manuel Bueno mártir/Cómo se hace una novela* (Madrid: Alianza Editorial, thirteenth edition, 1986), pp. 127, 128—129.

[20] See *Ibid.*, p. 132.

[21] We should turn our attention to M. Kundera — *La insoportable levedad del ser* (Barcelona: Tusquets, 1985), pp. 226—227. After asking himself, somewhat rhetorically: «Perhaps it is not certain that the author is unable to speak of more than himself?», he tells us that «the characters in my novel are my own possibilities which have never been realized. For this I love them all equally and all of them produce the same panic in me. Each one of them has crossed a frontier whose perimeters I could no more than pass by. It is this frontier (on the other side of which I cease to be me) which attracts me. It is on the other side of this that the secret which the novel deals with starts. A novel is not a confession of the author but an investigation of the human situation in the world.»

[22] Described fictitious situations, events, lifestyles, characters come out of the author and almost of necessity "immigrate" to the reader in such a way that the reader cannot

forget himself in the act of reading. Sabato has referred to «this Raskolnikov who we used to read about in the works of Maucci, who used to appear to wander — or perhaps better, to roam — about the Puerta del Sol, surrounded by wastrels and boneheads murmuring such incredible words as snacks and a quick bite» (E. Sabato, *Heterodoxia*, in *Hombres y engranajes. Heterodoxia* [Madrid: Alianza Editoria, Second edition 1980, Bolsillo, 445, p. 135). See Gullón, *op. cit.*, p. 41. He cites George Poulet: «To read is the act in which the subjective principle which I call 'I' is modified in such a way that I have no right, strictly speaking to consider it as myself.» According to Gullón such an affirmation is excessive and unaceptable, «man . . . has an irreducible base, an 'authenticity' . . . which remains. It projects itself into the reading and becomes part of it.» On the internal psychology of the reader (N. Holland and D. Bleich), see R. Selden, *La teoría literaria contemporánea* (Barcelona: Ariel, 1987), Letras e Ideas, pp. 146—151.

[23] See M. de Unamuno, *op. cit.*, p. 93.

[24] As was done by «Dante with those condemned to the inferno, Shakespeare with his historical figures, and Kafka with his nightmares.» (Sábato, *op. cit.*, p. 184).

[25] Y. M. Lotman, *Estructura del texto artístico* (Madrid: Istmo, 1982), Fundamentos, 58, p. 89. See Bonati, *op. cit.*, p. 205: «If the truth of opinions should have, as Ingarden believes . . . a reserve or absence of 'seriousness' in the way in which these phrases are realized by the reader then they should not have either transference or alienation or, for that matter, the plenitude of the literary object. There are certainly no such reservations nor lack of 'reliable' affirmation of the opinions which form part of the work because the opinions are themselves imaginary and this makes their presentation as 'truth' unrestricted. The lack of 'reliability' occurs on another level, in an 'awareness of the game' which envelops the mechanism of the production of images. Such are poetic opinions and they are presented in all seriousness, but, they are imaginary and the reader knows this intuitively."

[26] With regard to the difference between phantasy and the modification of neutrality see, for example, Husserl, *Ideen* I, *op. cit.*, 109—111. On this problematic subject see A-T. Tymieniecka, "Imaginatio Creatix. The 'Creative' versus the 'Constitutive' Function of Man and the 'Possible Worlds,'" *Analecta Husserliana*, Vol. III (1974), pp. 18—23.

[27] Bonati, *op. cit.*, p. 162.

[28] See E. Husserl, *Zur Phänomenologie der Intersubjektivität* III. (henceforth, *ZPI*) ed. Iso Kern (The Hague: Martinus Nijhoff, 1973), Husserliana XV, Text 22 (1931): «Teleologie. Die Implikation des Eidos transzendentale Intersubjektivität im Eidos transzendentales Ich» (pp. 378—386).

[29] See Husserl, *ZPI* II, *op. cit.*, pp. 158—159 (Annex XXI, 1922).

[30] *Ibid.*, pp. 154—155 (Annex XX, 1922). See also pp. 138—139 (Text 7, 1921).

[31] Husserl, *ZPI* III, *op. cit.*, p. 417 (Text 24, 1931).

[32] *Ibid.*, p. 359 (Annex XXI, 1932).

[33] P. Ricoeur, "Hegel et Husserl sur l'intersubjectivité," in *Phénoménologies hégélienne et husserlienne. Les classes sociales selon Marx* (Paris: CNRS, 1981), p. 12. With regard to the ability of the "aesthetic truth" to enlighten various "a priori" aspects of anthropological understanding, see M. Dufrenne, *Phénoménologie de l'expérience esthétique* (Paris: PUF, 1967), p. 658 ("Justification anthropologique de la vérité esthétique").

[34] See E. Husserl, *Erfahrung und Urteil*, ed. and compiled by L. Landgrebe (Hamburg: Felix Meiner, Sixth edition 1985), §86, e, f.
[35] Kundera, *El arte de la novela, op. cit.*, p. 55.
[36] See *Ibid.*, p. 16.
[37] See César Moreno Marquez, *La intención comunicativa. Ontología e intersubjetividad en la fenomenología de E. Husserl* (Seville: Thémata, 1989) Suplementos, serie mayor 1.

JERZY SWIECIMSKI

TRUTH AND UNTRUTH IN THE MUSEUM EXHIBITION

Museological dissertations often try to define the function of museum exhibitions. Museum exhibitions are conceived of there as 1) those which *present* objects of artistic or scientific documentary value, or 2) those which *transmit* knowledge (mostly fundamental, elementary and, in every case "ready" knowledge) to various categories of visitors (mostly unprofessional ones).

According to the first interpretation, exhibitions are conceived as a realm for scientific research (in this regard no essential distinction is made between the exhibition and museum laboratories or museum storage)[1] or, as a place where works of art through their being given to the recipient's perception, feeling, contemplation, and emotion, become — as it is sometimes formulated[2] — "objects of aesthetic experience," intentionally provided (i.e., in the visitor's consciousness) with specific, aesthetically valid qualities.[3]

According to the second interpretation, a museum exhibition is conceived as an instrument of informative-didactic value and is functionally parallel and structurally similar to a school or university manual. The role of the aesthetic factor is, in this interpretation, also recognized, although it is thought of as a factor subordinate to the exhibition's didactic purpose: the aesthetic form of the exhibition becomes thus "the form of the exhibition's content" and a means for its communication effectiveness.[4]

In museological papers we also find the thesis (formulated mostly as a normative statement, a postulate, or a program) that exhibitions, in order to fulfill any of their fundamental functions should become (in this "should" the normative character of the thesis is expressed) a presentation or a representation (transmission, message) of truth. Presenting or transmitting untruth becomes, thus, essentially contrary to the museum's function or, even more, to the museum's fundamental sense of being.

Are museums factually presentations (or transmissions) of truth?

Is truth (an absolute, indisputable truth) the single quality by which museum exhibitions are characterized?

A-T. Tymieniecka (ed.), Analecta Husserliana, Vol. XXXVII, 343—365.

Is truth that kind of message which — under certain conditions — it is generally possible to achieve in a "pure" form or, is it probably the case that, besides the truth, the content of every exhibition transmits some untruth which, because of its (unavoidable?) emergence is to be "neutralized" (i.e., intentionally postponed, transformulated, etc.) in every case of perception and understanding of the exhibition's content and meaning?

And, finally, may untruth, being probably one of the constituent ("essential") factors of the exhibition's content, become — under certain conditions — a means of transmitting truth? Expressing this a bit paradoxically, may it become a "form of truth"?

These questions are the subject matter of the present paper.

* * *

THE CATEGORIES OF TRUTH IN MUSEUM EXHIBITIONS

It should be noticed that although truth in the general sense of the notion is to be understood explicitly and is, as a rule, to be connected with assertions of appropriate, positive value, it may if considered in relation to museum exhibitions' content (to the message introduced into the exhibition by a scientific script, or to the meaning of the exhibition's form), be interpreted in various ways and from different aspects. Truth, as a factor or quality of the exhibition's content can also be concretized on various levels of this content.[5]

The following categories of the exhibition's truth should be distinguished:

Group A

1) Truth connected with the entity, or with the state of being of the exhibits (of single exhibits or of groups of exhibits)

2) Truth connected with a priori norms or definitions formulated as regards the exhibits and their features or as regards objects there designated (whether real or conceptual ones). The latter case occurs when exhibits are functioning as objects presented not "for themselves" but in order to symbolize or illustrate objects distinct from themselves; they then become representatives of some class, type, etc.

3) Truth connected with the content and value of assertions made about particular objects or their types (exhibits, their designata).

4) Truth connected with the artistic concept of the exhibition design, in particular with the expression of the exhibition as a whole, with its meaning. This truth is not connected with any assertions.

5) Truth connected with the appearances of the exhibits and with their aesthetic qualities.

Apart from that, one can distinguish two distinct situations:

Group B

1) When the truth of the exhibition (or of the exhibits taken as particular objects) is independent from any acts of consciousness (e.g., from our evaluation of the exhibit's content, meaning, etc. as regards its truthfulness).

2) When the truth of the exhibition (or of particular exhibits) is evidently a product of acts of cognizance and their expression.

The number of particular cases is, of course much larger than the number of the points systematizing them (A: 1—5; B: 1—2). And within some points, several subgroups and variants can be distinguished. Also the particular cases we usually meet in practice often result from a combination of these categories. Combinations of cases from Group A and cases from Group B may, in particular, occur. This makes the detailed image of particular cases rather complex. We shall limit ourselves, however, to the main end most typical ("pure") cases.

* * *

I

When truth is interpreted as a factor connected with the exhibit's entity or with its state of being, it is usually regarded as a quality essential to the exhibit's individual identity (to its "being itself"); it might even be identified with it.

This category of truth is typically of an ontological character: it immanently "belongs" to the given exhibit as it is independent of any acts of our cognizance in the sense that it remains definite, whether it is known to us or not, whether we formulate any statements about the given exhibit (as regards its identity) or not. To paraphrase: if the objects had a consciousness of their own, they "would know" what they are, no matter what we say about that, if anything.

Truth of this kind, being a factor which determines the object's individual identity, and "belonging" to a given object intrinsically may, however, be given to our cognizance, e.g., may be detected by us. This occurs when the given object becomes an object of our analysis, e.g., is taken up for its definition. However, irrespective of what result such analysis would give, the exhibit's "own" truth remains unchanged: the object's truth in this case is not a "product" of our cognizance. Therefore, in any exhibit one can distinguish its "own" truth and, on the other hand, information (with its content and value) which is ascribed to the given object. Such a definition may confirm (detect) the object's "own" truth, but it may also be erronous, hedged, or non-existent: e.g., it does happen sometimes that objects received by a museum remain undefined despite of all efforts to detect their identity.

Untruth – if it is considered analogously to the just referred to sense of the notion of truth (i.e., as a factor intrinsically "belonging" to the given object and defining it as an "untrue" one) – is essentially never possible. Its existence is never possible because, as we have mentioned above, any object, i.e., irrespective of what kind it would be, does "possess" an individual truth "of its own."[6] Untruth may, therefore, be in this case connected only with the content of the information added to the object. The circumstance that such "untrue" information is usually ascribed to the object itself is not only of no importance for the given object, but it also results in the just referred to category of the object's truth being given different interpretations. (We shall deal with further on.)[7]

An individual object's truth appears in objects whether they are in their origin the products of somebody's act of creation or not. In particular, it appears in all kinds of "ordinary things" in our everyday surroundings, in the objects that are elements of nature (e.g., geological, zoological specimens, etc.) and in the works of art. In each of these cases truth is founded on a different set of the object's features and qualities.

In particular, the object's individual truth may be founded a) in the object's morphology or b) in the object's origin.

In Case A the specific, individual (singular, unrepeatable) set of the object's features (e.g., those of its shape, of its physical substance, etc – even those which define the type of damage the object has sustained: as often occurs with museum exhibits, for instance in paleontology, archeology, etc.) is what determines its identity.

In Case B the object's morphology, although co-determining its individual identity, becomes a factor of secondary importance. For instance, a painting received by a museum is a "true Rembrandt" because, irrespective of all the features by which it may be recognized as an authentic Rembrandt, it is Rembrandt who was its author, irrespective of whether we know it or not, and, whether analysis confirms the object's identity or not. Another painting will "possess" its identity as a Rembrandt copy, because not Rembrandt but a copyist was its author. Even a forgery of a well-known masterpiece has "its own" truth indentity (exactly, that of "being a forgery"). The content or meaning of truth is, in all of these cases, singular, unrepeatable, and, therefore, individual.

II

Very close to the above is the interpretation, according to which an object's individual truth is understood as a value conditioned by a set of conventionally selected features of the object (those of a special category) or by a special kind of origin. In this way only some objects (those which fulfill the necessary conditions) are to be qualified as the "true" ones, while all the others remain in the category of "untrue" objects.

In a way similar to the previous case, the object's truth here (or, its untruth) is of an ontological character, i.e., it presents a factor which "belongs" to the object intrinsically (is immanent to it) and which may, at least be given to our cognizance, e.g., when an exhibit becomes an object for our analysis and determination. The analysis may, like in the previous case, "detect" the object's "own" truth. If it is wrong, it may negate it, or leave the object unevaluated. On the other hand, the object's truth (or untruth) is dependent on acts of cognizance insofar as it is constituted on the grounds of some a priori definition, norm, or standard, in many cases conventional or even arbitrary. A possible change in those assumptions does not influence, of course, the individual object's identity (that which it has in the sense covered in Section I), but it leads to the change in the object's value.

This case presents two main variants:

Variant A embraces cases where the object's truth is identified with its authenticity.

An a priori assumption determines in these cases all that should be understood as the object's authenticity, i.e., that the necessary condition the object must fulfill is the special category of its origin. The determination of this category in particular cases, is, as a rule, conventional.[8] For instance: if for objects of some special type, e.g., masterpieces taken to be the output of certain artists, the assumed a priori norm for their authenticity (and hence, for the truth of their factually being the works of the given artists) were: "to be completed entirely by the author himself" (i.e., without the help of the author's disciples), many works which have been only designed by their authors, or only finished by them — in which the main work was that of disciples — would be qualified as non-authentic. With a change in the qualifying standards, e.g., the assumption that for a work's authenticity it is absolutely enough that the work be designed by a certain artist, completed under his supervision, and finished and signed by him (irrespective of how much of the work was done by his disciples, by some craftsmen, etc.), then the number of works to be recognized as the artist's authentic output would be much greater.[9]

The status of being and that of the work's value (although the latter is based on conventional grounds) are intrinsically connected with the work and can, at least, be given to our cognizance, on their "detection."

Analysis may in fact "detect" which part of the given work is that of the author himself and which is that of his disciples. If the research is correct, the work's authenticity (or non-authenticity, partial authenticity) will be confirmed; if it is wrong, it will be negated or left undetermined.

Summing up: although in each of such cases the object's value is established on conventional grounds, it should always be distinguished from the assertions which are formulated about it and are, already, the result of some process of cognizance.

One can give yet another example: an original object received by a museum is modified in some of its features,[10] but due to its primary origin (for, after the introduction of changes, the object's origin becomes factually double) and to the preservation of the features regarded (conventionally!) as essential for its authenticity, it is — despite the secondary changes — still regarded as authentic and a "bearer" of truth.

For instance: if a work of art, as the result of the conservation process, becomes modified in some of its features (even to the extent that the whole of its individuality is changed), e.g., when its authentic, dark varnish has been replaced by a new one, when old canvas has been doubled, etc. then the work of art is still qualified as "authentic" (sometimes, paradoxically, even "more authentic" than it was before). Its authenticity is recognized, because the features of the object which are considered essential to it (e.g., the composition of forms, the configuration or setting of the objects represented in it, etc.) remained unchanged. The list of features is, of course, conventional: for instance, it does not matter that in such cases the objects after the conservation process "look" different, that they become physically different things.[11] In restored objects, e.g., works of art, this sense of authenticity is relatively less: one already distinguishes there the "authentic" and the "restored" (i.e., non authentic) parts. The final shape of such an object becomes therefore a combination of its primary and of secondary elements, the latter being a "concretization" of some concept. The object's structure may, in such cases, become extremely complex. We shall come back to this problem later (see Section III).

Variant B occurs when the object's truth is regarded as a value independent of the object's authenticity, or when the question of truth is limited exclusively to that of selected features, irrespective of the kind of physical *substratum* (substance, thing) which is their "bearer."

In such cases the whole physical substance of the given object can be replaced with an non-authentic one. The single condition which is necessary to preserve the truth is the precise reconstruction of the original's external features; sometimes the "list" of the features to be reproduced is determined by an appropriate convention. The authenticity of the material object is in such cases distinguished from the truth it contains and, being a museum exhibit, it transmits a scientific message to the audience.

For instance, the Hammurabi stele displayed in the Museum of the Near East (*Vorderasiatisches Museum*) in Berlin is a patinated plaster cast of the original. Although it is only a reproduction, it functions as a valuable exhibit, for it contains and transmits the same amount of scientifically essential data as the authentic object; its shape, size, engraved inscriptions, the bas-relief, the formulation of the content of Babylonian law (clear to anybody who can read Babylonian), even accidental damage of the original. The circumstance that its physical substance was made in plaster, and not in stone, by a contemporary museum technican, and not by a Mesopotamian sculptor is of no importance. The satisfactory concordance of some features with the original is the single factor which is essential for the exhibit's truth. Untruth would appear in such a case only if somebody tried to erase some parts of the inscription, to replace one text with another, etc., i.e., to essentially change the content of the exhibit or its artistic form.

In the case of the plaster cast of the Pleistocene Wooly Rhinoceros (*Coelodonta antiquitatis*) displayed in several museums, the sample of features essential for its truthfulness is remarkably smaller. The casts exposed in the Museum of the Earth (Muzeum Ziemi) in Warsaw and in the Natural History Museum (Haus der Natur) in Salzburg are painted black (although the original carcass when excavated was not black at all). The casts displayed in the Museum of Natural History in Cracow and in the Bieszczady Museum of Natural History (South-eastern Poland) are left in the natural color of plaster. The cast exposed in the British Museum (Natural History) in London is patinated and has the appearance of a bronze sculpture. All these exhibits show, that not only physical substance but also color are irrelevant to the truthfulness of the transmitted scientific message. All these reproductions despite their differences in color function as equally valuable exhibits, for they preserve the authentic shape and morphology of the original, which, n.b., no longer exists in its primary, authentic form.[12]

Many museums have exhibitions presenting reproductions. The sculpture galleries are especially characteristic in this regard. Features of the originals are replaced (e.g., the specific exterior of the marble by the features of plastic), but the differences are not regarded as something essential, although on the aesthetic level, the difference between a genuine marble sculpture with its typical lustre and irreplaceable texture of stone, and a cast, always dull in color, is obviously noticeable. The conventional selection of the features to be regarded as "important," and the depreciation of all other features (although significant for any work of art and its aesthetic qualities) is typical in these cases.

III

The features of an exhibit may be of heterogenous origin. This happens, where some of them are being introduced into the exhibit as a novelty.

The inspiration of such an operation may come from a scientific concept, formulated primarily as a statement or as a hypothesis, and presented visually by the media of the exhibit. The cognitive value of such a concept (its truth, untruth, or probability) consequently determines the value of the objects which take part in its "materialization" in a museum exhibition. In many cases the concept is of an evidently creative value. According to the content of the concept, the exhibits which are intended to help us visualize it are modified in their primary features (they undergo a certain technical process), sometimes to a large extent.

As in all the previous cases, the object's altered content "belongs" to it intrinsically and can, at least, be offered for our cognizance and evaluation. Analysis of the object may therefore "detect" its heterogenous structure, distinguish the elements which belonged to it primarily and those which have been secondarily introduced and may determine which ones convey truth, and which are hypothetical or evidently imaginary or untrue.[13] As regards singular objects, archeological reconstructions are the most typical in this regard: for example, the restored Palace of Knossos with all its restored parts (the truth of which was questioned by some authors), the Ishtar gate presented in the Museum of the Near East in Berlin, the Pergamon Altar in the Pergamon Museum in Berlin, etc.

Sometimes the changes introduced into the restorations, though inspired by a scientific concept, have, to some extent, their source in technical factors. In such cases the restorations have a more or less conventional character and their evident deformations, if they are to be properly decoded, have to be recognized not as errors resulting from a false assertion about the given object but as the result of limited possibilities for their presentation in the exhibition environment. The abovementioned Ishtar Gate, for instance, contains some untruth as regards the size and mutual proportions of its elements. These deformations resulted from the fact that the exhibit was installed in an interior room too small to allow a reproduction of the object in the original size and all to scale. Explanatory text added to the exhibit brings attention to these deformations and helps the visitor to understand the restoration properly. Technical limiting conditions become, anyway, an additional complicating factor in the heterogenous process of restoration.

The role of scientific concepts becomes especially distinct when the exhibits are intended to represent or illustrate not individual things, but a general (conceptual) object, a class of objects, a notion, etc. In such cases the objects which are represented, symbolized, or illustrated by the exhibits (sometimes by singular ones, sometimes by groups of exhibits) are the creations of scientific ideas and are determined, as such, in their ontological type, essence, degree of generalization, and defining features. The truth of the exhibits depends in such cases mostly on the satisfactory (conventionally accepted) concordance of their features with the features of their conceptual designata. Only some selected features that determine the exhibit as an individual thing are taken into consideration; all the rest, in particular, the features determining the singularity of the given exhibit, are dismissed and play no role in the presentation. The features which are selected in this regard, although factually individual, are (in a way corresponding to the definition of the designata) reformulated intentionally, i.e., they are intended to be understood as general features. Necessary concordance is already considered, therefore, on the level of this intentional generalization of the exhibit's features and of the general features of the exhibit's (conceptual) designata.

There are several variants within this type of case.

A. In some instances the role of the exhibit is limited to a representation of some general class of objects and the representation is not connected with any material changes in the exhibit.

In such cases the meaning of the exhibit is imposed on it by its representative function; the exhibit's structure and features are to be understood in the categories of the general object represented (concretized) by it.

a) The simplest case occurs when the object's meaning is limited to its denomination, e.g., when a natural-history specimen is exhibited as a representative of the species under which it is listed and consequently is identified *sub nomine* of its species; or when a work of art is presented as an example of a class to which it belongs (or, is supposed to belong), e.g., as a "Renaissance" painting, as "Mycenaean" pottery, etc.

In such cases the individual features of the presented exhibit, i.e., those which identify it as a singular thing are of lesser or no importance at all; they are intended to be ignored in the process of of understanding the exhibit's content (just as they are ignored in making the choice of specimens to put on display) or to be reformulated in such a way that they can be grasped as representations (concretizations) of general features (which of their essence, are never given directly in the perception of a singular, material thing, being always constructions of our reasoning). On this ground the choice of an exhibit within its class is free: no more a singular, definite object, but any object from the given class is in such cases needed: the single condition for its being chosen is that it factually belongs to its class and can represent it satisfactorily.[14]

When, for example, in a geological exhibition a piece of granite is presented as "granite," any piece of granite rock may be used for display irrespective of individual features (its size, shape etc.).[15] When in a zoological exhibition we illustrate "Crustaceans," then any crustacean, belonging to any family and to any species would be right for the purpose, for each of them will be the "bearer" of the same necessary amount of information.[16]

Similarly, in an archaeological or historical exhibition, any Roman sculpture will be a good illustration of Roman art, any example of Minoan pottery will serve to give the visitor an idea of what Minoan pottery looks like. Sometimes, when the type which is intended to be visualized comprises examples of a very diversified character, a selected group of exhibits (each of them representing a typical sub-group) may be chosen.

The naming of the exhibit (or of a group of exhibits) may be correct or erroneous on various levels: 1) the concept of the general object which is intended to be shown may be conceived well or erronously. 2) the classification of the given exhibit may be correct or wrong. The role of convention is, in such cases, double: for not only should the general object be conceived before its representation, but also a norm should be set as to which features of the exhibit are to be the representatives of the general features in the exhibit's conceptual designata. On each of these level the convention may contain truth, hypothesis, or untruth.

b) More complex cases occur when the general object which is represented in the exhibition does not have the ontological character of a thing, but belongs to the class of processes, events (in particular, when it is a general type of a process, event). In such a case the features of the exhibit (being necessarily the features of a thing, anyway) have to be intentionally reformulated into the features of an object ontologically different and approximately generalized.

For instance, when a geological specimen is used as an illustration of, say, the sedimentation process, nothing of this object ("sedimentation") is factually present in the exhibition. On the contrary, the exhibition presents exclusively an individual material thing and its features. Individual features of the factually presented exhibit have in such a case to be intentionally reformulated so that 1) they can achieve the character of general features and 2) they can be understood as the features of the process in question, i.e., that of sedimentation. The convention embraces therefore not only the naming of the represented general object (which can be conceived in many different ways, correctly or wrongly), but additionally the way in which the exhibit's features are expected (or, intended) to be understood as something completely different from what is seen.

c) The most complex cases embrace situations where the concept which is illustrated by the exhibits becomes an interpretation of a scientific idea concerning the origin, function, mutual relations, evolutions, etc. of particular objects, processes, or events. Such an interpretation may be in its origin scientific or non-scientific (though it is always presented as a scientific one or, at least under the aegis of science). Most of these cases involve a hypothesis (e.g., "how the Carpathians emerged"). Exhibits which are put on display "serve" as documents or proofs of scientific concepts: their content "in itself" (and, consequently, the truth "in itself") may in some cases not be in accord with the value of the idea they are intended to illustrate, e.g., "true" (original, well-determined) objects may become illustrations of a hypothesis which is completely wrong.

Untruth may in such cases appear on various levels and may, consequently, have different extent. In this class occur — fortunately rarely, however — conscious falsifications of scientific truth. These occur in museums which, instead of being institutions for scientific information, are converted into instruments of "ideological" indoctrination. the exhibit's identity there becomes intentionally falsified, sometimes to such an extent that the exhibits completely lose, in the understanding of them, their authentic truth. This happens even when "true" objects are used for such falsification, because the concept to which they are subordinated is a scientific lie. Where exhibitions of this type are arranged in a suggestive way and where because of this the untruth they present is hardly detectable (especially by a non-professional visitor), the informative-cultural and educational role of such exhibitions is especially destructive.

B. The representation of general objects (of all ontological types, but first of all of those which have the character of things and states of being) may involve changes being introduced into either the material substance or the objective structure of the exhibit. The role of a specific convention is in such cases much broader than in the ones previously considered, for not only must the norms regulating the selection of features which are to be regarded as defining ones be assumed a priori, but also the object which is to be represented must be explicitly defined ("conceived", scientifically "invented"): the representations become therefore visualizations of general scientific ideas of conceptual objects.

The most typical instance in this regard are paleontological restorations, especially restorations not of individual fossil organisms but of the species to which these organisms belong (or are supposed to belong). The general idea of a definite species becomes, therefore the main object of the representations. Documentary material things (fossil bones, etc.) are only used as material ("stuff") in such representations.

These restorations characteristically preserve some authentic, documentary elements in their primary state and of completing them with elements which, although provided with individual features (for all of them are individual, material things), are intended to be understood as generalizations of the elements they represent. The restorations possess in most cases an element of scientific creativity which is based on the data of research and on reasoning, but is sometimes inspired by free imagination.

Paleontological restorations are sometimes free of any novel elements: everything of which they are composed is authentic. In such cases, it is only the configuration of the documentary parts, that is the novelty (the way in which fossil bones are mounted into skeletons — correctly or wrongly, according to ideas of the restorer).

The truth and untruth of paleontological restorations is dependent on the results of research, on the probability (or non-probability) of the proposed hypothesis (all restorations are to some extent hypothetical!), as well as on the manner of their concretization — in their structure and in their features. The relation between documentary material and material of conceptual origin is especially significant in such restorations

This truth (or untruth) may be "de-coded" in our acts of cognizance, e.g., when a given restoration becomes an object of our analysis and revision. The information which is always added on in an exhibit may, therefore, confirm the exhibit's "own" content or may belie it, as is the case with restorations executed in the past and which appear to be false.

C. Representations of general objects (or, of their verbal definitions) may take the form of creations which, regarded as things, are materially and structurally different from the documentary material (of things also), on which they are based in their content.[17] These creations are executed mainly as works of art or objects formally similar to art: as models, sculptures, reliefs, paintings, or graphic works. Their content and, consequently, their truth (or untruth) depends on a satisfactory (and, as a rule, conventionally assumed) concordance between a selected set of their features and the general features of their designata. There is one rule in this regard: the more general is the represented object, the smaller is the number of features which are necessarily concordant in its representation.

For example, the "Mammal in general" (or the defintion of the group: "Mammalia") would need less of such features than a model which is intended to represent a certain species belonging to the Mammal family. The model of a whale exhibited in the British Museum is "naturalistic" to such an extent that it can illustrate a definite whale species. Features which would identify it with any particular specimen belonging to this species are already unnecessary in it, and even if they were included (e.g., because of the artist's accuracy in representing animals), they would have no importance for the correct understanding of the exhibit's sense. Untruth in such cases would appear only if the representation were provided with creatures which were not concordant with the definition of its conceptual designata (here, with a definition of a certain species), e.g., if the whale model were painted in red, or if it had three tails instead of one. Sculptures representing broader taxonomic units (with, therefore, broader specifications in their definitions) may be provided with individual features more freely.

The model of an Orthopteran from the Senckenbergian Museum of Natural History in Frankfurt am Main is made of brass. Although its color is that of metal and none of the Orthopterans are of that color, it is, as regards its content and truth, absolutely correct, because the morphological feature of color does not exist in the definition of Orthopters and, consequently, color can be ignored in its representation (interpreted *ad libitum*, just as the artist decides). In the same museum, graphic representations of Dinosaurs represent concrete species (Brachiosaurus, Iguanodon, etc). But although they are given an "abstract" color, a red, orange, blue, their content is correct too. Since nothing is known about the color of dinosaurs, our lack of knowledge can be expressed by a choice of color which evidently means nothing, but a graphic convention: almost "provocative" abstract color indicates that color here is void of any meaning.[18]

Misinformation (and therefore, untruth) may, therefore appear in works of art which are "too detailed" in their features, e.g., when the degree of their realism goes beyond the message they should transmit to the visitor. This is very often the case when a museum designer sculptor, or painter "fills in" the missing elements in the object he intends to represent with naturalistic, but completely imaginary features. The naturalistic convention of such representations, therefore, makes them very suggestive, but their "veracity" obviously misinforms. The restorations executed by Zdenek Burian or by Jay Matterness (Smithsonian Institution) are most typical in this regard.

Marginal cases among creations which represent general objects are those where the general object (the class of objects) is represented by a a single specimen, e.g., as was the case with the famous "living fossil," *Latimeria chalumnae* at the time when only one single specimen of this species was known. Representations (very naturalistic ones) were made and displayed in many museums around the world. Formally, they resembled a replica (partly a restoration) of single specimen that was given as a "model." In such a case representations with a high degree of veracity (high naturalistic) are fully acceptable. As regards their origin and function, they come very close to being representations of unique objects — just as happens, as a rule, in the case of reproductions of sculptural works of art.

The reason for the heterogenous structure of an exhibit's features may also lie in the didactic concept of the exhibition.

In its relation to the scientific concept, the didactic concept is always secondary; it presents a means for the interpretation of the primary exhibition's scientific idea. Didactic reasons influence, for instance, the formulation of the assertions to be visualized in the exhibition; sometimes they introduce a special structure into the exhibition regarded as a whole, e.g., the sentence-like organization of its elements, each of them becoming the bearer of a special meaning to be "de-coded" according to a special "key." The semantic structure of the exhibition resembles, in such cases, that of a handbook. Perception of it has to be realized by going in a determined direction: the exhibition has its beginning and its end, it is divided into chapters, etc. Although such exhibitions are intended to be conveyers of scientific information (and, therefore sometimes simplify and even primitivize, their primary content) they may, in some cases, become a source conveying untruth.

IV

Finally, the reason for the heterogenous structure of the exhibits or of the exhibition may lie in the artistic conception of it.

In many cases, the artistic conception of the exhibition becomes an expression of its scientific message (the exhibition's form becomes the form of the scientific content); in other cases, however, it is independent of it. Irrespective of this, the artistic conception may become a conveyer of information which is not programmed into the exhibition's script. Thus, information brings, as a rule, some untruth which needs "neutralization" in the process of understanding the exhibition.

For instance: Coptic columns displayed in the National Museum in Warsaw were received in fragments and had to be restored. The restoration was conceived schematically, i.e., the missing parts were suggested in their main shape, but not even in their size, and were deprived of any detail. The most characteristic feature was, however, the arrangement of the columns which were exposed standing in a "bunch" close one to another, i.e., as never occurs in any authentic arrangement of ancient architecture (such an arrangement is frequent only in medieval architecture where columns (precisely, semi-columns) are arranged around pillars or form a "bunch," being simultanously converted into slender shafts). The conventional, schematic arrangement of the Coptic columns found in the Warsaw museum is indisputably untrue; its schematicism, however, is so evident that it practically excludes any misunderstanding of the exhibit's idea: the exhibition presents merely a collection of restored columns and is not a restoration of their arrangement or of their architectural space (the space in which the columns are exposed is still that of a museum hall, not of a restored Coptic temple). Nevertheless, the "deciphering" of the idea of the exhibition is left to the visitor's knowledge and his capacity to understand museum conventions. Factual untruth is, in other words, easy to neutralize.

In other cases, however, the schematic or conventional way of presentation or restoration in exhibits leads to various misunderstanding. This happens when the visitor is not able to destinguish exhibit conventions (which are intended to be meaningless) from the scientific message. Factual untruth is then taken "for granted." One can give another example of artistic convention in the design of exhibitions. Many paleontogical exhibitions combine the principle of a diorama with that of a collection. In such cases quasi-surrealistic sceneries are created, where skeletons are situated against a painted (more or less realistic) landscape, and the restorations are incorporated into the scenery like models reduced in scale. Evident schematicism excludes, in such cases, any misunderstanding of the exhibition. However, the artistic conception can be misunderstood and be interpreted by the visitor (especially when the visitor is an art historian!) as an intention to create factually a surrealistic work of art (see the British Museum, the Natural History division, the hall of fossil mammals). Some influence of surrealism is, however, not excluded here.

Finally, numerous museum exhibitions are strongly influenced by abstract or by Pop Art. We already have mentioned the application of abstract color in the Senckenbergian Museum. In the Copehagen University Museum of Natural History (Universitetets Zoologisk Museum) there are no restorations, but all exhibition solutions are strongly influenced by abstract art (especially the portion of the collection mounted on the top floor — the showcases with birds). The color which is introduced there has, in some cases, the function of being an iconic sign (it is conventionally provided with some meaning). In other cases it is meaningless. Similar principles can be observed in numerous museums in the United States.

V

Truth or untruth in a museum exhibition may, finally, have their source in the exhibit's appearances, understood as phenomena given directly to our perception of things and resulting from the optical interrelation between exhibits and their surroundings (exhibition space, lighting, color, etc). Any real object we perceive is given to us under the form of its appearances: museum exhibits only present one special example in this regard.

The appearances of things are unconditioned by any content, intellectual concept, etc., which is, or which may be, "in-coded" into the objects we perceive. In the case of exhibits, it is unconditioned by any scientific message transmitted by them. Appearances belong exclusively to the sphere of "effects" — as some would probably call it. Hence they are to be distinguished from any kind of permanent features of the object (e.g., from morphological ones, like shape or permanent color). Appearances are always changing: they may last for a shorter or longer time, but irrespective of how long they last, one second, one hour, or one year, they are always ephemeral[19] — changes in the optical conditions in which the objects are perceived cause their simultanous changes.

The circumstance that objects which belong to museum exhibits are for the most part detached from their primary, natural surroundings[20] means that after their installation in the exhibitional milieu, they begin to "look different" in it. Aesthetic qualities (all of which are of typically phenomenal character) founded on the basis of the objects' appearances change correspondingly as well. The museum exhibit regarded as a presented object (or, more accurately, as a phenomenon of an object as a perceived "image" of a real thing) is therefore aesthetically different from what it was "before," in its natural environment or even in its pre-exhibition surroundings, even though, as a material thing, it is usually "the same."[21]

For instance, a painting, created in an artist's studio, when installed in the exhibition hall provided with different lighting, space, etc., presented against a background of different color, "looks" different sometimes so much that it often becomes an astonishing discovery, a "new" work of art: it reveals qualities which could not be seen in the conditions of the studio. Every painter who has seen his own works presented under various exhibitional conditions knows this phenomenon very well from his own experience.

A geological specimen, say a piece of rock which in natural surroundings looked very attractive and was taken for its color, after being transferred into a museum showcase (especially when the showcase is old-fashioned, deprived of artificial light), looks unattractive, grey and dull. We often scarcely recognize it as "one and the same" thing and feel disappointed when we now see it in a museum, especially when we remember its first, natural appearance.

A deep-sea fish is deprived of any appearance (and consequently, of any aesthetic qualities) at all when it is in its natural environment: living in complete darkness it is "nothing" as a phenomenon. Observed in daylight it astonishes us by its bright (often vividly red) color — "illogical" for the conditions in which it lives. In a museum display we see it with its "illogical" color.

A fragment of Mediterranean architecture which, in its natural surroundings, possesses aesthetic qualities connected with its natural lighting (for which it was designed!), when installed in a museum interior, under dispersed roof-lighting, becomes aesthetically neutral, dark and dull.

Summing up, one can say that: 1) any surrounding in which an object is given to our perception participates in the constitution of the object's appearance and, correspondingly, its aesthetic qualities. 2) The influence of the surroundings in which objects are observed acts on any kind of object, irrespective of whether the given object originated as a product of artistic creation (a work of art) or not (is an "ordinary thing," or a fragment of nature), with works of art, certainly, these changes of appearance are especially important as regards their aesthetic values, but the same rule and the same phenomena can be observed in all other kinds of exhibits, even those which in their origin, function, and meaning have nothing in common with aesthetics or art creation. In museums, where the aesthetic factor is neglected or deliberately ignored, the conditions in which particular exhibits are presented, are regularly provided with negative aesthetic qualities. 3) Regarding the changes of the object's appearance, especially those in works of art, it is often difficult to say which of these appearances should be qualified as the "true" one. In many cases, one can speak exclusively about different objects' appearances, all of which are equally valuable as regards their truth. One can speak, at least, about more or less favorable conditions for presenting exhibits, or about better or worse appearances as regards their effectiveness. And these differences are important enough to be looked after so that optimal design solutions may be found. 4) Modifications of the exhibits' appearances mostly occur as a result of a twofold set of relations: a) those between particular exhibits and b) those between exhibits and their "background." These two relationships never occur separately.

An object's appearance may, in some cases, bring some novelty to the sense of presented objects. This sense is more or less of emotive character and may sometimes become a strong factor modifying the meaning of the objects, to be precise, the meaning which is "programmed" in the exhibition script. Emotive expression may heighten the impression made by the "visual quality" of objects presented in a museum display: objects which are known from everyday's experience as "ordinary" ones, when installed in well-arranged exhibitional conditions become converted in their appearance to the extent that they begin to come across like authentic works of art. Irrespective of their being objects of cognizance, they become parallel objects of aesthetic experience given equally to our feeling and to our understanding. Many examples from artistically advanced archeological exhibitions, where items are presented as treasures, may be cited as typical instances. The double-leveled character of exhibitions (cognitive and emotively-aesthetic) is of great significance in contemporary culture. It is the role of museums, especially those presenting and interpreting cultural heritage, to function on both levels in a form of concretization.

* * *

SUMMARY

In speaking on truth and untruth in museum exhibitions one should not have in mind exclusively the values of the assertions encoded in exhibitions (in particular, into the exhibits) and transmitted to the visitors. Untruth especially may have a different origin and even different functions (not exclusively negative).

1) It does happen, of course, that museum exhibitions contain and transmit false assertions. Such assertions may result from erroneous research on particular objects and may be formulated before the exhibition is arranged. In such cases the exhibition only repeats the untruth, taking it as something ready, "for granted": one can describe such a case, paradoxically, as a "correct expression of some untruth." In some other cases, we are dealing with untruth resulting from the simplification of a true statement; the simplified version (and consequently, its expression by exhibitional means) is what should be qualified as something that is not true (whether valuable or not, is another thing!). One sometimes, encounters, fortunately rarely, untruth that is the result of conscious falsification of scientific truth. As was mentioned above, this case should be qualified as self-evidently negative with respect to the fundamental function of museums. It happens when museums lose their character of institutions devoted to objective research and the transmission of scientific knowledge. 2) Untruth may be connected with the conditions in which the exhibits are preserved and presented in a

museum display, as objects detached from their first surroundings. Objects which are unavoidably deformed (even in their physical features) are, in spite of their deformations shown as true (documentary, authentic) ones. 3) Untruth may be connected with the introduction of a creative concept (mostly of hypothetical value) into exhibits. Exhibits are restored and completed with elements which are novel to them, and may contain untruth if the concept in accord with which the restoration was completed was false. This is, to some extent, a special case of the first instance discussed above. And, finally, 4) untruth may appear on the grounds analogious to those which are essential to the constitution of any work of art. Since works of art (in particular realistic, representation art) are never strict "reproductions" of reality and always contain some amount of deformation that is characteristic — of, e.g., their composition and style — since they are factually — interpretations of the reality existing "outside" of them — exhibitions and their elements (first of all the exhibits) may contain an analogous amount of untruth. This untruth is mostly of an aesthetic or conventional character, and, if the exhibition is well designed, it may easily be "de-coded," even becoming a means of transmitting truth. This type of untruth may be, under some conditions, of extremely high value in that concretely it is this kind of untruth which may enable us to look at the represented reality creatively.

Museums regarded as institutions transmitting ideas about the world (the world of nature and that of culture) do not reproduce this world or its image in a strict mechanical sense. The moment of interpretation, in some cases, of evident creation is what makes museums manifestations of our human, contemporary outlook on the objects, events and processes of our times as well of the past. It is obvious that presenting only a given "image" of the world (scientific-cognitive or aesthetic-emotive) leads to some deformations of objective reality since it incorporates the elements of human subjectivity. This transformation, sometimes even deformation, belongs to the way we perceive and understand or feel the reality we live in, or the reality of past days. Therefore, museums, despite the untruth they convey, sometimes necessarily, unavoidably, play a significant role in our cultural life. They are an integral part of the worlds of science and art, a special means of their manifestation or presentation.

Polish Academy of Sciences
Krakow

NOTES

[1] Old, traditionally arranged exhibitions had entirely the function of storing scientifically valuable materials for research.
[2] See Roman Ingarden, *Studia z Estetyki* (Studies on Aesthetics), (Warsaw: 1958).
[3] *Ibid.* The book has been translated into English (in the USA).
[4] J. Swiecimski, "Scientific Information Function and Ingarden's Theory of Forms," *Analecta Husserliana* Vol. IV (Dordrecht/Boston: 1974).
[5] A similar diversity of truth (different "kinds of truth") can be observed in the content and form of works of art, where truth is not necessarily connected with any assertions, although it becomes a constituent factor of the work's meaning or even of its autonomous structure. It sometimes becomes a paradox to speak of the work of art's truth in cases where the work of art presents us a completely deformed image of reality. The popularly quoted artists' expressions: "as I see it," "that is the expression of my truth" clarify this problem fully.
[6] Therefore, the object's individual truth is, in the present interpretation, deprived of the character of value. It is exclusively a factor essential to the object's identity.
[7] See below, Section II.
[8] Very characteristic in this regard are the conventions for designating works of art to be recognized as "originals" (*l'épreuves d'artiste*) in the case, when the artist produces only the preparatory, negative form of the work (the mold, board, etc) and when exact works are executed as its reproductions (casts, prints, etc). Although all works which are made correctly, are identical in their form, only some of them (e.g., the authorized ones) are recognized as authentic.
[9] For different types of works of art, different norms are applied. Even with the works of one and the same type (e.g., for painting) different norms are applied depending on the historical period to which particular works belong.
[10] It can be disputed whether these features are, or should be regarded as primary ones. To be clear, I mean here concretely the features which are characteristic of the object at the time when it is received by a museum — irrespective of whether they are factually primary or secondary (e.g., have resulted by some damage, natural deterioration, etc.).
[11] For instance zoological specimens like "stuffed" birds in which only the skin and feathers are authentic and the whole body (essential for the living organism, certainly!) is replaced by cotton and wire are regarded (conventionally!) to be "authentic." This occurs, because only some (not all) external features are essential for the object's truth. For instance a change in shape due to bad preparation technicque is usually dismissed as something "of no importance" and plays no role in the object's qualification.

As we shall see below, it is on this basis exactly that the distinction between the value of a museum object and that of an exhibit is established. A valuable museum object may be worthless as an exhibit and, conversely, a valuable exhibit may be of no value as a museum specimen.
[12] The original does not exist: the authentic skin of the carcass was used to mount a restoration of the animal. The restoration (executed in a traditional form of a "stuffed" specimen) is in many of its features erronous and incomplete (the long fur with which the live animal was covered is lacking). Visitors, since they know about contemporary

rhinoceroses are not shocked by that, anyway; it would be different if it were, for instance, a restoration of a bear or a bison. The restoration (completed in 1928) is exposed in the Museum of Natural History in Cracow. Its cast is exposed in Salzburg at the Haus der Natur.

[13] Although in all such cases the guiding concept is based on documentary data, the creative character of the concept sometimes reaches beyond the given facts or a purely intellectual induction. This means that in the exhibit primary features (if not some of its material parts) will be removed and replaced by novel ones which have already been determined by the content of the conceptualization. The reasons for such deformations may be different in particular cases. It depends on the kind of the concept and on the role which hypothesis (and sometimes even "scientific imagination") plays in it.

[14] It is necessary to stress this additional condition so far that sometimes the specimens cannot fulfill their function satisfactorily because of their state, as when they are damaged, discolored, etc., when they present a rather "untypical" form within a given class.

[15] This is so with the stipulation that the specimen's shape is void of any meaning; a fragment of a granite sculpture, although it is a granite one, would not be suitable in this case simply because of its unnecessary content.

[16] We have to say: "necessary amount" because the number of features which is necessary depends in each case on the definition of the general object in question.

[17] In this way exhibits would be distinguished from research specimens by their illustrative value. Facsimile models are used instead of originals mostly because they illustrate a given species better than any authentic specimen could. This is true especially in the case of specimens preserved in alcohol which as the result are usually deformed and discolored. The secondary features of such specimens overlay primary ones, which excludes their use even though they are authentic. In their preserved state, such specimens would become transmitters of misinformation, of some untruth.

[18] It is, however interesting to analyze to what extent exhibit features which are intended to be void of information are factually conveyers of information (and of what kind of information). In exhibits such features are unavoidable. Sometimes it is difficult to distinguish the intentionally informative features from void ones. As a result, features which are intended to be deprived of any meaning, become factually the bearers of misinformation, or at least, of unnecessary information.

[19] The specific character of the exhibition situation is that the appearances of objects which are put on display are more or less stablized by the exhibition technique; it is, namely, the aim of exhibitions that objects which are presented should give one, selected (optimal) appearance. Therefore, the optical conditions in which objects are shown in museum exhibitions are to some extent artificially controlled.

[20] I say "for the most part" because there is a group of exhibits which are created in the exhibitional situation, to which they are intended to belong permanently. The museum exhibition is, therefore, their primary environment. They are adjusted to the conditions of their presentation from the first moment of their creation.

[21] I abstract here from cases in which, apart from changes on the level of appearences, objects detached from their natural surroundings undergo some physical changes, i.e., in their morphological features.

HERNÁN NEIRA

NIHILISM AND NOESIS: THE CONTRIBUTION OF PHENOMENOLOGY TO THE SARTREAN ANALYSIS OF FLAUBERT

I. SARTRE'S POINT OF DEPARTURE: HUSSERL, THE TRANSCENDENTAL EGO, AND THE ORIGIN OF MEANING

Based on principles held in common by Husserl and Sartre, namely, (1) the idea that the only origin of meaning is the activity of consciousness and (2) that consciousness *is* a relationship with the world, two distinct philosophical projects take form. Insofar as he is a phenomenologist, Sartre takes for granted the possibility of the existence of meaning proceeding from consciousness. However, the moment Sartre conceives of consciousness as trying to project its meaning beyond itself towards the world, he is forced to distance himself from Husserl. Sartre's distancing himself from his master takes place in two stages: the first is the *Transcendance de l'ego*, and the second is the *Cahiers pour une morale*.[1]

a) *The First Moment of Sartre's Distancing Himself from Husserl. The Transcendance of The Ego, The Impersonal Consciousness*

A structure like that of the cogito leads to a vision of consciousness as being double, that is, multiple. Each layer of consciousness would exist there without maintaining relations with the neighboring layer, or maintaining mechanical relations only, or even an opposition between them. This is a conception one finds in empiricism, but also in psychoanalysis and in Husserl. Sartre does not include Kant in his critique, because for the lattter the structure of the cogito exists only *de jure* and not *de facto*:

Le "Je pense" kantien est une condition de possibilité. Le cogito de Descartes et de Husserl est une constatation de failt. On a parlé de la 'necessité de fait' du Cogito et cette expression me paraît très juste. Or, il est indéniable que le Cogito est personnel. Dan le "*Je* pense" il y a un *Je* qui pense.[2] . . . L'ego apparaît à la réflexion comme un objet transcendant réalisant la synthèse permanente du psychique. L'ego est *du côté* du psychique . . . Il serait tentant de constituer l'Ego en "pôle-sujet" comme ce "pôle-objet" que Husserl place qu centre du noyau noématique. Ce pôle-objet est un X qui supporte les déterminations.[2]

A-T. Tymieniecka (ed.), Analecta Husserliana, Vol. XXXVII, 367–381.

L'Ego est l'unification transcendante spontanée de nos états et de nos actions . . . . L'Ego est créateur de ses états et soutient ses qualités dans l'existence par une sorte de spontanéité conservatrice . . . . Mais cette spontanéité ne doit pas être confondue avec celle de la conscience. En effet, l'Ego, étant objet, est *passif*, Il s'agit donc d'une pseudo-spontanéité qui trouverait ses symboles convenables dans le jaillissement d'une source, d'un geyser, etc. C'est-à-dire qu'il ne s'agit que d'une apparence. La véritable spontanéité doit être parfiatement claire: elle *est* ce qu'elle produit et ne peut être rien d'autre.[3]

If consciousness is not the site of the ego and if the ego is not a subject either, this is because there is no ego in consciousness. This "X" that Sartre speaks of is situated in the empirical world, among things; it is the ego of empirical psychology, in no way a structure of transcendental consciousness. The ego is nonetheless an object, but a transcendental and real object, it is not an intentional product, but a material product of a psychic nature or a psycho-physical nature with which consciousness maintains intentional rapports. No matter how powerful it may be, noesis can neither make the ego exist nor create it *ex nihilo* by means of consciousness' activity. Sartre thus concludes,

Le Champ transcendantal, purifié de toute structure égologique, recouvre sa limpidité première . . . . Nous pouvons donc formuler notre thèse: la conscience transcendante est une spontanéité impersonnelle. Elle se détermine à l'existence à chaque instant, sans que l'on puisse rien concevoir *avant elle*. Ainsi chaque instant de notre vie consciente nous révèle une création ex nihilo. Non pas un *arrangement* nouveau, mais une existence nouvelle.[4]

b) *The Second Moment of Sartre's Distancing Himself from Husserl. The Reconquest of The Empirical World*

The first stage of Sartre's distancing himself from Husserl is completed by the conquest of a consciousness that is pure spontaneity, by the conquest of liberty. The second stage is the recovery of the empirical world and its engagement with the activity of consciousness. In doing this, Sartre avoids the risk created by the egological structure of consciousness. Effectively, this structure threatens to create a split within itself and, moreover, makes the ego an atom, a monad opposed to worldly objects, themselves monads. The ties between the world and consciousness become problematic. Even if consciousness does not fall into solipsism, its relationship with the world can only be mechanical, and any question of the world's meaning is emptied of meaning. Consciousness, then, can never be sure of the world's truth or of its existence in the way that Descartes wished.

Sartre will surpass Husserlian phenomenology in the name of phenomenology itself, and this thanks to the instruments with which it provides him. He continues to assert the spontaneity of consciousness, while reattaching it to the world with certain and ontological relations. If one strips consciousness of its ego, it is reduced to pure intentionality, it becomes a *nothing*. The transcendant ego consists of acts and psychic states that subsist beyond the act that produces them, whereas consciousness, for Sartre, consists of pure acts deprived of permanence. Consciousness is not a substance; it does not subsist beyond its activity; it empties itself in its activity. It could not possibly enclose itself in the cogito because the cogito is not a structure that belongs to it. Consciousness is pure intentionality without ego — that is, it is *rapport* and nothing but rapport.

Since any relation implies two poles, the mere existence of consciousness implies the existence of a "something," of an exterior world to which it can relate. This implication does not mean that noesis could create the world; the existence of consciousness does not engender that of concrete things. If the activity of consciousness were to cease, consciousness itself would cease to exist, but no change would be brought to bear upon the world. Sartre thus conquers the certitude of existential, practical, and cognitive relations between consciousness and the world. However, these relations do not yet include semantic ones, Sartre remains convinced that the world does not create meaning; nor may consciousness impose its own meaning upon things without being modified by them. According to Sartre, the world — like the God of Racine and Pascal — does not answer; consciousness knows that the world exists, but does not know its meaning.

Sartre has conquered the empirical world, and the risk of solipsism seems to be parried by the affirmation of the contemporaneity of consciousness and the world. He has attained certitude of the concrete outside of consciousness in addition to the transcendental field that is highlighted by Husserl. Nevertheless, Sartre has yet to acquire the possibility of imparting meaning to this world. The vector of creative intentionality, which brings meaning from consciousness towards the concrete empirical, manages only with difficulty, in Sartre, to leave its imprint upon the world.

II. THE CERTITUDE OF THE EXISTENCE OF THE WORLD AS THE CERTITUDE OF NIHILISM

By placing the ego outside the transcendental field, the absence of meaning of the world touches the empirical life of man, inasmuch as he identifies himself with the psycho-physical ego. A transcendental ego has the same avatars as the rest of the world. The transcendance of the ego leads to a dead end road, insofar as one has no means of demonstrating that consciousness is able to give meaning to the concrete outside of it, to the empirical world, for certitude in this world does not suffice to answer the question of *what is the meaning of the world*?

A remarkable intuition of the young Nietzsche was to question not the certitude of the existence of the world, but that of the logos that would claim to seize it, and that of action that would claim to leave its mark on it.[5] What is tragic about Nietzschean thought is the rupture of identification between logos and the world. For Nietzsche there is no equivalence between the rational and the real; they remain two entities without any relationship. Modern reason, except perhaps in Kant, is founded upon a principle of reciprocity, according to which the vector of meaning proceeds from the world towards consciousness, or from consciousness towards the world. In the young Nietzsche, the essential principles of Western philosophy, individuation and causality, serve only to obscure the indistinct unity of the beings of the universe, including man. It would be futile to undertake any action there or to try to know it, because the universe is a nebula of forces against which man can do nothing.

The principles of causality and individuation constitute the a priori condition of possibility for noesis, praxis, and poeisis. What activity could one exert, and upon what would one act, if there were no difference between the Me and the Other, between consciousness and the world, between meaning and absence of meaning, and if one's actions always lead to unforeseen results. Indeed, the rejection of the principles of causality and individuation leads to the collapse of any hope of religion, cognitive or political redemption, and of any salvation by technical progress; it leads to the eternal return. In order to make the suffering provoked by the eternal return tolerable, noesis creates meanings, but these are only appearances that obscure a more profound metaphysical and, in fact, irremediable evil. Western Philosophy must choose between the Cartesian cogito that deprives philosophers of the world to save consciousness, and the certitude of the world which takes away the meaning of the concrete mundane.

III. FROM PURE CONSCIOUSNESS TO MAN AS CONSCIOUSNESS

It is not enough to assure noesis; it is equally important to assure that noesis may impose its seal upon the world. Sartre will always be haunted by the problem of the polarity that opposes Nietzschean tragedy to solipsism. He attempts to resolve the opposition of the logos and the real in three complementary stages. The first is a phenomenological description of the ontological structures of the being that is consciousness. This stage leads to a definition of man as being for-self, as liberty opposed to the blind concatenations of the world.[6] The transcendent ego "places" man and consciousness in the world. Thus the relations of consciousness with man imply the relations of consciousness with the ego, which acts in the universe of things.

The second stage is the description of the intermediate world between things and the individual, that is, a question of the history of the human being, of technique and of social groups in action.[7] Sartre concludes that praxis is only possible at the price of the uncertainty of the results which follow.

The third stage is the demonstration, in concrete examples, that the activity of consciousness may leave its imprint upon the world *de facto* and not only *de jure*. The possibility then that man has to leave his seal on things requires a prerequisite condition that this seal may be left on the psychic ego. This condition has been studied by Sartre, in abstract fashion, thanks to the description of the structure of man as being-in-the-world (*être-dans-le-monde*). Sartre did not, however, go on to prove that the activity of consciousness can go beyond its stage of simple possibility to become a fact. The examination of the poiesis of Baudelaire, Jean Genet, and Flaubert demonstrates this by showing that noeisis is capable of bringing meaning to the worldly ego. This attempt to prove the existence of meaning leads Sartre to undertake the study of concrete examples of consciousness-in-the-world, in particular that of Flaubert, in *L'Idiot de la Famille*.[8] This work will be analyzed in the next chapter. Sartrian phenomenological ontology thus evolves from the description of the field of purified consciousness to the description of the avatars of the intentional relations of this consciousness to the transcendent ego.

The evolution that takes Sartre from *La Transcendance de l'ego* to *L'Etre et le Néant* is realized essentially around two axes. The first is the finding that consciousness is not constitutive of transcendantal being. In Sartrian terms, "l'animation du noyau hylétique par les seules intentions qui peurvent trouver leur replissement (Erfüllung) dans cette hylé ne saurait suffire à nous faire sortir de la subjectivité."[9] This is because the world is assumed as a certain something that imposes its alien presence to consciousness in the intuition that consciousness has of itself. The second axis is the anthropologization of consciousness, a task carried out under the influence of Heidegger. After *L'Etre et le Néant*, man and consciousness constitute a single reality; their identification is complete. The transcendent ego confirms its role as an exterior object that inserts man into the world. The definition given in *La Transcendance de l'ego — consciousness is consciousness-of-something* — becomes, in EN, *man (consciousness) is relation with the world wherein man exists.* Thus, man (consciousness) relates to the world by means of his egological insertion. Two questions are then necessarily posed: "I° Quel est le rapport synthétique que nous nommons l'être-dans-le-monde? 2° Que doivent être l'homme et le monde pour que le rapport soit possible entre eux?[10]

IV. WHY FLAUBERT? WHY *L'IDIOT DE LA FAMILLE*?

In order to refute nihilism and solipsism, it is necessary to show that intentionality reaches the world in order to leave its mark there and all the while to take into account that the "subjectivité[de la conscience] ne saurait sortir de soi pour poser un objet transcendant en lui conférant le plénitude impresionnelle."[11] Sartre, in his analysis of Flaubert, adds two certainties to this constant of phenomenology. The first is the spontaneity of consciousness and of noesis; the second is that the certainty of the world leads to the certainity of its absence of meaning. However, this second conviction is not comparable to Nietzschean nihilism, "World" (*monde*) in Sartre signifies the organization of objects with respect to the center that constitutes each man, that is, the in-self (*en-soi*) organized by a human project. Each time that the question of leaving a human mark upon the world arises, the question of the in-self also appears as soon as man inhabits it and organizes it with respect to himself.

The meaning of the world is not immediately given the intuition that consciousness has of itself. Sartre needs only to describe *one* case in which noesis has left its imprint on the world in order to prove that science, praxis and poiesis are founded with respect to man. This process is based on the passage of the ontic towards the ontological. After this passage, every ontic manifestation of man reveals his ontological structure, his existence. It would thus suffice to show man leaving the seal of noesis in one of his empirical manifestaitons in order to have the right to draw conclusions concerning the being of the human being. If the being of a man is capable of bringing meaning to the world, then sense exists empirically, and the bridge that connects the meaning produced by the consciousness to the transcendent real is re-established for all humanity. Sartre nevertheless goes neither from the particular to the general nor from the a priori to the empirical. He moves instead from the ontic to the ontological and deduces the possibility of the meaning of the world from the meaning that an individual imparts to his own transcendent ego.

Why does Sartre choose to examine a writer and not just any man in order to realize his refutation of nihilism? It is because Sartre remains a prisoner of logocentrism, and thus for him writing is the *voie royale* to the production of meaning. The activity of writing has the goal of "saying," of creating a meaning; it is never simply a game or a diversion.[12] In literary activity, man is apparently more the master of the results than in any other form of praxis. Writing seems to be, by definition, the organization of signs and significations in a meta-signification. The objectivized noesis, which may be difficult to locate in other aspects of the transcendent ego, seems to show itself in all its purity in literature. A piece of writing is thus the location where meaning — if it exists — must manifest itself most explicitly. But, if it were discovered that it were possible to write about meaning, or that a work of art is not the product of the forces of its author, it would be necessary to completely renounce any intentional relationship of consciousness with the world. Nihilism and solipsism would dominate, once and for all, over poiesis. Moreover, insofar as poeisis is an ontical manifestation of consciousness, the whole ontological structure of man and noesis would collapse in impotence.

This explains Sartre's interest in showing noesis in an extreme case in which it has apparently disappeared completely: the case of Flaubert. Why this writer rather than another? Because Flaubert, given his psychic and his egological constitution, represents for Sartre the incapacity to signify. His worldly ego seems to be pure passivity; he does not manifest the presence of any intentionality; Flaubert seems to be the denial of the intentional vector that connects consciousness and world, a denial that lets him be dominated by nihilism.[13] More than a theoretician of the absence of meaning, Flaubert is himself this absence; his transcendent ego is the ideal of passivity, and consequently, the location where spontaneity seems to collapse.

V. BEING-IN-LANGUAGE, THE ANTINOMIES OF THE CASE OF FLAUBERT

a) *Noesis, Praxis, Poeisis, Cogito, Being-in-the-World as Being-in-Language*

A few distinctions are indispensable. We will call *poiesis* the noesis proper to the creation of literary works, and in a general fashion, to artistic production. We will thus distinguish poiesis from *praxis*, or the noesis proper to the enterprises carried out in the social, natural, or technical world. From here on *cogito*, signifies a poiesis that maintains the unity of consciousness by producing itself in time. These distinctions, which are ours and not Sartre's, do not concern the nature of noesis. Noesis, as the generic name of the activity of consciousness, is always one and the same, even if it focuses simultaneously on diverse objects.

An intentional object cannot become transcendent without running the risk of erasing the differences and all distance between consciousness and the world. To avoid this danger, artistic poiesis must not be conceived as simple romantic inspiration, nor as *creatio ex nihilo*. To avoid solipsism, it is indispensable that poiesis be an intention directed at transcendental material, at the in-self. Account must also be taken of the fact that the egological incarnation of consciousness, its being-in-the-world, is an incarnation in lanuguage, i.e., in psycho-linguistic and socio-linguistic conditions. Given that the significations of this language are social significations, trans-individual ones, literary poiesis is an intention of consciousness that is related to the socio-linguistic worldliness of the ego. By means of the linguistic incarnation of the ego, consciousness is thus related to an individual and, at the same time to social significations that this individual did not choose.

b) *The Being-in-the-Language of Flaubert as Passivity*

Relying on the testimony of Flaubert's sister and of his niece, Sartre writes that the author of *Mme. Bovary* has

de mauvais rapports avec les mots . . . l'enfant comprend sans pouvoir assimiler [. . .] il croit d'abord tout ce qu'on lui dit; par stupeur devant l'objet verbal, par amour dévot des adultes.[14]

[Flaubert] voil plutôt des impératifs que des affirmations: elles s'imposent d'elles-mêmes et puis il *faut y croire* puisqu'elles sont le don gracieux que lui font ses parents.[15]

Faute de la réciprocité [. . .] qu'établit une compréhension entière, avec toutes ses structures, la parole de l'Autre lui semble *parole donnée* [. . .] Dire [pour Flaubert] n'est pas *énoncer*.[16]

Le petit garçon [Gustave Flaubert] est mal visé [sic] dans l'univers du discours. Le mot n'est jamais *sien*, tantôt l'hébétude engloutit le verbe et tantôt celui-ci, tombé du ciel le tyrannise. Dans ce dernier cas, jusque dans l'intériorité profonde, il reste extérieur . . . . A l'âge où tout le monde parle, il lest encore *à imiter* les parleurs . . . . Il n'y a pas de commune mesure entre l'existence subjective de Gustave et l'univers des significations.[17]

Dehors et dedans, il voit les mots à l'envers, dans leur étrangeté sensuelle, il tient les lieux communs pour des impératifs gravés dans la matière verbale et que chaque individu a la mission de reproduire par les inflexions de sa voix; il persiste à penser que le verbe le ronge et ne pourra jamais le désigner tout à fait. Dans son cas, la difficulté d'apprendre à lire vient d'un touble géneral et plus ancien, la difficulté de parler.[18]

Culture, pour lui, c'est le vol [. . .] le mot est chose; introduit dans une âme, il la résorbe dans sa propre généralité; il s'agit d'une véritable métamorphose.[19]

c) *The Antinomies of The Case of Flaubert*

i) *The First Part of The Antinomy: Flaubert Confuses "La Langue" with "Le Langage"*

All in Flaubert is egological passivity. His consciousness seems to founder when faced with its being-in-language. Flaubert never manages to make the distinction made evident by Saussure between "langue" and "langage"; the author of *Mme. Bovary* uses the "langue" as if it were the "langage." Nowhere in his work does one find poiesis seeking transcendent signs to organize them humanly; nevertheless, he is spoken or as a great writer, as an artist. Flaubert seems to surrender to the nihilism of solipsism. The author of *Mme. Bovary* would be the ontic manifestation of an ontological structure of the human being. This structure would determine man in his incapacity to leave his mark upon things, not even on his own worldly ego, on his psyche. If it were true that following Flaubert that there is no difference between "langue" and "langage," nihilism and/or solipsism would be inscribed in the existence of man. These would constitute the possibility proper to being-in-language; they would be part of the ontological structure of man. In this case, Flaubert, misanthrope and nihilist — as his correspondance shows him to be from 1838[20] — would become the ontic manifestation of what has formerly been a mere possibility of man.

ii) *The Second Part of The Antinomy: Flaubert is a Great Creator*

A second fact opposes the first: Flaubert wrote during his entire life, from the age of eleven. He acted; he had a praxis that, as an artistic one, is a poiesis. To deny poeisis to him would be to contend that the works that are normally attributed to Flaubert should be attributed instead to the sickness of the author, to his social class, or to a language destitute of a subject — all of which would have determined that he write. Here is a contradiction that cannot be resolved at the empirical level and that demands the use of a transcendental phenomenological method. Indeed, it is impossible to accept that Flaubert, as a consciousness-in-the-world, one with serious linguistic troubles, both did not exercise poeisis *and* was also one the greatest French novelists. This is a Kantian problem, which can be referred to the third and the fourth antinomies of pure reason. It is not possible that a being be empirically free and not free at the same time; a choice must be made between natural causality and an extra-natural beginning called "liberty."

VI. GOING BEYOND THE ANTINOMIES: LIBERTY, "LANGUE," AND "LANGAGE"

Sartre goes beyond the antinomy by showing that Flaubertian poiesis consists of transforming an initial confusion, produced at the level of the transcendent ego, between "langue" and "langage," into a confusion *aimed* at by his consciousness. Sartre does not accept the recourse to Descartes' continual creation or to a *creatio ex nihilo* with respect to noesis. Since consciousness *is not*, since it is not an object, it is an empty interiority whose existence consists of objectivizing itself through praxis. That which consciousness objectivizes by means of its own activity cannot be the consciousness itself, because consciousness is not. Thus, it could not possibly objectivize the world, because the world is an object unconcerned with consciousness. What consciousness does objectivize is the lived (*vécu*), that is, not the world in its totality, but the world inasmuch as consciousness relates to it, inasmuch as consciousness goes beyond it. This is why Sartre, speaking of *L'Idiot de la Famille*, reminds that his goal in the book is "d'abord tout en dénombrant les conditions objectives et en les organisant, de les montrer maintenues et dépassées vers l'objectivation par le moment subjectif."[21]

This subjective moment is irreducible;[22] it is distinguished, then, from the simple transcendent material base where the poiesis of Flaubert is exercised. This moment, which introduces the mediation of consciousness in every rapport that man has with things, cannot be a simple copy of the psycho-linguistic structure that belongs to the transcendent world. In other words, subjectivization does not consist of a copy reproduced from the exterior and introduced into the consciousness. Literary poiesis is "langage," just as it manifests the impossibility of copying or reproducing the "langage" in its totality. However, if it is impossible to reproduce the "langue" in its totality, Flaubert's identification with the "langue" becomes equally impossible, because any linguistic performance must necessarily be "langage."

Is Flaubert's empirical and linguistic project condemned to failure? Yes, from the empirical point of view; no, from the transcendental one. Flaubert's identification with the "langue" reveals a transcendental intention of consciousness focused on the "langue," a vectorial relation of consciousness *vis-à-vis* the semantic world where it exists. Since the "langue" is a social ensemble of significations, it cannot be grasped in its entirety by an individual. The adult Flaubert knows that the "langue" exceeds any exhaustive apprenticeship, that it is something that an individual can use only partially and faultily. When a linguistic performance takes place, this faultiness makes "langue" become "langage." In other words, the "langue" becomes the individual's rapport with the semantic totality, which always remains partially alien to him.

Flaubert seeks, without ever attaining it, the *dégré zéro* of writing and "langage," where his work would be so "perfect" that it would not be differentiated from the "langue." All his effort and much of his greatness lie in a quest after the depersonalization of both himself and of art, so that nothing shows, in the texts, of the point of view of an individual's poiesis that created them. "L'art n'a rien à démêler avec l'artiste"; "Je veux qu'il n'y ait pas dans mon livre *un seul* mouvement, ni *une seule* réflexion de l'auteur," avows Flaubert to his mistress, Louise Colet, when he is writing *Mme. Bovary*.[23] Flaubertian poeisis is anti-poetic; it consists of hiding the poiesis of the artist and tends to situate it in the point of view of everyone and no-one. Flaubert wishes to attain what no artist has understood as well as he has, he wishes to accomplish what the aesthetic reason and the objective spirit of his period demand: to surpass (*dépasser*) romanticism.

Flaubert, the idiot incapable of situating himself in the "langage," embraces this "disease" as his own, to transform it into a virtue and to become himself the artist he always wanted to be. He will speak as others speak in order to distinguish himself from them by becoming the person whom no one can recognize as different; he will dissimulate his own "langage" under the veil of "langue."[24] The composition of his works must be so sublime that it be impossible to notice the work behind them, so subtle that his writings not seem written, that they seem the simple manifestations of an idea, a model. Flaubert wishes, according to Sartre, that his work become like "un *être naturel*, comme un arbre, comme un paysage, que les générations nouvelles acceptent [*Mme. Bovary*] au même titre que les *choses* du monde urbain ou rural et que les institutions."[25]

VII. CONCLUSION

Beginning with the certainty of the spontaneity of noesis and the certainty of the existence of the world, Sartre attempts to prove that noesis does not remain closed in the interiority of consciousness. To attain this goal is to prove that noesis can leave its imprint on the world. The analysis of Flaubert permits Sartre to show how this individual, who apparently embodies denial of any spontaneity, must maintain a constant *activity* of denial. Flaubert must construct, must invent a "langage" that resembles the "langue"; he must adopt the point of view of "langage," of consciousness, to become "langue," Sartre shows that for the author of *Mme. Bovary* the confusion of the "langue" with "langage" conceals a poeisis originating from the spontaneity of consciousness, an extra-empirical causality. In Flaubert, a transcendental personalization of his empirical impersonality takes place. This personalization, which is the product of poeisis, consists of pushing the will-to-be-"langue" to the point that such an enterprise manifests its impossibility. This impossibility is the idiocy of Flaubert, and it manifests the seal of the singular consciousness of the artist on his ego, a writer by vocation if not by profession.

It is useful to remember that the etymology of the word "idiot" stems from *ιδιωτεσ* (*idiotes*) which signifies both the private singular, as opposed to the universal of the state, and ignorance.[26] What characterizes the semantic point of view of Flaubert's infancy is his empirical ignorance of his own linguistic singularity. What characterizes his literary life, his poeisis, is the quest after impersonality that raises him above his social class and distinguishes his aesthetic from the romantic one. Beside the anthropological and moral thrust in *L'Idiot de la famille*, one can find therein the outline of theory of the semantic-being-in-the-world. This involves questions of the rapport between language and liberty, and that between languages and noesis — rapports that are little analyzed in the earlier works of Sartre. There remains, however, an ambiguity that is damaging to Sartre's work on Flaubert. The limits between what is termed "consciousness" (*conscience*) and what constitutes the "lived" (*vécu*) are unclear. At times these terms seem synonymous; at others the psychologizing aspect of the word "lived" prevents it from being understood as being identical with consciousness.

The Flaubert study confirms the Husserlian influence in Sartre beyond the enriching influences of Marxism and psychoanalysis. These latter theories show Sartre the necessity of realizing the phenomenology of a concrete example of liberty in the world by going beyond the description of the ontological structures of man to examine the ontic manifestation of this liberty. The problem of the meaning or sense of the world is Nietzschean in origin; that of action is Kantian and Marxist. By moving beyond noesis towards praxis, Sartre uses the phenomenological method to concern himself with questions that were not explicitly addressed by Husserl. In doing this, the theory of consciousness is opened upon morality. The latter opens the door towards a literary aesthetic where the examination of the writing — that is to say, the work itself — becomes as important as the examination of the will-to-be of the artist's consciousness.

We have not evoked the progressive-regressive method, nor the existential psychoanalysis that Sartre uses to analyze Flaubert. These would each merit an entire article. The method of *L'Idiot de la famille* could not be examined without the complement of its semantic and its ontology. Also, we have considered the author of *Mme. Bovary* in his infancy, and then, without transition, at the age of thirty-one. Our analysis lacks the historical and social aspects of phenomenology that transform the idiot into an artist. We do not deny their importance. In this article, we have emphasized phenomenological semantics and aesthetics, as well as the phenomenological foundations of the being-in-the-world-of-language. These three aspects of Sartre's thought have only rarely been discussed by the specialist and the collaborators of the philosopher.

University of Paris VIII

NOTES

[1] Cf. J. P. Sartre, *La Transcendance de l'ego* (Paris: Jean Vrin, 1985) (hereinafter, TE), and J. P. Sartre, *Cahiers pour une morale* (Paris: Gallimard, 1985).
[2] Cf. Sartre, *La Transcendance de l'ego, op. cit.*, p. 26.
[3] Cf. *Ibid.*, pp. 59 ff.
[4] Cf. *Ibid.*, p. 79.
[5] Cf. F. Nietzsche, *La Naissance de la tragédie* (Paris: Gallimard, 1978); and on Nietzsche, cf. Jacques Derrida, *De la grammatologie* (Paris: Minuit, 1985).
[6] Cf. J. P. Sartre, *L'Etre et le Néant* (hereinafter, EN) (Paris: Gallimard, 1980) (hereinafter EN).
[7] Cf. Sartre, *Cahiers pour une morale, op. cit.*, and *La Critique de la Raison Dialectique* (Paris: Gallimard, 1980).
[8] Cf. J. P. Sartre, *Baudelaire* (Paris: Gallimard, 1980), *Saint-Genet, commédien et martyr* (Paris: Gallimard, 1985); and *L'Idiot de la Famille* (Paris: Gallimard 1970, 1973) (hereinafter, LIF).
[9] Cf. Sartre, EN, *op. cit.*, p. 27.
[10] Cf. *Ibid.*, p. 38.
[11] Cf. *Ibid.*, p. 27.
[12] Cf. Simone de Beauvoir, *La Cérémonie des adieux*, pp. 289—294 (Paris: Gallimard, 1987); and *Autoportrait à soixante-dix-ans*, in *Situations X* (Paris: Gallimard, 1976).
[13] Nietzsche never read Flaubert's correspondence where Flaubert develops his aesthetics and his pessimistic conception of the world. There are surprising analogies between these two men who hated their century and democracy, who considered scientific progress to be an illusion and who worked out a theory of the impersonalism of art. This analogy is limited, however; the aesthetics of Flaubert and that of Nietzsche are very different from one another.
[14] Cf. Sartre, LIF, *op. cit.*, pp. 23—24.
[15] Cf. *Ibid.*, p. 24.
[16] Cf. *Idem.*
[17] Cf. *Ibid.*, p. 26.

[18] Cf. *Ibid.*, p. 27.
[19] Cf. *Ibid.*, p. 40.
[20] Cf. G. Flaubert, *Correspondance*, Vols. I and II (Paris: Gallimard, Bibliothèque de la Pléiade, 1973).
[21] Cf. Sartre, LIF, *op. cit.*, p. 1787.
[22] Cf. *Ibid.*, p. 1787.
[23] Cf. Flaubert, *op. cit.*, letters of 26—7—52 and 8—2—52.
[24] Cf. G. Flaubert, *Dictionnaire d'idées récues, Bouvard et Pécuchet*, in *Oeuvres de Flaubert*, Vol. II (Paris: Gallimard, la Pléaide, 1968).
[25] Cf. Sartre, LIF, *op. cit.*, Vol. III, p. 29.
[26] Cf. Manfred Franck, *Idiot ou individu, L'intuition embryonnaire du "Flaubert", Etudes Sartriennes*, II—III, in *Cahiers de Sémiotique textuelle*, Université de Paris X, Nanterre, 1986.

WOLFGANG WITTKOWSKI

GOETHE AND SCHOPENHAUER: A PHENOMENOLOGY OF THE FINAL VISION IN *FAUST II*

The climax of *Faust* is its dying hero's vision of the future. Revered as Goethe's legacy to the German people — indeed, to all of mankind — these words have also been abhorred as the essence of the Faustian element in human nature that has led our technological civilization to the brink of self-destruction. A third way to read them is as Goethe's prophetic warning against our modern Faustian era, with its exploitation of man and nature. But in this case, why does Goethe allow Faust, the devil's associate, to enter heaven? Surely that would compound a theological provocation by moral and political affronts as well; such a sanction would undermine or trivialize the text's previous condemnation of Faust.

The essence of my argument is a phenomenology of Faust's final vision. Surprisingly, only rudiments of this exist to date. I shall also take context into account, particularly the "salvation," scenes leading up to the vision, and other scenes (especially that opening Act IV) that were among the last Goethe wrote.[1] Likewise I shall consider the implications of the author's own words as well as a number of parallels in the writings of Schopenhauer that have previously gone unnoticed.

Schopenhauer's conversations with Goethe during the winter of 1813—14 significantly influenced his conception and writing of *Die Welt als Wille und Vorstellung* (Part I, December, 1818). As Arthur Hübscher notes, this work was produced "directly in discourse with Goethe and with Goethe in mind."[2] Goethe read this work not over an entire year, as he originally anticipated, but in only two months (January and February, 1819), for he read "with an enthusiasm which Ottilie had never before seen in him."[3] Indeed, throughout the work he repeatedly encountered his own thoughts "which he had perhaps once felt, but never consciously formulated," arriving "at a clearer understanding of his own ideas." (Hübscher pp. 80, 78) Consequently their final conversation in 1819 became "a session of mutual instruction."[4]

Schopenhauer knew that he was in accord with Goethe when he defined his philosophical method as "phenomenology" and wrote: "All

A-T. Tymieniecka (ed.), Analecta Husserliana, Vol. XXXVII, 383—408.

primal, that is, original, evidence is a perception" (§14). Admittedly, his method of linking empiricism with ideas did lead him (even more than Goethe) into the metaphysical speculations both men so deeply despised in principle. Perhaps the philosopher was inspired to this by an artistic "intuition" (Hübscher p. 42) that is, of course, entirely legitimate in symbolic art itself.

1. SALVATION AND INTENDED VALUE INTENTIONALITY

> Whoever strives with all his power,
> We are allowed to save him
> (11936f.)

The angels can and do save Faust because "[heaven's] love from upon high has sympathized with him" (11934—400): Faust is a "loving one" himself (11751, 11902). The love Faust expresses is through his ceaseless striving, something Goethe expressly mentions as a precondition for his salvation (June 6, 1831 to Eckermann). In fact, Faust's final words express both that striving and the pleasure he takes in it (what I later shall put into the terminological straight-jacket of "intended value intentionality"):

> Then to the moment I might say:
> Abide, you are so fair!
> . . . .
> As I anticipate a bliss so high
> I now enjoy the highest moment.
> (11518ff.)

These words recall the wager Faust made in the pact with the devil: "If to the moment I should say: Abide, you are so fair!" (1699ff.). In that case, Faust would immediately surrender to the devil his life and soul (which he has already pledged at any rate). But things never come to that: Faust dies at a hundred years old, never desiring to eternalize the final moment.

Instead, the dying Faust celebrates the present moment as the "highest" because he anticipates in it the final moment of future fulfillment that he might want to eternalize. He "anticipates," Schopenhauer would say, the "ideal" in a purity that may occur in a work of art but seldom is seen in real life.

In the *Annalen*, Goethe spoke of "a dark presaging of the great truth" when he began *Wilhelm Meister*, claiming that through anticipation the poet "foreshadows the world." He wrote these words "only a few days after he started to read *Die Welt als Wille und Vorstellung*" (Hübscher p. 78) On pages 320 to 321 (§45) of that text, Goethe marked the following lines:

> Thus alone, i.e., in the fact that *we ourselves* are the will, have we in fact an anticipation of that which nature (which is just the will that constitutes our own being) strives to express. And in the true genius, this anticipation is accompanied by so great a degree of circumspection that he recognizes the Idea in the particular thing, thus, as it were, *understanding the half-uttered speech of nature.*

Immediately following this passage, Schopenhauer writes of "anticipation" of the Ideal, of the "Idea, so far as it is known a priori, at least half." Here he is referring to the genius, the artist and his work, and the critic, but this can be extended by analogy to all work. Man and idea, intention and intended ideal are but "different ascending grades of distinctness and completeness" of the one will to life, a point mentioned frequently in these pages. The dying Faust derives infinite pleasure that his "will to do deeds" is directed toward a value, an ideal. Put in Schopenhauer's terms, he takes pleasure in anticipating transforming the Idea into reality. In phenomenological terms, Faust takes pleasure in intending his value-oriented intentionality. Stated in Platonic terms — and in substance these concepts owe much to (Neo-)Platonism — Faust is filled with *Eros* for the Ideal.[5] In Goethe's terms he is a "loving one": his intentions derive from love.

For this reason and because of its extraordinarily intense value intentionality, Faust's entelechy can be reabsorbed into nature's household. At the end of Act III, the choir girls are granted — or rather, condemned to — the same fate, but on a more superficial, passive, and archaic level. In contrast, the choir leader, Helena, and Euphorion may descend to Hades. Thus, in Goethe's philosophy of life, Faust's re-acceptance into nature constitutes a symbolic and poetic "as if" meant to reconcile us to mortality and death: this too was a sort of "world as will and representation" or imagination. The same is true of the transmigration of souls, which Schopenhauer admittedly 'represented' as being objectively scientific.

Furthermore, in spite of his fundamental pessimism, the philosopher devoted what can only be called a 'loving' interest in many phenomena, for example to art and to the theory of colors. On the latter subject, he even defended Goethe's views against Newton's. It was Goethe the scientist who maintained that "everything which happens among men in the higher sense must be considered, examined and judged from an ethical point of view" (1832).[6] Schopenhauer fully agreed with this. He concluded his 1836 study *Über den Willen in der Natur* with a "Reference to Ethics." Despite the scientific foundation of his metaphysical system, he insisted that ethics must remain the touchstone of his — indeed, of all — philosophy, since "what is most important, nay, alone essential, in our whole existence, that on which everything depends, the real meaning, pivot, point (*sit venia verbo*) of it, lies in the morality of human actions." And in *Die Welt als Wille und Vorstellung* of 1818, he wrote that "the cardinal fact in the life of man is his ethical worth, that worth which counts for eternity" (§66).

According to the salvation scene, the case of Faust, ethical and non-value are combined: a "strong spirit" and its energetic value intentionality with the "distressing," immoral "earth remains" of which he must be purged (11954—65). Therefore, we should particularly resist any inclination to appropriate specific phrases for sole application to the 'credit' side of Faust's ethical account. The final lines, quoted above, particularly tempt us to do this for they express the ethos of indefatigable and energetically striving value intentionality as the work's intended theme. However, these lines embrace the couplet:

> The traces of my earthly day
> No eons can impair
> (11583ff.)

Thus the 'loving' Faust's intended value intentionality remains linked to his last plan and to the way in which he has hitherto sought to put it into action. Conversely, from the first line on, the final aspect is imbued with the impassioned ethos of 'loving' value intentionality — before this finally becomes itself the object of his intention. In order to prevent one aspect from obscuring the other, we must consider each separately. We have done this with the inspired and inspiring "how" of the language and the 'loving' value intentionality, and will now proceed to do the same with the "what," the values of the plan intended by the final vision.

2. FAUST'S FINAL VISION IN CONTEXT

> A swamp still skirts the mountain chain
> And poisons all the land retrieved;

I hope the marshland yet to drain,
And thus surpass what I achieved.
(11559ff.)

Faust wishes his venture to

[B]ringing the earth back to itself,
Setting the waves a boundary,
Putting a bond around the sea.
(11541ff.)

Instead of respecting and using the mountains as a natural boundary, Faust has pushed the "bond," the dike, far out towards the sea. Consequently a marsh, which he overlooked or actually created on his newly won land, now threatens to render the already "teeming land" (11106) uninhabitable. He plans a "groove" to save "all the land retrieved" — a new venture to salvage and justify the earlier effort. In this, Faust is "analogous to the essence of modern man," as perhaps already implied in the *Ankündigung der Helena* (1826). Moreover, Faust leaves no room to doubt how he intends to remove the dangerous situation he has himself created:

Feel free to use finesse;
Get throngs and throngs of laborers here,
Spur on with pleasure or severe,
Use ample pay, allure or press!
And every day inform me how the throng
Has pushed the groove we undertook along.
(11551ff.)

Veiled echoes of this recur in the vision. Faust praises

the edges of the hill
That has been raised by bold men's zealous will.

And we know how this work was accomplished:

Human sacrifices bled,
Tortured yells would pierce the night,
And where blazes seaward sped
A canal would greet the light.
(11127ff.)

The strange and uncanny affair was the work of the devil. Now Faust no longer speaks "sorcery" (11423). This is part of his last transformation on earth. "Human without guile" he stands before Nature (11406). From now on, not Mephisto but "the master's word alone has real might" (11502). Enjoying the sound of (as he believes) the "throng which slaves for me" (11540), Faust has become the exploiter and oppressor of his own. There is no reason to assume that this attitude has changed in his final vision, even if he no longer expresses it directly. That final vision, however, contains revealing details. Speaking of the intended work, Faust ignores the individuals and their happiness: it is the masses he needs as an instrument and for whom he intends to provide a paradise:

> Green are the meadows, fertile; and in mirth
> Both men and herds live on this newest earth,
> . . . .
> A veritable paradise inside,
> Then let the dams be licked by raging tide.
> (11565ff.)

A beautiful vision, indeed. But let us compare Helena's Arcadia in Act III. It possesses woods and mountains, a hierarchy of nature, society, and institutions. It has traditions from a fatherland and yet is part of the "first world" in history. The "youthful vigor" of the land makes every individual "free" to achieve his own happiness in the context of the whole. With "leaping waves on each side" (9442—9573), Arcadia's "shore and sea" have truly "ended their dispute," that is, they have been brought back to each other, as Mephisto ironically says of Faust's pirate expedition (11222). But Faust triumphantly affirms that he only brings the earth "back to itself" (see above). That is, he reconciles the earth with itself, but not with the sea. In fact, he forcibly separates them, only to have them clash forcefully at the border separating them.

Faust's venture ends in total catastrophe, in the kind of localized destruction of the world and mankind presented in the "Klassische Walpurgisnacht" as an event constantly recurring throughout history. In those variations, the work explores the cause: the use of evil means. Force is displayed not only in the recruitment of workers and 'resettlement' of Philemon and Baucis, but even in the related violence inherent in the abstract artificiality and revolutionary arbitrariness of this design of a "society in its earliest stages,"[7] unscrupulously isolated from any historical roots.

This entire concept is partly motivated by a commendable disgust with courtly profligacy (see the conversation with Mephisto at the beginning of Act IV), but now Faust also desires power over man and nature. He speaks as an omnipotent savior of the world and benefactor of mankind: "For many millions I shall open regions" (11565) and he wants to "join" them:

> At such a throng I happily would stare,
> Would with free people on free ground their freedom share.
> (11580f.)

These lines have reconciled even otherwise critical interpreters. Is Faust not envisioning an egalitarian society such as that of the French Revolution? But with the word "throng," he glosses over reference to the slaving men – people he has no qualms about putting under the yoke for his own ends. The ground is "free" because the envisaged land has become private or collective property; but this property would be taxed by the Emperor or the Church (11037f., 11115f.), if not by Faust as well. People are "free" because they are not directly subject to the Emperor. They are, however subject to the Emperor's vassal Faust who, as the self-appointed neighbor, resorts ever more quickly and unscrupulously to the right of the stronger, to the principle of force, wherever people do not "crouch" and "cringe" as subjects (11134).

We shall look at this in greater detail in relation to Schopenhauer's writings. At this point, the question is: are our reservations assuaged at all by the core of the final vision? Faust links it to its dramatic climax, the struggle against the sea:

> And as it gnaws to rush inside with force,
> A common impulse speeds to fill the gaps.
> Yes, this is the highest wisdom that we own,
> The wisdom I am totally devoted to:
> Freedom and Life, they are deserved alone
> By those who every day must conquer them anew.
> (11571ff.)

Exhilarating words! And "common impulse" can indicate spontaneous solidarity. But is this any more than the mere force of necessity? These people have no choice but to comply and are doomed to struggle against the sea, albeit under Faust's leadership.

It is at this very point that Faust displays the "demonic" element, which "rarely recommending itself to our affections by goodness of heart," exercises enormous force and power "even over the elements, and who can say how much farther such influence may extend?" Such people are "deceived," if not "deceivers" and may be overcome by nothing but the universe itself, with which they are in conflict.[8]

Indeed, Faust resembles a "war demon" (as Goethe called Napoleon), a god-like constructor in whose "theatre of war"[9] human beings must every day win anew their "Freedom and Life," that is, must struggle for their barest survival.

An image truly worthy of Schopenhauer of the suffering inflicted by the all-embracing life will. Faust is clearly like a man who "far surpassed others by extraordinary intellectual powers, and who inflicted unspeakable sufferings upon millions of others — for example, as a conqueror" (§64).

Conquerors of every sort pretend to themselves and to others that they benefit the world, and Faust indulges in such 'love' intentions (see above). While Goethe exposes the deep, problematic nature of such self-delusion he also makes it seem humanly and psychologically plausible. What Faust presents is nothing less than a pure, laboratory-type model — projected onto a totalitarian collective — of the same lifestyle he himself enjoys and praises as "the highest wisdom":

> For many millions I shall open regions
> To dwell, not safe, but actively and free.
> . . . .
> Surrounded by such danger, each one thrives;
> Childhood, manhood, age lead active lives.
> (11565—11578)

Year in and year out, from generation to generation, people are ever prepared to hasten back and forth, to and fro, between field and dike. It is a man-made circle performed by men (where, might one ask, are the women?): an institutionalized vicious circle that remains static, never moving from one spot and certainly not progressing beyond the "beginnings" of society to any kind of culture.

In contrast, it was precisely in the knowledge bestowed by art, religion, and philosophy that Schopenhauer saw the possibility of salvation from the vicious circle of life will and suffering. Goethe himself found refuge from the wretched "business of war and flight" of the French campaign[10] in culture, science, and art. At the end of the campaign, this refuge was found in the domesticity of his own home; up to that time, he found it in the far more elegant residences of his hosts.

Such havens, alternatives to Faust's "utopia," are represented by the aged couple on the hill with the chapel and linden trees, a symbol for justice and dance, that is, for happy yet disciplined sensibility. Here the old people used to rescue and nurse shipwreck victims, performing on earth a task that the good spirits continue in heaven. For Faust's part, he plans a permanent demonstration of the precise opposite and he even destroys this particular area of culture.

Admittedly, when this happens Faust "feels annoyed" (11341). He knows no better; he has acted carelessly and irresponsibly. He is unable to solve the dilemma of ends and means and to fulfill the law of action, combined in the two maxims Goethe summarizes as the true "highest wisdom": "The doer is always without a conscience; no one has a conscience except he who contemplates."[11] The synthesis sought is, of course, the doer who also contemplates:

> Thinking and doing and thinking that is the sum of all wisdom, acknowledged and put into practice from the earliest times, though not perceived by everybody. Both must move back and forth in life continually, like breathing out and breathing in; like question and answer, the one ought not to take place without the other. He who makes it a rule to assess action by thought and thought by action, something that the spirit of human sense secretly whispers into each newborn child's ear, cannot go wrong, and if he does go wrong, he will soon find his way back to the right path.[12]

Faust, however, makes no effort to follow this law any more. Earlier in his "dark aspiration," he repeatedly remembered the right road (329) — namely that he had strayed from it. At the end, he is not aware even of this much. Apart from the climax of the Helena adventure, Faust "ever errs as long as he will strive" (317). When it comes to the 'love intentions' of the final vision, which reflect his self-gratified enjoyment of his own greatness and striving, Faust violates the ideal of the ruler: "Ruling and enjoyment are not compatible. Enjoyment means belonging to oneself and others in gaiety; ruling means being of service to oneself and others in the most earnest sense."[13]

Faust enjoys (as he predicts he will at the end of Act IV [10181—88, 10252—59]) "commanding" (*Befehlen*) inasmuch as it relates to his "strength" (*Kraft*) to achieve his "deed" (*Tat*). For that reason, he fails to achieve the object of ruling and the responsibility it entails. His loss of eyesight then becomes both symbol and cause of his failure — of his crimes against mankind — as well as of his unawareness of it all.

This entire situation is reminiscent of Schopenhauer's theory of the deeper, generally unrecognized common nature of all suffering:

> . . . the difference between him who inflicts the suffering and him who must bear it is only the phenomenon, and does not concern the thing-in-itself, for this is the will living in both, which here deceived by the knowledge which is bound to its service, does not recognize itself, and suffering in another, and thus, in the presence of excitement, buries its teeth in its own flesh. . . . The inflicter of suffering and the sufferer are one. The former errs in that he believes he is not a partaker in the guilt. If the eyes of both were opened, the inflicter of suffering would see that he lives in all that suffers pain in the wide world. . . . And the sufferer would see that all the wickedness which is or ever was committed in the world proceeds from that will which constitutes his own nature also. (§63)

The basic situations are similar, although with his metaphysical speculations the philosopher transcends the phenomena that capture the poet's attention. Goethe endows his hero with the very ingredients Schopenhauer sees in the tragedy of life itself: "the never-satisfied wishes, the frustrated efforts, the hopes unmercifully crushed by fate, the unfortunate errors of the whole life" (§58). Admittedly, he does deny Faust the self-knowledge that Schopenhauer would prefer to see as the climax of life and tragedy (§51). On the contrary, Faust's blindness and death are perfectly timed to spare him the terrible realization of his tragic bankruptcy. Furthermore, Goethe redeems him and sends him to heaven, which will purge him more benevolently than purgatory and "lead him to clarity" (309).

At the same time, however, Faust's lack of insight represents a cruelly ironic mockery of the hero with which Schopenhauer would be in agreement. Of course, this is no help to the exploited workers or to the thousands who will be engulfed by the sea as it surges back to reclaim the land. Goethe does nothing to mitigate the earthly disaster.[14] On the other hand, he also spares us direct sight of the total catastrophe. All of this is part of the higher poetic justice with its effect, the catharsis on stage and in the auditorium. It serves to project the perspective into the cosmic, the "bird's eye view." As is frequently the case in *Faust II*, the catastrophe results from a hubristic transgression of human limitations, a transgression that is punished, expiated, and compensated for. Thus poetic justice and catharsis help to "moderate" not only our "pain" but also our petty "desire" to make condemning judgments self-righteously and unobjectively — a tendency found in Schopenhauer.[15]

Let us reconsider some of these points as we move forward. Thanks to his 'love intentions,' Faust's constant striving dominates the final vision. The problematic nature of this vision derives from its intended object and the manner of the intention: Faust's "love" emerges in such a "pure" form because it is here that he finally perfects his individuality, surpassing and stylizing himself to become the very incarnation of the aspiring and active man: a man without conscience or responsible consideration for his fellow human beings.

The hero of this tragedy arouses our sympathy, admiration, and fascination. And it is for this reason alone that he also is able to arouse in us pity, awe, and shock. Goethe takes great care to develop both sides of this spectrum of feelings. Faust fascinates and impresses us "esthetically" (Schiller).[16] This should not let us forget the dubious nature of his ethics (Schiller), but neither should this awareness let us forget the indefatigable aspiration that he so magnificently exemplifies. While this is grandiose in itself, it is in pragmatic terms indispensable whenever major difficulties and obstacles must be overcome. It is only from this kind of delicate balance between the phenomena and our reactions to them that the desired third element can result: the fruitful synthesis of doing and thinking, of aspiration and care, of action and conscience.

Writing as a philosopher, Schopenhauer does not offer the reader a picture of human experience that seizes our whole being; that is the province of the poet. But neither does he refer us to the active life. Instead he recommends that we seek refuge in meditation and asceticism, in the pietistic ideals of his own youth.

3. FAUST AND THE FOUR GRAY WOMEN

It is in Faust's fight with the "four gray women," Want, Guilt, Need, and Worry (or Care), that Goethe carefully develops his hero's final transformation into a character blindly devoted to nothing but action, without the worries of conscience, careful consideration, and responsible prudence. These four women are reminiscent of the witches in *Macbeth* and of the allegories of the late medieval morality plays in the *Everyman* style. Goethe was familiar with similar treatments of the Worry motif found in writing since antiquity. However, he never expressly mentioned a point that was apparently much more present in his thinking: the way in which Schopenhauer linked the forms (or masks) of Worry with the main Faustian motif of striving.

According to Schopenhauer, the moving force behind the life will is constant alternation between the two polar principles of want and its passionate appeasement. The experience of "want, need, worry" means suffering and pain.

> If we succeed, which is very difficult, in removing pain in this form, it immediately assumes a thousand others, varying according to age and circumstances, such as lust, . . . jealousy, envy, hatred, anxiety, ambition, covetousness, sickness, etc. If at last it can find entrance in no other form it comes in the sad gray garments of tediousness and ennui, against which we then strive in various ways. If finally we succeed in driving this away, we shall hardly do so without letting pain enter in one of its earlier forms, . . . the whole material of which was already there before, yet could not come into consciousness as worry because there was no capacity left for it, and therefore this material of worry remained indistinct and unobserved in a cloudy form on the farthest horizon of consciousness. But now that there is room, this prepared material at once comes foward and occupies the throne of the reigning worry of the day. (§57)

Goethe's groups of scenes — a peak in world drama — follows Schopenhauer's pattern, which he very likely recalled. In the appearance of the four crones, as well as in the scenes immediately preceding and following, Faust struggles with the "various forms" of Worry (or Care). With each succeeding challenge Faust musters increasing sovereignty to resist their attacks on his 'life will.'[17]

After a difficult inner battle, Faust relieves his spirit of the worry 'Want' by appropriating the homestead of his elderly neighbors, for it is "the most cruel rack" (11257) to have to feel — especially "in riches" — this "want" (11252f.) What begins as a forcible relocation of the elderly couple quickly turns to murder and destruction of the very thing he covets, thus attracting the worry 'Guilt.' Its mastery demands greater power and 'art': Faust rationalizes the murder down to a "robbery," the crime down to an "ill-considered, savage blow." He allows that the deed was "commanded fast" but maintains that it was, above all, "too fast obeyed!" (11371—83). He shifts the blame to others: "I curse it; share it and leave!" (11373).

Nevertheless Faust is still susceptible to certain types of worry at this stage. Such susceptibility vanishes completely when Worry and her various masks appear in the form of the four gray women. The most extreme version of the worry of "Need" (which is the worry of "Death") challenges him to consummate his existence, bringing himself and his work to completion. From this moment on, Faust renounces the devil's magic and his own sorcery: from this point he relies solely on his own human potential, intensifying and stylizing his individuality to become the incarnation of the principle of active striving.

It was probably not only Schopenhauer's description of the forms of Worry but also his theory of "acquired character" (§55) that enables Goethe to present this process with such refinement. In Goethe's *Urworte Orphisch*, written in 1817, the accumulation in the fourth stanza of the words "will," "willing," "free will," and the "iron must" as the basis and consequence of voluntarily relinquished "freedom" may contain echoes of his conversations with Schopenhauer in 1812 and 1814. In that text, Goethe's concept of freedom of will already seems to be as restricted as Schopenhauer's. Moreover, Goethe's own comments on the poem dating from 1820 combine the philosophers published thoughts and vocabulary with his own:

The fact that "innate strength and individuality determine the fate of the human being more than everything else" justifies the notion of the "unchangeable nature of the individual." His *demon* (Emphasis added.) . . . survives everything, and this is then the individual's true nature, the old Adam, or whatever one likes to call it, which, however often it is driven out, returns again and again, ever more relentlessly." It can "certainly be fragmented." The "demon, . . . independent and egoistic, which intervened in the world with absolute will and was only vexed to find that . . . its first and original character was restricted in its effects and hindered in its inclinations," becomes "aware that it can determine itself" —

— that is, "embraces a second being, which is possessed of the same external indestructible will as itself." This and then "hope" are the subjects of the poem — impossible in Schopenhauer.

For his part, Schopenhauer sees man as the arena for a struggle between his Christ-like nature and his Adam-like nature. In the passages Goethe quotes to help reflect his own thoughts, the philosopher frequently mentions an individual "demon." In agreement with Goethe, but much more explicitly, Schopenhauer rejects the ethical and anthropological voluntarism of the Enlightenment tradition. As Schopenhauer notes, "According to that doctrine, the individual wills what he knows, and according to mine, he knows what he wills." A person's motives determine him and mark his "intelligible," "original character," which is reflected in his "empirical character" through the "deed."

The second passage of which Goethe takes particular note in the book maintains that this "knowledge," together with the growing awareness of what we are "capable of doing," enables us to "carry out the unchangeable role of our own person . . . deliberately and methodically as if we had learned it." This is precisely what Faust means when he explains to Worry that up into his old age his life has been a restless "progression" from one desire to another. This ceaseless embracing of one wish while discarding another is, in fact, the basic structure of Schopenhauer's concept of the life drive. Faust's character, in developing itself, becomes more and more identical with the life drive, or will to live. Therefore the most interesting aspect here is that in Schopenhauer's view, when character is acquired in this way, the "core" of individuality develops in an ever purer form because

> if we desire to grasp and possess one talent, we must *renounce* [*entsagen*] . . . innumerable others . . . . [W]e will endeavor to cultivate those talents which are naturally prominent in us. We will always turn to those occupations in which they are valuable and to the purpose and entirely avoid, even with *self-renunciation*, those pursuits for which we have naturally little aptitude. (§55) (Emphasis added.)

Nowhere does Goethe express this dramatic structural law more purely than in the figure of the aging Faust. Here he allows — to use Schopenhauer's words — "what is bad in the character to come out more strongly with time." Schopenhauer further notes that if "we could form a good resolution only after a long conflict with a bad inclination, this conflict will pass by . . .: it is in the resolve which follows that we shall see what kind of men we are." Our "deeds" are the "expression of character, the mirror of the will," in which we recognize "our inmost self, the kernel of our will."

Faust's palace is his inner fortress, the kernel of his personality and will. This is why the concerted attack mounted by the four gray women founders at his doorstep. Granted, the maskless principle of "Worry" (Care) can penetrate into the palace but then it, too, is repelled. In this scene, activity and Worry (the creative and the obstructive principles of anticipatory intentionality) clash head-on. (Compare Schopenhauer, §57: "Excessive joy and very keen suffering . . . are produced not by what is really present, but the anticipation of the future.") Faust perfects (that is, radicalizes) the form of his existence by conquering Worry swiftly, though he manages this by an act of sublimation, which satisfies every possible claim to existential absurdity.

One cannot break the spirits' iron ties;

And yet your power, Worry, creeping, great,
I shall refuse to recognize.
(11492ff.)

Faust refuses to recognize what he cannot avoid recognizing. He does the impossible and pays dearly for it: "pure" activity triumphs in the ruins of his exhausted physical strength. As Worry departs, it makes him blind. That is, when Worry leaves him, so do foresight, careful consideration, conscience. There are no longer any restraints on his will to act, and Faust takes a conscious pleasure in this (as he, in his final moment, will take pleasure in his continued striving):

Night seems now to fall more deeply –
Yet inside me there shines a brilliant light;
What I have thought I hasten to fulfill:
The master's word alone has real might.
Up from your straw, you servants! Every man!
Let happy eyes behold my daring plan.
(11499ff.)

Here the principle of constantly active aspiration is depicted in its purest form as a splendid inner light. However, this aspiration is also blind, lacking any orientation in reality. It lights without enlightening; burning it burns itself out, dragging others down with it to destruction. Moreover, the shadows cast by this light also are perfected: unprincipled inconsiderateness and the illusory design in Faust's final vision of utopia. Goethe also addresses this earlier through individual psychology in *Part I* (1808) and later, even retrospectively, through social psychology in Act IV of *Part II.*

4. THE UTOPIAN SOCIETY AS AN INSTITUTION FOR COMBATING WORRY

The moment in which Faust dies is not one he wishes to grasp and hold ever unchanging but one he cherishes precisely because it is the very moment in which he anticipates an energetically intended future: the realization of Faustian striving per se. That is why God already nullified Faust's bargain with Mephisto in the Prologue.

But why does Faust enter the pact in the first place? It is not to quench his thirst for knowledge, although frustration on this point does lead him to consider taking his own life on Easter Night. Faust strikes this bargain because his life will is consistently disappointed in its pursuit of satisfaction and finally paralyzed by the melancholia of Worry. Alone in his den Faust muses, "Since joy on joy crashed on the rocks of time":

> Deep in the heart there dwells relentless worry
> And secretly infects us with despair;
> Restless, she sways and poisons peace and joy.
> She always finds new masks she can employ:
> She may appear as house and home, as child and wife,
> As fire, water, poison, knife —
> What does not strike, still makes you quail,
> And what you never lose, for that you always wail.
>
> (644–651)

Thus the concept Goethe borrowed from Schopenhauer — Worry ever in "new masks" — is in fact one the young philosopher originally took from Goethe and developed further. A similar give and take evolves when the two address the perpetual frustration of the life will. Schopenhauer expresses it thus: "the wish, i.e., want, is the condition which precedes every pleasure. But with the satisfaction the wish and therefore the pleasure cease" (§58). Faust's despair is even more pessimistic when he thinks every morning that the day "will scorn my *every* wish, fulfill not *one*." This too is already the frustrated confidence of the potential man of total action who torments himself with his inability to translate one single ideal into reality:

> The god that dwells within my heart
> . . . .
> He cannot move the world outside;
> And thus existence burdens me,
> Death is desired, I hate life.
>
> (1566–71)

Faust's awareness of the inevitable disappointment of all earthly desires (Schopenhauer considers this the fundamental experience of "existence" as it is "cast into time") makes it impossible for any moment ever to satisfy him. This is the point of the wager. The pact has no other function than to oblige Mephisto to provide practical, concrete confirmation of Faust's enraged despair in creation and in God.

This metaphysical experiment in nihilism backfires on the devil. At first he wants to ruin Faust's soul through the conventional route of earthly sin ("Dust he shall eat, and that with zest." [334]). However, from the *Fragment* of 1790, Goethe intends Faust to "perish" from the "insatiability" of his desire for enjoyment, thus deflecting his "spirit," which "sweeps ever onward with unbridled might," from its "hasty striving" that "leaps over the earth's delights" (1856ff.). This now becomes Faust's own desire, except that he refuses to succumb, to allow the process of torment and accusation come to an end. In fact he wants never to be content, because that proves the legitimacy of his protest. In this way, the wager itself becomes the real test; the trial in the sense of process and judgment becomes more important than the technical and bureaucratic monstrosity of the pact. Each wish becomes disappointment and disillusionment, and each disillusionment is succeeded by yet another wish and disillusionment so that "enjoyment, distress, [annoyance and success] [p]ursue each other" without end. As Faust concludes: "*Nothing but restless activity proves a man*" (1756ff.). (Emphasis added.)

Can we take this at face value? In the final analysis, the justification for Faust's salvation, the much-praised incessant aspiration, is nothing but compensation, suppression of worry and existential disappointment, desperate flight from the tormeting life will into its very jaws. Schopenhauer offers the same lesson, although his conclusions are not identical to the ones Goethe comes to in *Faust.* According to Schopenhauer, the continual pursuit of what does not exist in life, "lasting satisfaction of the harassing desires and worries, which are constantly breeding new ones" makes us hasten "into the infinite"

> or, what is more rare and presupposes a certain strength of character, till we reach a wish which is not satisfied and yet cannot be given up. In that case we have, as it were, found what we sought, something that we can always blame instead of our own nature as the source of our suffering. And thus, although we are now at variance with our fate, we are reconciled to our existence .... The result ... is a somewhat melancholy disposition, the constant endurance of a single great pain, and the contempt for all lesser sorrows or joys that proceeds from it; consequently an already nobler phenomenon than that constant seizing upon ever-new forms of illusion. (§57)

The philosopher nevertheless extends the traditional model of the sufferer of melancholy, which here appears to be quite intact, to apply to the languishers in *Weltschmerz* of his own age and our own modern apostles of existential-theological philosophy, the "ever-famished yearners for the out-of-reach" (8204ff.), who appear to be trying to stage a comeback. Goethe takes a different approach, drawing a line from that model directly to our modern obsession with activity. We can note this in Faust's final vision and its context.

The new world Faust creates out of the ocean is a rival to creation; thus Faust is a rival to God. However, the scene reads differently when one bears in mind the model of melancholy (complete with the melancholic's heroic pride and guilty conscience) and views the sequence as part of a Luciferian experiment in nihilism designed to expose the world — in true Schopenhauerean manner — as the worst of all possible worlds.

Even when Faust's focus is on exposing the world in this way, he has already abandoned his desire for knowledge at any price. He no longer wants to know "what secret force hides in the world and rules its course" (383). Now, as ruler of a new land, he is interested only in what can be achieved on earth. He is no longer tormented by the fact that the universal law of the life will and suffering can be mastered (or suppressed) only by unbridled activity. In fact, his very nature has become unbridled activity. He embodies this life form and defines it in the face of Worry, the life principle diametrically opposed to his own:

> Through all the world I only raced:
> Whatever I might crave, I laid my hand on,
> What would not do, I would abandon,
> And what escaped I would let go.
> I only would desire and attain,
> And wish for more, and thus with might and main
> I stormed through life . . .
>
> (11433—39)

Whether Faust's life was or was not as he describes it, this is what he means his life to have been like. Only "in his progression" did he find "agony and joy, / At every moment still dissatisfied" (11451—52). This is Schopenhauer's aimlessness of the life will but imbued with the new value of permanently and actively striving intentionality. And it is expressed with the self-confidence of a man dedicated to pure action, without the pathos of melancholic despair and metaphysical nihilism, without any religious or worldly worry.

At this point, Faust appears to be immune to any kind of worry or care whatsoever. Yet it is precisely Faust's new pathos of perfectly worry-free active intentionality that reveals echoes of his arch-opponent. Were this not the case, it is unlikely that Faust would mandate his own life form as a permanent institution for his new utopian society. Man must secure his life over and over again, day after day: the aimless, eternal circle of the life will is simplified and domesticated by its regular recurrence. This method of combating one form of worry suppresses every other form of worry as well. In fact, the potential for anxiety inherent in this struggle is anesthetized by man's ceaseless anticipation of and preparation for an active confrontation with worry.

Intense activity that prevents one from worrying can often be essential in life. Goethe even recommends it as a way of dealing with worry and melancholy — albeit in doses which ensure that worries arising from reasonable prudence, from foresight and hindsight are not crushed. "Doing and thinking and doing" should alternate in a life-sustaining balance.

Faust demonstrates just how dangerous it can be to adopt the extreme, radical form of this struggle, to become completely immune to worry. Governments sometimes resort to this technique — for example, by proclaiming a war or war-like state of emergency as the permanent and normal state of affairs — so that the people leave them undisturbed to govern as they wish.[18] The Jacobins did this. Bismark, "the white revolutionary," spread the fear of a permanent crisis against which politics had to erect "dams."[19] Napoleon later appeared to Goethe to be a war demon, the campaign in France a theater of war — and Faust's utopia, as we have already said, the staging of one.

In France, the poet was seized at the power of the element of water and by the necessity to restrict all his thoughts to the elementary needs of the moment. Admittedly, he did not blame this on evil intention, but on the negligence of the warring leaders.

Fault also brings about evil, but this does not result from deliberately evil intentions but from sheer thoughtlessness. At first he is even reluctant to allow it to happen, although eventually he becomes indifferent to its quality. The *liberté, egalité, fraternité* of his utopian society become as questionable as his supreme desire to stand "on free ground among free people." We have seen that. And Faust, harbinger of happiness for mankind and a "revolutionary from above," does not even want 'progress' since in his view enlightenment and education would only serve to "raise rebels" (Act IV, 10158f.). Like the revolutionary theoreticians of natural law, Faust the ruler wants to reconstruct the "virtues of primitive human society" (p. 606) by force, arbitrarily and artificially, replacing the characteristics of the present society as it has evolved historically. Faust's new constitution lends itself all the better to the mechanics of ruling and is a projection of the lifestyle of its architect: in it, combating worry through restless activity has become an institution.

5. FAUST AND THE SEA: "ELEMENTS" OF THE LIFE WILL

After the completion of Act V Goethe took pains to shift Faust's struggle with the sea in Act IV even more clearly into the double perspective of the modern power and work psychology. Here Faust reveals his plans to Mephisto, who creates the opportunity for his success (p. 606): the victory of the emperor, who will reward Faust with the seashore. To summarize,[20] when Faust describes the sea, it is not the rich source of life presented in Act II, but a destructive force such as that seen in the *Witterungslehre* (1825). Its introduction exhorts us to see and read symbolically:

That which is true, identical with the divine, we can never recognize directly, but can only see in the reflection (*Abglanz*), in the example, symbol, in isolated and related phenomena . . . . The elements must be called the height of despotism . . . , colossal opponents . . . , which we must eternally fight and can only master through the greatest strength of spirit, through courage and cunning in the individual case . . . . The ultimate . . . is to become aware of what laws and rules nature bears within herself in order to impress that unbridled, lawless creature.

When the tidal flood threatens, Goethe thinks of mountains, the natural border, and from viewing the workings between sea and shore Faust immediately extracts this principle for his planned venture:

Pressing as the flood may be,
It has to creep around the hill.
It may move forward with despotic arrogance,
Slight elevations yet resist with ease and pride.
(10222ff.)

But Faust also uses this principle as an expressive metaphor for his compulsion to seize his neighbor's property. In this case, his "greatest spirit strength," "cunning and courage" acquire quite a special character. Seeing the homestead, Faust cries out in helpless torment:

> The almighty grip of my despotic will
> Breaks down before that sandy hill.
> (11255ff.)

The parallels with the previous, much earlier quotation reveal that Faust has become like the element. When Faust attributes despotism to the sea, he is unaware that those words can also be read as describing the later Faust himself. He calls the sea "imperious," "unbridled," and "imbued with power," but Faust will become all of these things as well, especially when he undertakes to "lock the imperious ocean from the coast" (10202, 10229, 10215).

The identity and affinity of the elemental in Faust and the sea is highly reminiscent of Schopenhauer's one will to life that is present in every natural being and phenomenon, "which here, failing to recognize itself, turns its weapons against itself, and, by seeking increased happiness in one of its phenomena, imposes the greatest suffering upon another." "[T]he bad man . . . is himself the whole will; . . . consequently he is not only the inflicter of pain but also the endurer of it, from whose suffering he is only separated by an illusive dream" (§65).

Goethe does not allow himself to embark on such metaphysical speculations, restricting himself to phenomena that are symbolic. He prefers to offer "a particular phenomenon" in which "the general is also conveyed without one's being aware of it, or not until later" (HA 12, 471). So he wrote: "In the works of man as in those of nature it is really the intentions which most deserve our attention."[21]

In this sense, and expressed as anthropomorphically as poetic fiction permits, Faust compares his intentions with those of the sea. In contrast to the alleged "imperiousness" and "sterility" of the element, he invokes "the liberal mind which cherishes all rights" (10219, 10203f.). Good intentions — but what becomes of them? Later Mephisto reveals to us that "the ocean's freedom frees the mind." In practical terms, however, what this irony means is that Faust's "naval enterprise" is emancipated to become "piracy" (11177).

As mentioned earlier, Faust's covetousness crashes around the hill with the rage of a tidal wave until in "self-disgust" he tires of honoring the rights of others (11271f.). He attacks his neighbor's property and in so doing submits himself to the law of the strongest and of violence. However, already much earlier in the text than this, he unwittingly undermines his commendable intentions when he reproaches the sea (in language similar to that discussed here) and then dedicates himself to the trinity of power (*Herrschaft*), estate (*Eigentum*), and the deed (*Tat*) (10187f.). He says that in "commanding" he wants to find "his bliss." The "god that dwells within his heart" will then (and, in fact, does) "move" "the world outside" as he wants (see above, 1566—69). What he whose "soul is fraught with high design" whispers "to those most closely in his confidence" is "brought about and everybody stands astounded" (10252ff.). Indeed Lynkeus laments the resulting destruction:

What for many centuries
Pleased all eyes — now is gone.
(11336f.)

But Faust's new land with its population, inasmuch as they already exist, will also vanish: victims of his hubristic commands. Meanwhile, he sacrifices the worry of his conscience on the altar of violence, as the chorus of the devil and his three men of might declare:

The ancient word still makes good sense:
Accept the rule of violence!
If you are bold and don't give in,
Then risk your house and home — yourself!
(11374ff.)

Yourself. Your very own soul. Faust gives his soul away now, not in the devil's pact. Now perhaps we can understand his reaction when, at the beginning of that last stage of his life, that of the ruler, he observes the sea exercising its sterile rule with thundering force: he experiences "passionately unbridled blood," "wry perversity of temper" (10204ff.). Despair, anxiety grasp him from the void. He realizes that *tat twam asi* = "that is you" (Schopenhauer §40) *tua res agitur*. The play within the play mirrors Faust. It is a projection of Faust's self onto the element. Against this force and against his fear of it he musters all his strength — everything within him that resembles this violent element. Precisely this makes him become increasingly like the element he battles. After his triumph over Worry, his incessant striving becomes, quite literally, "purely" that which he earlier claims it to be in mirroring his opponent: force taking up the fight against the force of nature just as it would against any other force.

> There wave on wave imbued with force has heaved,
> But to withdraw — and nothing is achieved;
> Which drives me near to desperate distress!
> . . . .
> Such elemental might unharnessed, purposeless!
> There dares my spirit soar past all it knew;
> Here I would struggle, this I would subdue.
> (10216ff.)

Faust's will reacts 'sublimely.' Like Kant and Schiller, Schopenhauer explains the sublime in terms of the confrontation of the individual with the "absolute desert," (§39) "the desolation left in its tracks," (Kant),[22] of the victory of the dam-builder over the sea, "the most important of the elements." The "relative greatness outside him is the mirror wherein the individual perceives the absolute greatness within himself" (Schiller).[23] It is certainly "strength of spirit," "a power of resistance . . . which gives us courage to be able to measure ourselves against the seeming omnipotence of nature" (Kant). But Faust's is not the esthetic/ethical "freedom" of "reason," its "independence of natural laws" (Schiller).

Certainly Faust no longer cares or worries strongly about what "we are wont to be solicitous (worldly goods, health and life)" (Kant). However, this is not identical to Schopenhauer's "forcible detachment from the life will." In Faust, "the pure subject of knowing" is far from gazing "through that strife of nature, through that picture of the broken will" and "unshaken and 'unconcerned' comprehending the Ideas even in those objects which are threatening and terrible to the will" (Schopenhauer, §39).

That may be the perspective of someone on the receiving end of this force, but in fact Faust goes only half way. He looks at the sea and "he sees clearly our dependence, our strife with hostile nature, our will broken in the conflict" (§39). But what he sees is a mirror image, against which his own proportion and share of the universal life is rebelling — and in this act of revolt consolidates itself as that very same will.

Faust's active striving derives from his desperate flight from a desolate feeling of futility and worry. This he indicates expressly when he makes the pact with Mephisto and unconsciously in the final vision. Between both stations, confronted by the sea, futility and worry escalate to anxiety and despair, indeed, to a bleak sense of spiritual loss. Though unconsciously, he senses the whole spectacle as the fate shared by the perfectly "Faustian" human being and the sea, a symbol of the "element," both embraced by the common primordial element, that of the life will.

Faust's meeting with the sea in Act IV provides us with Goethe's final commentary on the Faust of Act V. The strange and apparently unmotivated abruptness with which Faust's anxiety and despair appear reveal these emotions to be the deeper motives — as does his heroic but hubristic reaction to them. This reaction is meant to immunize him against future occurrences of fear and despair. And his tragically vain struggle is linked — in a final reckoning, as it were — with the death of Euphorion or, to be more precise, with the massive lament that the Chorus strikes up over the body of Faust's dead son. The final strophes in particular also apply to Faust himself: erring into the elemental; ceaseless striving for the absolute; the failure; the awful life reckoning which, in its general validity, lacks none of Schopenhauer's pessimism.

But whereas the philosopher accepts only "salvation" into nirvana and that only through meditation and high art, Goethe's Choir, starting with the strangely sudden "yet" of the fourth line from the end, calls arbitrarily — in spite of everything — for valiant faith in the wonder of the constant renewal of human existence (such as we see in the salvation at the end). And this happens, as the Choir claims, in precisely that sudden and sovereign artistic hypothesis, in two ways: through "good," "motherly," "creative Nature" (which is, more for Goethe than Schopenhauer, part of the life will, despite its destructive tendencies) and its product, art: the high and also the popular and ritual art of song.

Goethe shared the same indomitable, indeed Promethean, will to creation and life that Faust continues to display in the face of all the onslaughts of worry. And Faust's doubts, his despair at life, at its cruelty and futility, overcame Goethe, too. At times they affected him more terribly than the philosopher of pessimism. While my many comments on this point offered above and the passages from *Faust I* cited here support his point, it is most clearly documented in the final lines of this lament from the Chorus:

But you rushed most vehemently,
Heedless, into tangling flaw,
Fell foul incontinently
Of propriety and law;
Yet at last exalted yearning
Lent to native courage weight,
Splendid laurels to be earning
You were minded — not so Fate.

Who succeeds? That query dismal
Muffled Destiny leaves moot,
When misfortune most abysmal
Strikes the bloodied people mute.
Yet sing afresh new hymns and anthems.
Stand no longer bowed in pain:
For, as it has done through the ages,
The earth will grow new songs again.
(9923–38)

State University of New York at Albany

NOTES

* Translated by Penny Willard.

** The author is grateful to Carol Newhouse for her editing of this article.

[1] Cf. Wolfgang Wittkowski, "Faust und der Kaiser. Goethes letztes Wort zum *Faust*," *Deutsche Vierteljahresschrift für Literaturwissenschaft und Geistesgeschichte* Vol. **43** (1969), pp. 631–651.

[2] Arthur Hübscher, *Denker gegen den Strom. Schopenhauer: Gestern — Heute — Morgen* (Bonn: Bouvier/Grundmann, 1973), p. 80.

[3] Adele Schopenhauer to her brother Arthur, January 1819. *Goethe, Gespräche*. Gedenkausgabe, Vol. **23** (Zürich: Artemis, 1949), pp. 43f. *Die Welt als Wille und Vorstellung* quoted according to the first edition with the since the second edition. *Faust*, text and commentary references quoted according to *Goethes Werke*. Hamburger Ausgabe (hereinafter, HA), ed. Erich Trunz, Vol. **3** (Munich: Beck, 1982). The translations follow in part those of *Faust. A Tragedy*, trans. Walter Arndt, ed. Cyrus Hamlin (New York: Norton, 1976) and especially *Goethe's Faust*. The original German and a New Translation and Introduction by Walter Kaufmann. Part 1 and Sections from Part 2 (New York: Doubleday, 1961). Partly, however, they have been adapted by myself and Penny Willard, the translator of this essay from the original German.

[4] J. W. v. Goethe, *Tag- und Jahreshefte*, on 1819. Quoted by Hübscher, *op. cit.*, p. 78.

[5] See the classical philologist Dieter Bremer in the discussion of my paper "Vielfalt der

Autonomien im Kunstwerk: Fauste Ends," in W. Wittkowski, (ed.) *Revolution und Autonomie. Deutsche Autonomieästhetik im Zeitalter der Französischen Revolution. Ein Symposium* (Tübingen: Niemeyer, 1990), pp. 364f. Cf. also the somewhat extended version of this essay: "Goethe, Schopenhauer und Fausts Schlußvision," *Goethe Yearbook* Vol. **5** (1990), pp. 233—268.

[6] J. W. v. Goethe, "Principles de philosophie zoologique" (1830). HA 13, p. 224.

[7] J. W. v. Goethe, Outline of Act IV. of 16th May, 1831. HA 3, p. 606.

[8] J. W. v. Goethe, *Dichtung und Wahrheit*, HA 10, p. 177.

[9] J. W. v. Goethe, *Campagne in Frankreich 1792.* HA 10, p. 188.

[10] *Ibid.*, p. 344. See Richard Fischer, "'Dichter' und 'Geschichte': Goethe's 'Campagne in Frankreich,'" *Goethe Yearbook* Vol. **4** (1988), pp. 235—274. This article encouraged me to examine Goethe's symbolic analogies between natural phenomena and human behavior.

[11] J. W. v. Goethe, Maximen und Reflexionen. HA 12, 399, n. 251.

[12] J. W. v. Goethe, *Wilhelm Meisters Wanderjahre* II, 9. HA 8, p. 263.

[13] Goethe, Maximen und Reflektionen, HA 12, 368, no. 101.

[14] W. Wittkowski, "Irrestorable Destruction and Tragic Reconciliation in Goethe's 'Faust'" in *Goethe Proceedings*, ed. Clifford A. Bernd *et al.* (Columbia, S. C.: Camden House, 1984), pp. 93—106.

[15] Goethe, *Dichtung und Wahrheit.* HA 9, 581. Cf. W. Wittkowski, "Katharsis. Goethe, Aristoteles und der Streit der Philologen," *Goethe-Jahrbuch* 104 (1987), pp. 113—127.

[16] F. Schiller, Preface to *Die Räuber*, 1782.

[17] W. Wittkowski, "'Gedenke zu leben!' Schuld und Sorge in Goethes 'Faust'," *Publications of the English Goethe Society*, Vol. **38** (1968), pp. 114—145.

[18] See Rudolf zur Lippe, "Objektiver Faktor Subjektivität," *Kursbuch* 35 (1974), pp. 1—36, here p. 6; Thomas Leithäuser, Birgit Volmberg *et al.*, *Entwurf einer Empirie des Alltagsbewußtseins* (Frankfurt: Suhrkamp, 1981), p. 49.

[19] Lothar Gall, *Bismarck. Der weiße Revolutionär* (Frankfurt/Berlin/Vienna: Ullstein Propyläen, 1980), p. 688.

[20] See W. Wittkowski, "Faust and the Sea. Goethe's Verdict on his Hero," *Analecta Husserliana* XIX (1985), A-T. Tymieniecka, ed., pp. 433—445.

[21] Goethe, Maximen und Reflexionen, HA 12, 517, no. 1078.

[22] I. Kant, "Analytik des Erhabenen," *Kritik der Urteilskraft*, §28.

[23] F. Schiller, "Über das Erhabene," Schiller NA 21, pp. 47, 49.

SITANSU RAY

THE TAGOREAN INTERPRETATION OF "AMI": MAN'S SELF-ESTEEM

The poem "Ami"[1] ("I") was written by Rabindranath Tagore (1861–1941) on May 29, 1936 and was included in his poetical work *Shyamali* (1936). *Shyamali* is the feminine form of *Shyamal*, which means green. Tagore used the word *Shyamali* as a proper noun to name not only this poetical work but also the mud house he lived in from time to time in his old age. *Shyamali* is also the title of the concluding poem of the collection. This poem was composed on August 6, 1936 and is most affectionately associated with the earthen house called *Shyamali*.

Let us come to "Ami" again, with which we are more concerned. "Ami" is the Bengali synonym of the Sanskrit *Aham*, the German *Ich*, and English "I". Any corresponding synonym in any language denoting the first person singular form of the personal pronoun may be thought of. Transcending its day to day worldly usage that every *Ami* (I) is the principle instrument of epistemology, the main laboratory of existentialism, and the boundless realm of phenomenology. The poem "Ami" is one of the best poetical expositions of the phenomenology of the human mind. Conversely, the phenomenology of the human mind finds its best intrgrated yet concise poetical manifestation in the poem "Ami".

Quite metaphorically, viz., using a rhetorical expression of transferred epithet, Tagore says that it is the encolouration of man's consciousness which tinges the emerald as green and the ruby as red. It is not the sun alone, but it is the very light of man's visual perception, which illumines the sky. The rose is beautiful only because man perceives it as beautiful.

If this whole idea be accused of being mere philosophical doctrine and not be accepted as a passage from a poem, Tagore's answer is — it is the essential truth and that is why it is poetry. Philosophy and poetry are not opposed to each other. They are complementary and very often identical. Tagore is proud of realizing this truth. But, his pride should not be taken as self-centred fancifulness. He is proud on behalf of each and every man; or rather on behalf of the whole of mankind.

A-T. Tymieniecka (ed.), Analecta Husserliana, Vol. XXXVII, 409–412.

Man is not a passive component of creation. He himself is the nominative of all sorts of creativity including the hues of precious stone, the lustre of the sky, and the beauty of the rose. That is why he asserts in full confidence:

Amari chetanar range panna holo sabuj,
  chuni uthlo ranga hoye.
Ami chokh mellum akashe,
  jwale uthlo alo
    pube paschime.
Golaper dike cheye ballum "sundar,"
  sundar holo se.

This very beginning of the free-verse-style poem "Ami" declares his thought regarding man's self-esteem.

In keeping with the nihilistic fashion the ontologist may declare that there is no emerald, no ruby, no lustre; no rose, and even no "I" and no "you." On the other hand, the Infinite Omnipresent has always been endeavouring to know himself through the limit of man's self. That very apparent limit is called "Ami" (I). In the depth of "Ami", i.e., my self, light and shade came into union, forms evolved enlivened with feelings. All the so-called theoretical nullifications turned into vital affirmations by virtue of an illusive magic as it were, by virtue of lines and colours combined with pleasure and pain.

By virtue of creative imagination, man's self is all-encompassing. He has the brush and colour in his hands, as it were. He composes, draws, and paints the world anew. Moreover, he has, combined with this sense of beauty, the most precious wealth of his (her, also) heart, and that is love.

Suppose, a total annihiliation takes place by means of a universal cataclysm, natural or man-made. Suppose, there is no trace of human civilization or human being any more. Matter and energy will remain in some form or other in the solar system, but there will be no feeling, no sense of beauty, and no love. There will be nobody to say, "You are beautiful, I love you." If that stage should come, the Creator will have to sit again for one more age-long *Sadhana* (prayer) for the renewal of man, because it is only man who will utter, "You are beautiful, I love you." The illuminating, dynamic and loving nature of man's self makes him greater even than his creator, the master of all matter, cosmic energy, and life.

There is another of Tagore's poems bearing the title "Ami,"[2] one composed by him on February 11, 1931. This poem is included in the poetical work entitled *Parishesh* (The End).

This poem is also most important from the point of view of man's self-interpretation in existence. The gist of the poem is as follows:

The inner "I" within my self is the vital source of all my words, movements, music, and all manner of creativity. The firsthand understanding of my own self may suggest that I am the be-all and end-all of my own self. But actually it is not so. That "I" extends beyond my self. That is why, I know myself better when I recollect the memory of my beloved, when I identify myself with the heroes of history and the mythology I read. Very often that inner "I" is overshadowed by my constant daily chores. But, in reality the realm of my inner self transcends my matter-of-fact living. My inner "I" is omnipresent in the past, present, and future. "I" can move everywhere and whenever. My apparent "I" perceives that inner "I".

One more poem still with the title "Ami"[3] is to be found in the "Addenda" to the collection of poems titled *Shesh Saptak* (i.e., the concluding octave), published in 1935. The central idea of the poem is the delight of the poet's realization of his own existence.

"Achhi achhi ei ye ami achhi."

This means, how fine it is that I am present, whether in the midst of my natural and social surroundings or in the midst of the vast extra-terrestrial world.

There are other excellent songs of Tagore expressing this very wonder, this very delight, that I have won my place in the midst of so vast a creation.

In various prose writings also, this basic philosophy of the human self is the key tone. One such is the essay entitled "Amar Jagat"[4] (My World) in the book *Sanchay* (Accumulated Wealth). Here, Tagore wants to establish the truth that the world is not merely the play of impersonal radioactivity; the world is after all mine. This is the basic claim of what is designated as phenomenology.

Just a few pieces from Tagore's works have been cited here. The vast range of Tagore's works can supply immense data regarding man's self-esteem. These are not necessarily confined to the Bengali ethos alone, rather these could be tested, verified, discussed and interacted with in Hegelian, Husserlian, and Peircean terms. The East and the West may come to unity with a view to reach a universal realm of man's creative imagination which is the generating ground of science, philosophy, literature, music, the arts, and all manner of intellectual and aesthetic creativity.

Visva-Bharati University

NOTES

[1] Rabindranath Tagore, *Rabindra Rachanavali* (Tagore's Works), Birthday Centenary Edition (Government of West Bengal), Volume 3, pp. 392–393. Henceforth abbreviated as *RR* and followed by the number of the volume.

[2] *RR* 2, pp. 880–881.

[3] *RR* 3, pp. 235–237.

[4] *RR* 12, pp. 559–565.

MARIO A. PRESAS

THE MAGIC OF ART IN THE MAGIC-LESS WORLD

> To write a poem, is to work a minor magic.
> Language, the tool of such magic, is so mysterious a thing, and we know nothing of its source.
> — Jorge Luis Borges*

I

When man ponders his own being, he cannot but feel the pull of the factual constraints to which he is subject in his *there-being*, in his *Dasein*("human existence"), and the counter-pull of the dream-plans he dwells in by inalienable right, a strain which is related to the measure in which his future makes sense of his present. Furthermore, it is bewildering to discover that things possible, once achieved and transmuted into indestructible facts, are no longer possible as they once were. But man also discovers that there are ways and means of cheating the restrictions of the *here-and-now* ("*Da*"), of the facts to which he is condemned, inasmuch as being also implies getting outside and beyond and transcending oneself — that is, precisely, freeing oneself from the *here-and-now*.

Such liberation, perhaps, cannot take many forms — maybe only the four classic ones: action, cognition, religious experience, and art. A ceaseless *activity* seeks escape from all restraint in the transformation of the world; this is the attitude of the Faustian man; *theoria* achieves the infinite in the vision of all existence as ordained *sub specie aeternitatis; religion* senses that limitation flows from the subordination of the finite to the divine, and thus forges new links with the absolute in which being is rooted; and lastly — and to a certain extent in a contrast with these forms — liberation through *art* reconciles the free-ranging spirit of man with the natural and factual world which surrounds him, and keeps limitless possibility in existence. This is, precisely, what distinguishes the aesthetic from all other approaches, which, inasmuch as they operate within the framework of the real, find existence to lie, as we were saying, between the pull and counter-pull of the factual and the

A-T. Tymieniecka (ed.), Analecta Husserliana, Vol. XXXVII, 413—421.

non-factual, as a possibility that cannot be realized at the present time. "What man cannot achieve in himself, he achieves in the medium of art, which thus represents a necessary release from the shackles which real life place upon his freedom."[1]

This process of *liberation* must not be seen as an *evasion* of reality. Evasion simply ignores limitations which, notwithstanding, continue to exist; but art brings about a release from the ontic confusion of chance and chaos, and points the way to a new order of things which sheds light upon and gives meaning to possible actions — in the imaginary context of the unreal, of course.

II

This luminous intelligibility of imaginary things is, perhaps, what Aristotle speaks of when he affirms that art is above chronicle: "Poetry is a higher and more philosophic thing than history." (*Poet.* 1451 b) Indeed, inasmuch as the latter deals with the particular and the accidental, and "consists in telling us, for example, what Alcibiades did and what he suffered," it simply tells us what has happened, whilst poetry, on the other hand, tells *what could happen*. It rises above the confusion of factual detail. Poetry, art, in its selective re-arrangement of acts, highlights how the actor in a drama is inevitably ruled by destiny, within the limit of what is probable and necessary. By re-arranging the successive events within the rigid structure of a myth or a fable — that is, by means of what Ricoeur likes to call its *mise-en-intrigue*[2] — art (in this case, the dramatist) puts on the stage a truth (a "world") as a possibility. We could put it in another way: If we hold to the idea that the sense of our actions is only made clear backwards, as it were, from a finale to which they were leading up without our knowing it, then we might sometimes think that in real life "nothing ever happens." As Sartre puts it: All that happens is that "the stage-decor changes: people come on and go off stage, and that is all. There are no beginnings and there is no ending . . . . Such is life. But when we tell the story of our lives, all that changes."[3]

Why? Because a narrative order cuts out all unnecessary detail, everything that is not pertinent to the end of the story — to the meaning of the whole thing — and dwells only upon such episodes as go to make up the aristotelian ideal of "an action whole and complete, with a beginning, a middle, and an end." In real life, it is only *a posteriori* that any given event begins a story; rarely do we understand that middle part which brings about a change of fortune, with the clarity which an artist gives to things in his work; and finally, we almost never feel the end of a series of events as intensely as we do when experiencing an artwork disposition — inasmuch as "no action, taken by itself, is an end (in the sense of a conclusion), unless it be the conclusion of the whole course of events in a narrative, the unravelling of a knot, the recognition which comes as the reward of previous reversal, the sealing of the hero's destiny in an ultimate event which explains everything that has gone before, and provokes in the spectator a *catharsis* of compassion and terror."[4]

As our lives unfold, they grow into shapeless things, our existence comes to no definite conclusion, but rather becomes enmeshed in a web of fortuitous and valueless circumstances, a chaos wherein it is utterly confounded with things foreign to itself. What is wrought by art, is the drawing out of the darkness of factual circumstance, the crucial moments of a man's life, although in so doing it inevitably attributes them to an unreal personage, in whom a real man may see himself mirrored and clarified. Hence, in such a work, "all useless details disappear, whilst all that necessarily flows from the living logic of character is brought together, concentrated, united in a being who might be less 'real' but is nearer to the truth." Thus Pirandello, whom I have just quoted, comes to anticipate the Sartrean dilemma: meaningless life *versus* narrative, since — as he used to confess — "the truth is, that life must be lived, or must be written about."[5]

Endowed with a richer variety of expression, the same idea is at the root of the theory of art explicitly sustained by Proust: "True life, life at length stripped of all its veils and clearly perceived, that is, the only life which is truly lived, is literature; that life which, in a sense, is ever-present in each one of us, just as it is in the artist."[6]

III

But a dramatic work puts before us a course of action, a theme, a human character as *possibility*, as scheme, as it were, for the guidance of the reader who is to build it up and fulfill it. Such character, be it Hamlet or Sherlock Holmes, as Borges says, "it is a thing of hazard, shaped directly or nearly so by the shifting moods of inconstant readers."[7]

The intrusion of the reader effects the *concretization* of the plot that the work proposes as an *Idea* or a Scheme (Ingarden). The work of art proposes an imaginary world, but one that could be dwelt in: or to say it better, "an imaginary variation of the world," to use a Husserlian expression which Ricoeur enlists.[8] "World" could here be taken to mean, the whole intelligible and comprehensible panorama of passions, actions, and expectations that the work unveils before us through its *mise-en-intrigue.*

The appropriation of this proposed world by the reader presupposes, as Kant used to say, a first moment of "disinterest" ("*ohne alles Interesse*", *Kritik der Urteilskraft*, §2) — "a willing suspension of disbelief", Coleridge called it — on the part of the reader, who, thus stripped of his proper self, enters upon the imaginary world in such a way as to experience, in a second moment, himself in "the imaginary variation of his own *ego*,"[9] in response to the metamorphosis of reality into the fantasy enacted by art.

This sloughing-off of the *ego* and of the factual world gives him, as we have seen, a truer understanding of himself, inasmuch as every man is, in the last resort, when he reads, "a reader of his own self," as Proust points out.

Well then, such absorption in what we are reading, consciously or unconsciously provokes in our minds so decisive a conflict between the possible and the real world, that we are inhibited from reifying — hypostasizing, factualizing — a purely fortuitous state of things in any limited or finite form; and thus art is allowed to play its game afresh, pursuing its serious purposes of liberation within the realm of such reading. Would not this be what Hegel had in mind, at the end of his *Vorlesungen ueber Aesthetik* where, in spite of having insinuated the apparent "death of art," he wrote: "Where art is involved, then, we are not simply dealing with a pleasant or useful toy, but with the liberation of the spirit from the content and the forms of the finite; with the presence of the Absolute and its reconciliation with the world of sense and appearance; with an unfolding of the truth, not confined within the limits of natural history, but revealed as world-history, of which art is the fairest aspect — that art which is our best reward for hard labour in the real world, and souring advances in cognition."?[10]

IV

Now the Idea — the truth, the world, the scheme, or whatever we care to call it — propounded by art, is expressed in the *indirect language of symbolism*, and cannot be translated into our directly descriptive language. To claim that it cannot be so reduced, does not imply any abdication of our rational faculty but simply takes into account the limitations of reason itself. The indirect language of aesthetic symbolism speaks to us in terms of the senses, of that which is beyond the senses, and beyond the reach of sensitive intuition. The artist, however, creates "a spiritual equivalent" (Proust), something which evokes an *aesthetic Idea*. In the words of Kant, the aesthetic idea is "an inexpressible representation of the reason, which is the source of much thought, but which no single thought, or rather concept, can satisfactorily enshrine, and hence no language render comprehensible."[11] Thus, we may very well admit, with Wittgenstein, that "we cannot speak" of that which lies beyond the borders of possible experience, and hence of descriptive language, but that it does not follow that we "must be silent about it," if it be true that there is another form of speech, another "language-game," which consists precisely in propounding those "representations of imagination," the study of which — as Kant concludes — "gives life to our cognitive faculties, and spiritualizes the language of plain words."

V

Let us take a look at the same thing from another angle. The language of art differs from that of common usage; and this is true not only of literary art, but (with some reservations) of the plastic arts also. It rings the changes upon language in such a way as to make it lose its ordinary meaning, and this — for example, a "semantic impertinence" — disconcerts at first the reader or spectator. Such deviation from common usage, however, which is an essential part of metaphor, does not make of the poet's language a thing apart and irrelevant, without any reference, but rather conjures up something semantically new out of the ruins of direct speech, something which reflects the new things we are able to perceive, just as, in the words of Aristotle, the right use of metaphor lies in the perception of similarities, the bringing together of "things" which in ordinary life, and in the language which tells of it, would be separate and distinct but for the vision drawn from the imagination and depicted in a work of art.

At this point we can again see our way to distinguish liberation from mere evasion. The latter allows us to escape from reality to a land of words or colors without reference; but art, on the other hand, frees us from the bonds of commonly accepted reality, not that we may flee from it, but that we may re-describe it by means of deviation, through indirect language. In this sense — following Proust's line of thought, referred to earlier — we could say that metaphor (or art in general) operates as an instrument of discovery, wrought by the imagination, to open our eyes to a new dimension to which the spontaneous direction and the acquired habits of our daily lives have blinded us.

Moreover, much the same purpose is served, in epistemology, by the scientific *model*, as Max Black (mentioned by Ricoeur) explains in his *Models and Archetypes*, for the scientific model is also an instrument of discovery, a fiction by means of which we discard an inadquate interpretation of reality and find our way to a new one which we feel to be more satisfactory. Mary Hesse, following in the footsteps of Max Black, speaks of the model as being a tool of re-description. As has been shown, metaphor — poetic language, art in general — pursues this very strategy: on the one hand it suspends for us that current interpretation of things, whose inadequacy we are aware of, and on the other hand, it sets us free to embark upon a re-description of reality.[12]

VI

Nevertheless, the paths which the scientific model and the poetic metaphor lay before our feet are stubbornly divergent. In the first case, we have a re-description of reality which necessarily excludes any explanation based upon "powers" or "forces" outside the realm of instrumental reason: a *prosaic* vision of the world, as Hegel put it, or a magic-less, a *disenchanted world* (*entzauberte Welt*), as Max Weber felicitously expresses it. In the second case, however, the emphasis is on events and ways of communication with reality, which, though true, are not susceptible of explanation in the general conceptual terms of reason alone. What the poetic vision can sometimes allow us to perceive, for example, is the full depth and breadth of love, or friendship, the full range of human communication, man's response to the mystery of death, etc. Experiences such as these that are not to be translated into the terms of common enunciative language.

The disenchanted world threatens to obliterate all such other powers, which are just as much of the essence of man as is his intellect, and to thus engender a mechanical society stripped of all mystery. The machine would be the symbol of such a de-humanized world — the machine which we can imagine refusing to submit to a rational plan, and arrogantly assuming the right to sit in judgement upon, and preserve or destroy everything that mankind has achieved in thousands of years of history. The machine substitutes the smooth-cut stones of the skyscraper for the tremulously hand-wrought work of the artisan. Speaking metaphorically, the poet Rainer Maria Rilke arrays against each other that disenchanted world and the enchanted world of Orpheus — the machine and poetry. Imagination is the key to both of them: the artist's mission is to keep mankind aware of this intimate participation in life and reality — not that he may deny the achievements of scientific interpretation, but that he may resist scientistic reductionism.

It is strange to observe that both the descriptions given here — the one embodied in Weber's statement, and the other in Rilke's poem[13] — should be based upon the same German verb *zaubern* (to enchant), in one case prefixed with *ent-* (dis-) and in the other with the augmentative *ver-*. The poet celebrates the *verzaubertes Dasein*, i.e., the enchanted existence, as compared with the *entzauberte Welt*, i.e., the dis-enchanted world. It is almost as if we were now, so many centuries later, reading an upside-down version of Plato, who warned us of the poisonous effects of art, which only ceases to be dangerous if one has the antidote of cognition ready to hand. Today, we seek to recover the magic of art as an antidote to a one-sided and totalitarian interpretation of the world and of life, which is largely built upon inferences unjustifiably drawn from the scientific approach, and from a boundless confidence in technical solutions.

Such is the synopsis of Rilke's "argument" in his tenth *Sonnet to Orpheus* (Part II), of which the following is an approximate translation:

The machine threatens everything we have achieved,
Claims to enter our very selves and be no longer our slave.
Smooth-cuts the stones of the solid building,
That the noblest human hand tremble not one fair moment there-on.

It never looses its grip, that we may escape it,
Or banish it to the grease of its silent factory.
It is life — which it believes itself to understand better
    than any other —
Itself, with the same will to ordain, to create, and to destroy.

But for us, existence is still a thing of magic,
It still springs up in a hundred different places, in
countless pure powers,
That none may come nigh to, unless he kneel in admiration.

Words are softly lost in the ineffable . . .
And music, ever new, tremulously, and stone by stone,
Builds up its divine house in the space not needed.

National University of La Plata
University of Buenos Aires

NOTES

* "Escribir un poema es ensayar una magia menor. El instrumento de esa magia, el lenguaje, es asaz misterioso. Nada sabemos de su origen."

[1] Emilio Estiú, "Arte y liberación," *Humanidades* **XXXIV** (1954), La Plata, p. 58.

[2] Paul Ricoeur, *Temps et Récit* (Paris: 1983), Vol. I, p. 11.

[3] Jean-Paul Sartre, *La náusea* (Buenos Aires: 1953), p. 54.

[4] Paul Ricoeur, "De l'interprétation" in *Du Texte à l'action* (Paris: 1986), p. 14.

[5] Cf. Mario A. Presas, "Vida y arte en Luiggi Pirandello," *Criterio* No. 1707 (1975), Buenos Aires, p. 23.

[6] Marcel Proust, *En busca del tiempo perdido*, Vol. 7: *El tiempo recobrado*, Spanish trans. (Madrid: 1969), p. 246. Besides, the idea that effect of the artist's work is the "exact opposite" of that wrought upon us by the practical purposes of life, intelligence, and habit especially, is notably similar to that which I shall expound in connection with metaphor, which can be briefly expressed in passages such as the following: "All that our self-esteem, our passions, our imitative capacity, our abstract intelligence, our habits, have done, *is what art will undo in us*, impelling us in the opposite direction, towards those deep places where lies buried, unbeknown to us, what has truly been." *Ibid.*, p. 247.

[7] Jorge Luis Borges, "Sherlock Holmes" in *Los conjurados* (Madrid: 1985), p. 49: " . . . está hecho de azar. Inmediato o cercano / lo rigen los vaivenes de variables lectores."

[8] Paul Ricoeur, "La fonction herméneutique de la distanciation" in *Du texte à l'action, op. cit.*, p. 114.

[9] *Ibid.*, p. 117.

[10] Cf. Mario A. Presas, "La muerte del arte y la experiencia estética," *Diálogos* **40** (1982), Univ. Puerto Rico, p. 69.

[11] I. Kant, *Kritik der Urteilskraft*, §49.

[12] Cf. Mario A. Presas, "La verdad de la ficción. Estudio crítico sobre Paul Ricoeur," *Revista Latinoamericana de Filosofía* Vol. **XIV**, No. 2 (July, 1988) Buenos Aires.

[13] Thus do I make bold to translate (I will not say, poetically), Rilke's Sonnet. The original, in his *Gesammelte Gedichte* (Frankfurt am Main: Insel Verlag, 1962), p. 513, runs as follows:

Alles Erworbne bedroht die Maschine, solange
sie sich erdreistet, im Geist, statt im Gehorchen, zu sein.
Dass nicht der herrlichen Hand schöneres Zögern mehr prange,
zu dem entschlossenern Bau schneidet sie steifter den Stein.

Nirgends bleibt sie zurück, dass wir ihr ein Mal entrönnen
und sie in stiller Fabrik ölend sich selber gehört.
Sie ist das Leben, — sie meint es am besten zu können,
die mit dem gleichen Entschluss ordnet und schafft und zerstört.

Aber noch ist uns das Dasein verzaubert; an hundert
Stellen ist es noch Ursprung. Ein Spielen von reinen
Kräften, die keiner berührt, der nicht kniet und bewundert.

Worte gehen noch zart am Unsäglichen aus . . .
Und die Musik, immer neu, aus den bebendsten Steinen,
baut im unbrauchbaren Raum ihr vergöttlichtes Haus.

RUBEN CARRASCO DE LA VEGA

EL PROBLEMA EINAILOGICO

I. LA ONTOLOGIA

Analizar las preguntas *¿qué es el ente*? y *¿qué es el ser*? como preguntas, constituye el propósito principal de este breve ensayo, que, por una parte, no aspira a ser completo, sino simplemente el preludio de una investigación de más amplios alcances, y, por otra, no sigue ni se atiene en sus resultados a ningún filósofo o corriente filosófica, sino que pretende, más bien, siguiendo la máxima de Edmundo Husserl "a las cosas mismasi", enfrentar en sí mismo el tema propuesto, como quien descubre por primera vez un domino, y lo explora, en busca de la verdad, indagada por todos, pero no poseída por nadie.

Ahora bien, desde los griegos la ontología ha pretendido contestar estas dos preguntas. Pero ¿cómo ha entendido la ontología su cuestión fundamental? La ontología debe preguntarse por el ente y también por el ser. Pero ¿cuál de las dos interrogaciones constituye propiamente *la pregunta ontológica fundamental*? ¿Cuál es, en rigor, la relación entre estas dos interrogaciones? ¿Son distintas? ¿O, más bien, al ser detenidamente analizadas, se explicitan como una y la misma pregunta? ¿Qué busca propiamente la ontología?

Desde Aristóteles la ontología ha sido definida como *ciencia del ente* y también como *ciencia del ser*. Tanto el ente como el ser han caído dentro del dominio ontológico. No es posible tratar el uno sin que el otro entre en consideración. Pero, en rigor, ¿cuál es el objeto propiamente dicho de la ontología?, ¿el ente o el ser?, o, por lo menos, ¿en qué medida y en qué sentido la ontología puede ser a la vez la ciencia del ente y del ser?

En rigor, la palabra ontología — acuñada no por los griegos, creadores de esta ciencia, sino recién en el siglo XVII y como simple interpretación de la concepción aristotélica — etimológicamente quiere decir *estudio o ciencia del ente*. El vocablo griego *ón, óntos*, significa ente. Este es el origen y el sentido literal de la palabra. Pero entonces ¿de qué manera surge *el ser* como objeto de la ontología? La filosofía occidental, desde Aristóteles, ha entendido *la cuestión ontológica* como

A-T. Tymieniecka (ed.), Analecta Husserliana, Vol. XXXVII, 423—445.

la cuestión del ser, y la cuestión del ser como *la cuestión del ente en cuanto ente*. La situación parece confusa y requiere aclaración rigurosa.

¿Qué quiere decir propiamente ente y qué significa ser? ¿Estos dos términos mientan lo mismo? ¿En qué sentido el *problema* del ser es *el problema del ente como ente*?

II. SER Y ENTE

Los términos ser y ente propiamente no significan lo mismo; y, por lo tanto, deben ser fijados y mantenidos como distintos, sin que esto implique perder de vista la relación fundamental y *sui generis* que los liga.

La ontología, en efecto, si quiere fijar con rigor y claridad su objeto, ante todo, tiene que distinguir ser y ente. Esta distinción es fundamental y constituye una de las primeras exigencias de la indagación.

Veamos esta distinción. Desde luego advertimos que *hay entes*, pero no *seres*, sino pura y simplemente *ser*. En español, tanto en el lenguaje corriente como en el filosófico, solemos hablar también de seres, en casos en que ontológicamente en rigor deberíamos decir *entes*. Y aún en singular acostumbramos decir *un ser, este ser* para significar *un ente, este ente*. Esta confusión terminológica debe ser evitada en la filosofía. Por eso, nosotros, para extirpar toda imprecisión y confusión, vamos a eliminar la expresión *seres* en el sentido de *entes*, y sólo emplearemos los términos *ser* y *entes* como completamente distintos. De acuerdo a esta distinción, el término *ente* podrá ser usado en singular o en plural, pero *ser* sólo en singular.

Ahora bien, es palmario que *el ser no es un ente* ni *menos los entes*. El ser se presenta a nuestra consideración por lo pronto como *aquello* — y decimos *aquello* porque en rigor no podemos decir *qué es* — que determina a los entes en cuanto entes, como *aquello* que hace que los entes sean antes. Ente soy yo, ente es la mesa sobre la que escribo, la tinta, el tintero que utilizo en este momento; ente es la casa en la que habito, el ruído producido por un automóvil que corre por la calle, y el automóvil mismo, y también la calle; ente es el aire que respiro, la luz del sol, la oscuridad de una noche tempestuosa; entes son los árboles de un bosque, los animales que caminan por la tierra, las estrellas, la luna, el sol. En fin, todo lo que hay y encuentro en el mundo es ente; *todo lo que es*. En sentido riguroso nuestra palabra española *ente* significa eso: *lo que es*. Algo que es significa algo que tiene ser; pues, si no tuviese ser no sería, no sería ente. El ser hace que un ente sea ente.

El vocablo *ente* procede de la expresión latina *ens, entis*, participio presente del verbo *esse* que significa *ser*. Pues, así como sonriente significa en rigor el que sonríe, cantante el que canta, ente — como participio presente de ser — quiere decir *lo que es*.

Pues bien, si evidentemente hay multiplicidad de entes, el ser, presente en todos los entes, es uno y el mismo en todos ellos; y, por lo tanto, él mismo no es un ente ni todos los entes juntos. Por eso, no se puede definir el ser como se definen los entes, no se puede descubrir y determinar el ser como se descubren y determinan los entes.

Esta distinción, que no es separación, debe guiar la meditación ontológica. Si los entes constituyen, por así decirlo, una dimensión, un orden muy peculiar, el ser conforma otra dimensión, otro orden totalmente distinto. El ser nunca puede ser un ente; ni un ente o los entes, el ser.

La distinción entre *ente* y *ser*, tradicionalmente confundidos, hasta el punto de que en la mayoría de las lenguas sólo se usa una palabra — *être* en francés, *being* en inglés — es pues clara y terminante. Podría intentarse, empero, un desarrollo más riguroso. Sin embargo para nuestros propósitos basta con lo explanado.

III. LA CUESTION ONTOLOGICA FUNDAMENTAL

En el pensamieno occidental, que arranca de los griegos, el problema ontológico fundamental, hemos apuntado ya, ha sido entendido como la cuestión del ser, y esta cuestión, a su vez, como el problema del "*ente en cuanto ente*", según la fórmula clásica de Aristóteles.

El problema ontológico no es, pues, nuevo. Se remonta a los comienzos mismos de la filosofía de Occidente, esto es, a aquellos tiempos memorables en que nace, se constituye y desarrolla la filosofía griega. Los filósofos griegos plantearon la cuestión ontológica como problema fundamental de la filosofía, como cuestión de radical primacía en la meditación filosófica. Desde entonces los filósofos de Occidente han luchado con este problema sin llegar a una comprensión adecuada y satisfactoria, sin alcanzar una solución definitiva. A través de los 26 siglos de la filosofía occidental la cuestión ontológica ha mostrado su rostro enigmático de problema perenne.

Sin embargo, las peripecias que ha experimentado la ontología no son pocas ni insignificantes. El problema del ser ha tenido sus adversarios, sobre todo, en el transcurso de la Edad Moderna. Ya desde el Renacimiento aflora ciert a hostilidad, que conduce lentamente al abandono y al rechazo de este problema como problema. Se llegó a discutir y negar con razones poderosas su legitimidad. La preocupación primaria por el problema del conocimiento, consustancial a la filosofía de la Edad Moderna, con descuido y menoscabo del problema del ser, ha constituído el motivo principal de la pérdida de este problema, el horizonte de investigación que hizo retroceder cada vez más la cuestión del "ente en cuento ente", hasta acabar rechazándola como de todo punto vana y sin contenido. Así, por ejemplo, en el siglo XIX resultó habitual este modo de pensar, y se creó así una tradición cuya fuerza sigue gravitando en nuestros días.

Pero, los problemas cuando son problemas genuinos y no pseudoaporías, sean descubiertos o abandonados — y esto último naturalmente no porque un día queden resueltos o porque se desvanezca su problematicidad — siguen siendo problemas. Vuelven a surgir, toda vez que la situación histórica lo permite, con toda su ineluctabilidad como enigmas insolubles. Y esto ha acontecido con el problema del ser. El abandono, cada vez más decisivo, operado en la Edad Moderna, ha sido seguido en nuestro siglo del inevitable retorno. Y en medio de este retorno nos hallamos los hombres de hoy. El problema ontológico ha recobrado o por lo menos lucha en nuestros días por recobrar sus derechos.

Hay que reivindicar la cuestión del ser. He ahí la tarea primera de la filosofía de nuestros días y he ahí la cuestión primaria de tantos esfuerzos de hoy. Y esto, naturalmente, no por afán de plantear problemas insolubles. Suena un poco y mucho a ingenuidad y a falta de sentido espiritual que alguien niegue y reniegue de la filosofía porque no ha alcanzado la solución de los postreros enigmas de la realidad, si sabemos que la misma ciencia llamada positiva y experimental, después de tanto saber acumulado, tampoco ha llegado a resolver sus problemas fundamentales ni ha permitido suprimir tantos males, flagelos y enfermedades que hieren y acosan cruelmente la existencia humana.

El hombre, anclado en este mundo que es un enigma, escarcuña, a veces dolorosamente, las entrañas de lo vivo y profundo. No está en su voluntad eludir o no las dificultades allí donde la contradicción es evidente, allí donde el problema es ineluctable. Porque está en este mundo, así y así constituído, y no por el mero afán de hacer difícil lo fácil, le salen al paso inevitablemente los problemas. Los problemas fundamentales de la filosofía no son mera invención inane del hombre, sino contenidos lancinantes, entrañas vivas de la realidad. Son cuestiones cuya solución no depende del deseo del hombre; problemas, en fin, con rostro de esfinge. Las soluciones, aparentemente definitivas y cuando en verdad parecen anunciarse como tales, sólo son rastros muertos de un movimiento que no es visto desde sí mismo sino desde fuera. El filosofar humano — no conocemos otro — es esencialmente movimiento, y el movimiento, en su genuina palpitación estremecida, sólo puede ser vivido desde dentro. Visto desde fuera deja de ser movimiento, esto es, deja de ser lo que es.

Pero si queremos volver a hacer ontología, plantear de nuevo la cuestión ontológica e intentar su solución, necesitamos previamente elaborar y desarrollar con el mayor rigor posible la pregunta misma, determinar y precisar qué interrogamos cuando preguntamos por el ser. Se alza ante nosotros, como problema previo aún no resuelto satisfactoriamente, la aporía que podemos caracterizar como *la pregunta por la pregunta del ser.*

IV. LA PREGUNTA POR LA PREGUNTA DEL SER

Hay, indudablemente, un problema del ser. Pero ¿en qué consiste propiamente este problema? Todo problema puede ser formulado en forma de pregunta. Es ciertamente una pregunta. Ahora bien, ¿cuál es la pregunta fundamental relativa al ser que la filosofía intenta resolver? Antes de resolver un problema es menester comprender con plena transparencia el contenido mismo del problema. No podemos pretender responder una pregunta si en verdad no sabemos cuál es propiamente esta pregunta. Emerge aquí, diríamos, la cuestión de la cuestión. Aporía en verdad previa, pero importante si se quiere seria y conscientemente volver a despertar la comprensión adecuada de la cuestión del ser, plantear en forma precisa el problema, para sólo luego intentar su solución.

¿Cuál es propiamente el problema del ser? ¿La pregunta por el ser reza simplemente: *qué es el ser*? ¿Podemos dar por supuesto que la interrogación *¿qué es el ser*? constituye la adecuada formulación de la cuestión fundamental del ser?

La pregunta ontológica es tan fundamental y ha tenido una historia tan llena de peripecias, que ha sonado la hora de saber qué preguntamos propiamente con ella. ¿Por qué muchas veces se ha rechazado o se ha creído necesario prescindir y dejar de lado problema al parecer, según una tradición veneranda, tan magno? ¿Se trata verdaderamente de un problema y además fundamental? Pues, si la respuesta o respuestas que encontramos en el pensamiento occidental, no satisfacen, es menester volver la mirada a la pregunta misma en cuanto tal. Ante todo, se hace necesario proyectar la luz a la pregunta misma y dejar que se nos muestre en toda su transparencia. Hemos llegado a tal punto que necesitamos de una vez dilucidar *la interrogación por la interrogación.* Si la pregunta por el ser es una pregunta. Hemos llegado a tal punto que necesitamos de una vez dilucidar *la interrogación por la interrogación.* Si la pregunta por el ser es una pregunta, ¿qué es propiamente una pregunta? ¿Qué es en general preguntar? ¿Cuáles son los elementos esenciales de toda pregunta? Sólo cuando sepamos qué es una pregunta en general, podremos mostrar cuál es propiamente la pregunta ontológica y si es y por qué una pregunta preeminente.

Nos enfrentamos con *la pregunta por la pregunta del ser,* con el problema previo de cómo plantear el problema. Y de este modo se nos convierte en problema el problema mismo del ser en cuanto problema.

V. ANALISIS DE UNA PREGUNTA EN GENERAL

¿Qué es una pregunta?, he ahí la pregunta por la pregunta, que ahora vamos a intentar responder.

Al preguntar, indudablemente, preguntamos algo, pero si decimos simplemente que preguntar es preguntar algo, no determinamos en forma precisa y completa la estructura esencial de la pregunta. La esencia de la pregunta exige afirmar: *preguntar es preguntar algo de algo.* Este rasgo esencial se completa y precisa con este otro: todo preguntar implica una combinación *sui generis* de saber y no saber. Si pregunto sé lo que no sé y sé lo que sé.

Veamos en qué consisten estos dos rasgos característicos y cómo constituyen la esencia del preguntar. El preguntar, en efecto, no surge en la ignorancia absoluta, sino en medio del saber, pero de un saber que no es completo, perfecto, sino circundado de no saber y que pugna por rebasar la frontera que lo separa del no saber. El conocimiento humano siempre se halla rodeado de un horizonte de no saber, y, por tanto, siempre frente a la tarea de resolver nuevos problemas que son otras tantas interrogaciones. Un ente que poseyese el saber total y completo no preguntaría, pero igualmente un ente sumido en total ignorancia. El hombre no está ni en un estado ni en el otro, por eso pregunta, y esto acontece de tal modo, que el preguntar constituye un rasgo esencial suyo, un carácter que lo distingue de todos los demás entes. El hombre es el ente que pregunta. El hombre desde que es hombre se erige en el único ente que pregunta o puede preguntar. El preguntar humano en cuanto tal constituye una forma de conocimiento, y tal vez la forma más alta que al hombre le es posible alcanzar con relación a los enigmas postreros de la realidad.

De aquí resulta la importancia decisiva en el ámbito de la filosofía de saber plantear los problemas, de saber formular las preguntas.

El hombre es el ente que sabe y no sabe; siempre hay, por una parte, algo que sabe, y, por otra, algo — y en realidad mucho — que no sabe. Pero el límite entre el saber y el no saber es móivil; el conocimiento humano progresa, esto es, el saber va penetrando cada vez más en la zona no conocida. Ahora bien, el primer paso de este progreso es el problema, la pregunta, conciencia peculiar del saber y del no saber, o dicho de otro modo, saber del saber y a la vez saber del no saber. En el preguntar, en efecto, el saber y el no saber que representan una determinada situación alcanzada por el conocimiento humano, no están simple y estáticamente separados por una línea divisoria, sino que entran en un juego muy peculiar. Cuando pregunto no sólo no sé algo de algo que sé, sino que *sé algo que no sé de algo que sé.* En el conocimiento ya establecido y fijo hay un saber circundado de no saber, algo conocido e inmediatamente más allá algo no conocido. En el preguntar lo no conocido entra en la zona del saber, aunque, claro está, como no conocido, y así entonces hay en él un saber del no saber, pero de tal modo que este último emerge de un fondo de saber aprehendido como saber. El "yo sé que no sé", implícito en todo preguntar, se sostiene, por así decirlo, en un "sé que sé". Por eso, la formulación precisa de esta situación, exige esta enunciación completa: "*yo sé que no sé algo de algo que sé*". Podríamos también decir: en el conocer el saber está circundado de no saber; en el preguntar, en cambio, es el no saber aprehendido como tal el que se halla rodeado del saber. Pues bien, esto resulta del hecho de que todo saber es esencialmente *saber algo*. Constituye un contrasentido *saber nada*; y el mismo saber socrático, el "yo sólo sé que no sé nada", por la misma razón, resulta una enunciación contradictoria en sí misma. Una pregunta en cuanto tal es ya un saber, pero un saber que se bifurca, por así decirlo, hacia dos "algos": el no saber y el saber que se extiende más allá de este no saber. Hay, por una parte, el no saber y el saber, y, por otra, el preguntar mismo, esto es, el saber de ese no saber y de ese saber. El saber peculiar que es el preguntar abraza un no saber rodeado de saber. Por eso, al preguntar pregunto algo de algo, algo (no conocido) de algo (conocido); pregunto, es decir, sé, aprehendo la relación peculiar de algo no sabido dentro de algo sabido. El conocimiento del último algo puede moverse en los límites más amplios posibles, hasta el punto de consistir en el mero saber el nombre con que este algo es designado.

En la pregunta en cuanto tal, esto es, en cuanto *pensamiento* contenido como el sentido de una proposición interrogativa, hay, pues, la referencia a dos instancias, a dos "*algos*", como nos hemos visto obligados a decir para mantener el plano del preguntar en general en que nos movemos. Preferimos contravenir ciertas reglas habituales de la lengua antes que sacrificar el rigor del análisis. En la interrogación ¿qué es el hombre?, por ejemplo, mediante la palabra "qué" nos referimos a algo (no conocido) y mediante el término "*hombre*" a algo (conocido). Pero el primer algo es siempre algo del otro algo. Ahora bien, el preguntar propiamente tal, esto es, el producir y aprehender el sentido contenido en la proposición interrogativa, es, como todo aprehender, un saber. Es un saber que aprehende los dos "*algos*": el algo no conocido en cuanto no conocido y el algo conocido en cuanto conocido, y la relación peculiar entre ambos. Por eso, al preguntar *sé lo que no sé y sé lo que sé.*

La pregunta, pues, *se refiere de un modo peculiar a algo de algo*, y en esto se parece al juicio, y en realidad viene a ser una forma correlativa de éste. Pero el preguntar mismo — naturalmente no la mera pregunta fijada en una proposición — implica un movimiento de búsqueda, una tendencia a contestar la pregunta, a convertir lo no sabido en algo sabido en cuanto tal. El verdadero preguntar no es sólo un preguntar por preguntar, sino en realidad un proceso de búsqueda, un buscar algo de algo. Por eso, en todo preguntar hay un alguien que pregunta y *un algo a que se pregunta*, un alguien que busca y un algo en donde se busca algo.

No se vea en la expresión "*un algo a que se pregunta*" un mero descuido gramatical que vulnera el genio habitual de nuestra lengua, sin una intención precisa y un sentido especial. No se crea, por ejemplo, que si una mesa viene a ser la cosa a la que preguntamos, sea la mesa el ente del que esperamos la respuesta. El análisis que sigue nos ha de mostrar el sentido peculiar y nada arbitrario de esta expresión y cómo mienta en verdad una situación habitual en la investigación filosófica y científica.

Pues bien, la nueva caracterización apuntada, nos permite ahora afirmar que en todo preguntar concurren los siguientes factores esenciales: 1) *El que pregunta*; 2) *Aquello de que se pregunta*; 3) *Aquello que se pregunta*; 4) *Aquello a que se pregunta.*

Para abreviar y hacer más ágil la exposición, vamos a designar también los tres últimos factores con las siguientes expresiones: "*el de que*", "*lo que*" o "*el que*" y "*el a que*" se pregunta.

Veamos como funcionan en el proceso de interrogar estos diversos elementos.

Pongamos un ejemplo sencillo. ¿Qué sabor tiene la naranja? Preguntamos aquí algo de algo. Tenemos mentalmente algo, la naranja, de la que preguntamos algo, el sabor de la naranja. La naranja constituye "el de que" o "aquello de que", y su sabor, "lo que" se pregunta. Dirijamos, por ejemplo, la atención a preguntas que para un análisis superficial parecen fundir ambos elementos en uno solo: ¿qué es la naranja?, ¿qué es sabor? Ni aún aquí hay, o sólo la naranja o sólo el sabor. No tenemos mentalmente presente la naranja sola o el sabor solo; preguntamos algo de la naranja y algo del sabor. ¿Qué es lo que hace que la naranja sea naranja?, ¿qué es lo que hace que el sabor sea sabor? Preguntamos, para decirlo ontológicamente, por el ser de la naranja como naranja y por el ser del sabor como sabor. Por eso, las preguntas anteriores pueden formularse también así: ¿cuál es el ser de la naranja como naranja?, ¿cuál es la esencia de la naranja?, y en la misma forma para el sabor.

Esta distinción es esencial y aparece en toda pregunta. Pues, como ya hemos dicho, *preguntar es siempre preguntar algo de algo*. ¿Por qué? Porque nunca un ente se nos presenta como completamente simple; un ente siempre muestra diversos aspectos. De la naranja, por ejemplo, podemos preguntar por su sabor, color, forma, tamaño, propiedades nutritivas, etc. Nunca un ente da lugar a una sola pregunta, sino a varias. En todas estas preguntas relativas a un mismo ente el "*aquello de que*" es el mismo, pero el "*aquello que*" cambia con la pregunta. Esto desde un punto de vista, pero desde otro, hay que reconocer que si no hubiesen los dos "algos", tendríamos que admitir que un ente es completamente conocido, y, por tanto, no da lugar al algo desconocido y peculiar de la pregunta. Pues si el hombre poseyese el saber total y perfecto, no tendría nada que preguntar.

El último de los componentes de todo preguntar, "*aquello a que se pregunta*", requiere, como anunciamos, una aclaración especial. En la vida diaria solemos hacer preguntas dirigidas a *alguien* en espera de su respuesta. Esto parece mostrar que habría que hablar en general, más bien, de "*alguien a quien se pregunta*". Pero en la ciencia y la filosofía, concebidas como investigación, como verdadera búsqueda y no como conocimientos ya logrados y apropiables en un libro o asimilables de los labios de una persona que expone, no cabe preguntar a alguien para que conteste. El mismo que pregunta tiene que encontrar la respuesta. Aquí surge el preguntar como un verdadero buscar, indagar, investigar; y naturalmente el que investiga intenta encontrar por sí mismo la respuesta y no simplemente recibirla ya hecha. ¿Significa esto que no habiendo persona alguna a quien se pregunta no hay propiamente el cuarto elemento de la pregunta? Ni siquiera resulta justo decir que en este caso la misma persona que interroga constituye también la persona que es preguntada. Pues, aún en el caso corriente de la vida cotidiana en que se pregunta a alguien, la situación real cambia de fisonomía si nos colocamos en la nueva perspectiva, que intentamos iluminar con el propósito de destacar el peculiar cuarto elemento de la pregunta. Esta nueva perspectiva, en efecto, muestra que en la misma pregunta dirigida a alguien, la persona a quien se pregunta no puede ser considerada propiamente como aquello a que se dirige la interrogación. La persona propiamente no es la *preguntada*, sino que desempeña simplemente un papel intermediario — no esencial para la subsistencia del verdadero preguntar — entre *el que pregunta y aquello a que propiamente se interroga*. Pues, aquel a quien se pregunta y responde, ¿propiamente de dónde saca la respuesta? No se diga de sí mismo o que la recibe ya hecha de otro, porque entonces cabe seguir indagando ¿ y este otro a su vez de dónde la saca? Si interrogo, por ejemplo, ¿de qué color es la naranja?, si en verdad hago esta pregunta porque no conozco el color de esta fruta, y respondo yo mismo u otra persona por mí después de una búsqueda, la respuesta sólo puede ser obtenida, leída, por decirlo así, en la naranja misma; el color de la naranja está en la naranja, y sólo en la naranja misma puedo leer la respuesta a la interrogación ¿de qué color es la naranja? Entendido de este modo el preguntar y el proceso de encontrar la respuesta, propiamente no hay la persona a quien se interroga, sino más bien *algo a que se pregunta*. No un algo que es un alguien racional y dotado de lenguaje, capaz de decir por sí mismo la respuesta, sino en general un algo en que la misma persona que pregunta puede encontrar la respuesta.

Concebido de este modo el preguntar, se pone al descubierto un hecho notable: cuando yo pregunto y yo mismo respondo después de buscar — y este es naturalmente el único caso que se presenta en la investigación científica o filosófica — "*el aquello a que*" se desplaza al "*aquello de que*" o en cierto modo se identifica con él. La naranja, que es "aquello de que se pregunta", funciona también como "aquello a que se pregunta". Por eso, si nos estuviese cerrado todo acceso a la aprehensión del color de la naranja, nunca podríamos responder la pregunta que interroga por el color de esta fruta.

Hay, pues, una cierta identificación del "*aquello de que*" con el "*aquello a que*" se pregunta. Así como el "*aquello que*" se pregunta reside en el "*aquello de que*", "*el que*" en "*el de que*" — el color de la naranja está en la naranja — , "*el a que*" a su vez reside o se identifica con el mismo "*de que*". Sin embargo, esto de ningún modo significa la abolición de la distinción existente entre estos tres componentes. Los tres elementos en cuestión son siempre distintos en cuanto cumplen una función diferente en la pregunta. El verdadero preguntar es un buscar, y este buscar es un movimiento en el que los tres elementos son claramente discernibles. Preguntar es *buscar algo de algo*, algo no conocido dentro de algo conocido; es buscar "*el que*" en el mismo "*de que*", pero en cuanto iniciamos la busca, "*el de que*" empieza a funcionar también como "*el a que*" se pregunta.

Pues bien, a la luz de los elementos esenciales de toda pregunta, descubiertos en las consideraciones anteriores, podemos ahora elucidar las interrogaciones tradicionales de la ontología: *¿qué es el ente?* y *¿qué es el ser?*

VI. LA PREGUNTA ¿QUE ES EL ENTE?

Preguntar ¿qué es el ente? o ¿qué es ente?, equivale a interrogar ¿cuál es el ser del ente?, ¿qué es lo que hace que el ente sea ente?, ¿cuál es la esencia del ente? La interrogación no pide determinar este o aquel ente individual, esta o aquella especie de entes, sino el ente en general, todos los entes. Es *la pregunta por el ser del ente.*

De los entes preguntamos por su ser. Interrogamos por el ser de los entes. Los entes son el "*aquello de que*"; el ser, "*aquello que*" preguntamos. ¿Dónde buscaremos la respuesta? ¿Dónde leeremos, apresaremos el ser de los entes? En los mismos entes. Los entes son también, por lo tanto, el "*aquello a que*" de la pregunta.

¿Puede ser ésta la pregunta ontológica fundamental? ¿No satisface la exigencia de ser la pregunta por el ser, por el ser como "*aquello que*" propiamente preguntamos? ¿La pregunta ontológica no ha de ser acaso la interrogación en la que el ser tiene que funcionar como "*el que*" de la pregunta, como aquello que propiamente busca la ontología?

Podrá objetarse contra esto diciendo que el ser y no los entes constituye el tema fundamental de la ontología. Concedido. Pero ¿cuál es entonces la forma estricta que debe asumir la formulación de la cuestión ontológica fundamental? Exigimos satisfacer la necesidad de formular la pregunta como pregunta. ¿La interrogación ¿qué es el ente? no satisface esta exigencia? Así parece. La ontología pregunta por el ser, busca el ser, pero a través del ente, en el ente. Pero entonces no debe preguntar *¿qué es el ser?*, sino más bien *¿qué es el ente?* Parece que podemos decir, en este sentido, con Nicolai Hartmann: Aristóteles tenía plenamente razón al afirmar que la *próte philosophía* es la ciencia del *on hê ón, del ente en cuanto ente.* (Cf. *Nicolai Hartmann, Zur Grundlegung der Ontologie*, 3a. ed., 1948, Westkulturverlag Anton Hain, Meisenheim am Glan, págs. 41 y 42. Hay versión española de José Gaos: *Ontología.* I: *Fundamentos*; Fondo de Cultura Económica, México-Buenos Aires, 1954.) La formulación aristotélica de la cuestión ontológica fundamental es a su modo, sin duda, insuperable. Como sólo mienta el ente en tanto ente, o sea, sólo en su nota más general, mienta mediatamente, a través del ente, a pesar de todo, el ser. Pues, lo único común a todos los entes es el ser. Preguntar por el ente en cuanto ente es preguntar por aquello que hace que el ente sea ente, por el ser.

Sin embargo, creemos necesario penetrar más a fondo en el contenido de la pregunta en cuestión.

¿Qué significa propiamente la expresión *el ente*? ¿A qué se refiere? Desde luego, *el ente* puede significar en general los entes o también un ente particular. Por eso, se aplica a todos y también a cada uno de los entes. Decimos, por ejemplo, "este caballo es ente", "mi escritorio es ente". O también, "*el ente no es el ser*", es decir, "los entes no son el ser", en el mismo sentido en que enunciamos, por ejemplo, "el hombre es mortal", esto es, "los hombres son mortales". Pero, por otra parte, la expresión *el ente* puede tomar un sentido muy singular. "*El ente*" puede mentar no este o aquel ente determinado, ni los entes, sino *el ente en general, la idea de ente, la esencia de ente, lo que hace que el ente sea ente.* Esto es, en este sentido, "*el ente*", propiamente, ya no es un ente o los entes. ¿Qué es entonces? El ser. Es decir nuestro idioma, como también otros, no hace sino aportar otro término para designar lo mismo. Pero entonces ¿las preguntas ¿qué es el ente? y ¿qué es el ser?

no resultan equivalentes? En cierto sentido, sí. Volveremos sobre esta cuestión y arrojaremos más luz, en el próximo capítulo, al elucidar la estructura de la pregunta ¿qué es el ser?

El mismo fenómeno encontramos en interrogaciones relativas al contenido objetivo mentado por un concepto universal. Por ejemplo, en la pregunta ¿qué es el triángulo?, la expresión *el triángulo* puede mencionar los triángulos, todos y cada uno de los triángulos *reales*, trazados o imaginados en algún lugar del universo o que han sido trazados o van a ser trazados en el futuro. Entonces la interrogación ¿qué es el triángulo?, pide que se precise y determine que es lo que hace que los triángulos sean triángulos. De inmediato viene la respuesta precisa del geómetra: "el triángulo es el polígono de tres ángulos". La contestación es inobjetable. El ser "polígono de tres ángulos" constituye, decimos, el ser del triángulo, lo que hace que los triángulos sean triángulos.

Ahora bien, el término "*triángulo*" puede tener otro sentido y con éste emerge una aporía muy singular. "*El triángulo*" puede también mentar no un determinado triángulo o los triángulos, sino "*el triángulo en general*", "*la idea de triángulo*", "*la esencia de triángulo*", "*el ser de los triángulos*". Entonces, "*el triángulo*", en rigor, y aunque parezca paradójico, no es triángulo, no es ya un triángulo. Porque, por ejemplo, en cuanto tal, en cuanto triángulo en general o idea de triángulo, no tiene ángulos, ni siquiera exactamente tres. "*El triángulo*", en este sentido, es, más bien, lo que podríamos llamar "*la triangularidad*". Pero "la triangularidad" justamente no es triángulo; es el ser de los triángulos. Los triángulos tienen tres ángulos, pero no su ser.

¿No ocurre lo mismo con expresiones como "el caballo" o "el hombre", por ejemplo? "El caballo en general", es decir, lo que hace que un caballo real sea caballo, no es, a su vez, un caballo. Es puramente lo que habría que denominar "*la caballocidad*". "El hombre en general" no es un hombre; no camina, no piensa, por ejemplo. Es, propiamente, lo que tal vez podríamos llamar "*la hombredad*". En este sentido, y en forma aparentemente paradójica, podemos declarar: "el caballo" no es caballo, "el hombre" no es hombre; los caballos no son "el caballo", los hombres no son "el hombre".

Pero, entonces, ¿la pregunta ¿qué es "el triángulo"?, en este nuevo sentido, qué significa propiamente? Pide el ser de "la triangularidad", esto es, "el ser del ser del triángulo". "La triangularidad" es, propiamente, lo que se dice un *ente ideal.* La interrogación ¿qué es "el triángulo"?, viene a ser entonces la pregunta por el ser de este ente ideal. Pero esta pregunta ya no puede ser contestada con la respuesta del geómetra: "es el polígono de tres ángulos", a no ser que esta nueva expresión "polígono de tres ángulos" se tome también en un nuevo sentido, esto es, simplemente como equivalente a "triangularidad". En este caso, podemos decir: "el triángulo" es "el polígono de tres ángulos". Pero, en el sentido en que decimos, por ejemplo, "este triángulo es un polígono de tres lados", ya no podemos enunciar "*el triángulo es un polígono de tres lados*, esto es, no podemos decir: "*la triangularidad*" *es un polígono de tres lados*". Este, al parecer un simple juego de palabras, responde a una realidad efectiva y desaparece como tal juego cuando el pensamiento intuye con claridad y se apropia con precisión de la distinción establecida.

Normalmente las preguntas del tipo indicado, ¿qué es el triángulo?, ¿qué es el caballo?, ¿qué es el hombre? y otras semejantes, relativas a clases de entes ordenadas de acuerdo a la relación de especie y género, son contestadas con arreglo a la norma de la lógica clásica que dice: la definición se hace por el género próximo y la diferencia específica. ¿Qué es el triángulo? La respuesta reza: el triángulo es el polígono de tres ángulos. Y el triángulo, decimos, queda definido. De esta manera pueden ser determinadas todas las clases de entes. Una clase de entes es incluída en otra clase superior, ésta a su vez en otra, y así sucesivamente. Siempre detrás de una clase de entes, aparece otra. Pero con este procedimiento el ser de los entes en cuestión, propiamente, se nos escapa constantemente, retrocede, por así decirlo, cada vez más, hasta que encuentra un límite: el ente, el ente en general. El ente ya no puede ser subordinado a una clase superior, porque abraza a todos los entes. La pregunta ¿qué es el ente?, entonces, ya no puede ser contestada. El ente resulta indefinible. *Al llegar al ente, damos con el ser.* "El ente" en verdad mienta el "ser". Por eso, podemos decir "*el ente*" *no es un ente*, o directamente "*el ente*" *no es ente*, en la misma forma que declaramos: *el ser no es un ente* o *el ser no es el ente.* "El ente" no es los entes, no es el conjunto de los entes. Por eso, no puede ser descubierto, por ejemplo, apuntando simplemente a la totalidad de los entes. "El ente" mienta lo que más bien podríamos llamar, para evitar la equivocidad, "*la entidad*", que equivale a decir "*el ser del ente*", "*el ser de los entes*". Mientras el ser retrocedía, la definición era posible, pero en cuanto aparece el ente como ente, el ser, la definición se hace imposible. La definición, en el sentido tradicional del término, que pretende exhibir el ser, más bien lo encubre.

Resumamos y precisemos la distinción fundamental que nuestras reflexiones anteriores han logrado poner en claro. Expresiones, dijimos, como "el triángulo", "el caballo", "el hombre", "el ente", tienen dos sentidos distintos y fundamentales, que es menester destacarlos y mantenerlos en la indagación ontológica, rigurosa y constantemente, como diferentes, a tal punto que resulta lamentable que la lengua no tuviese dos vocablos distintos y utilizables, en cada caso, sólo para cada una de las dos significaciones. En un primer sentido significan: *los triángulos, los caballos, los hombres, los entes.* En un segundo sentido lo que hemos llamado: *la triangularidad, la caballocidad, la hombredad, la entidad.* Estas últimas expresiones equivalen a: *el ser de los triángulos, el ser de los caballos, el ser de los hombres, el ser de los entes.* Las expresiones *la entidad* y *el ser de los entes* pueden sustituirse por: *el ser*, pura y simplemente.

Correlativamente, las preguntas correspondientes tienen también dos sentidos. ¿Qué es el triángulo?, puede significar ¿ qué son los triángulos? o ¿qué es la triangularidad?. *La definición de la lógica clásica sólo sirve para contestar la primera de estas interrogaciones.*

Sin embargo, la última afirmación requiere una aclaración más precisa. La tesis de que la definición de la lógica tradicional sólo es adecuada para contestar la primera de las interrogaciones tiene que comprenderse en rigurosa conexión con esta otra: *la definición de la lógica clásica sólo sirve para definir especies de entes.* La pregunta ¿qué es el triángulo? puede ser entendida en dos sentidos que en realidad constituyen dos interrogaciones distintas. La respuesta: "el triángulo es el polígono de tres ángulos", sólo contesta la que hemos dado en llamar primera pregunta. Cuando hacemos interrogaciones de este tipo, normalmente sólo las pensamos en el primer sentido, y, por eso, no consideramos ni vemos dos preguntas. Nos basta la primera. Con este modo normal de entender estas preguntas, el ser propiamente tal se nos escapa como el agua de la red, y no sólo *el ser en general* sino también el ser de cada una de las especies de entes. La definición "el triángulo es el polígono de tres ángulos" no apresa propiamente el ser del triángulo, lo mienta simplemente con otra expresión. La ontología debe iluminar plenamente las dos preguntas y mantener rigurosamente su distinción, si no quiere dejar que el ser se le diluya y escape. Ahora bien, en rigor, la segunda de las preguntas — y con esto llegamos a la anunciada aclaración precisa — puede también ser tomada como una pregunta del mismo tipo que la primera, no que se identifique con ésta, sino como una pregunta relativa a entes. Esto es, en la pregunta ¿qué es la triangularidad?, la triangularidad puede ser concebida y de hecho es concebida, como ente. La triangularidad, en efecto, se dice, es una esencia, esto es, un ente, un ente ideal. Y cabe, entonces, preguntar por su ser, exactamente como interrogamos, en el primer sentido, ¿qué es el triángulo? La pregunta, entonces, podrá ser contestada con arreglo a la regla del género próximo y la diferencia específica. Por consiguiente, nuestra tesis de que la definición de la lógica clásica sólo sirve para contestar la primera de las preguntas, quiere decir que las respuestas del tipo "el triángulo es el polígono de tres ángulos" o "el hombre es el animal racional", únicamente responden la primera de las interrogaciones. La segunda pregunta podrá también ser contestada con arreglo a la misma regla del género próximo y la diferencia específica, pero la respuesta en todo caso será otra. Las preguntas del tipo *¿qué es . . .?* sólo son aplicables a entes, a especies de entes. De ahí que, no sólo hay que afirmar que la interrogación ¿qué es la triangularidad?, puede ser respondida como la otra, sino que ya la pregunta misma exige esta clase de respuesta, implica que su "*aquello de que*" se pregunta es ente. La triangularidad, hemos dicho, es el ser de los triángulos, pero ya la pregunta misma encubre, por así decirlo, este ser, al tratarlo y concebirlo como ente.

Ahora bien, volvamos la mirada a la pregunta ¿qué es el ente? y apliquemos la distinción establecida. La pregunta, naturalmente, resulta tener dos sentidos; en el fondo, constituye la forma de dos preguntas distintas. La primera podríamos formularla así: *¿qué son los entes?* La segunda, que de acuerdo a nuestra terminología sería ¿qué es la entidad?, equivale a *¿qué es el ser?* Cuando más arriba afirmamos, de acuerdo con Aristóteles y Nicolai Hartmann, que la pregunta ontológica, al parecer, debería asumir la forma ¿qué es el ente?, esta pregunta, palmariamente, fue tomada en su primer sentido.

Pues bien, cuando preguntamos normalmente ¿qué es el hombre?, ¿qué es el caballo?, ¿qué es el triángulo?, sabemos lo que interrogamos y sabemos cómo podemos responder. Se trata de interrogaciones relativas a especies determinadas de entes. La pregunta *¿qué es . . .?*, aplicada a estos entes, tiene un sentido preciso, determinado. Pero cuando llegamos al *ente* ¿es justo seguir preguntando de la misma manera? Cuando llegamos al ente pura y simplemente ¿no será menester tomar precauciones con relación a este, al parecer ineludible, *¿qué es . . .?*, o, mejor, evitarlo, eliminarlo? Pero entonces ¿cómo preguntar? Si *el ente* no es un ente, una especie de entes, ¿ es justo seguir preguntando *¿qué es el ente*? Esta pregunta pide el ser, la definición del ser de los entes. Pero sabemos que sólo son definibles las especies de entes, no el ente.

Creemos necesario afirmar, de modo fundamental y riguroso, que así como *la definición de la lógica tradicional sólo es aplicable a las especies de entes y no al ente ni al ser*, igualmente *la pregunta de la forma ¿qué es . . . ? sólo es aplicable a entes particulares o a especies de entes, pero no al ente ni al ser.* No sólo basta proclamar, en este caso, la inutilidad de la definición tradicional, sino también, recíprocamente, la de la pregunta coordinada con ella.

Dijimos, parece que la pregunta ontológica fundamental debería asumir la forma ¿qué es el ente? Y ahora resulta que, como toda pregunta presupone o determina una forma adecuada de contestación, parece también que con esta formulación la ontología difícilmente puede seguir su marcha adelante.

Veamos, sin embargo, para iluminar mejor y de un modo más completo esta dificultad, que pasa, propiamente, con la pregunta ¿qué es el ser?

VII. LA PREGUNTA ¿QUE ES EL SER?

La interrogación ¿qué es el ser? pide "*la esencia del ser*", o, por así decirlo, "*el ser del ser*". Del ser preguntamos ¿qué es?, esto es, ¿cuál es *su ser*? El ser es "aquello de que" preguntamos; y *el ser de este ser*, "*aquello que*" interrogamos.

Pero, si en verdad interrogamos así y entendemos de este modo la pregunta, resulta palmaria su inoperancia y su inutilidad. Nos enredamos en un proceso interminable de interrogaciones y más interrogaciones. No damos con *la pregunta última* que es como, en efecto, debe ser la pregunta ontológica. Si puedo *preguntar por el ser del ser*, entonces también debo seguir interrogando *por el ser del ser del ser*, y así *ad infinitum*.

¿Puede ser ésta la pregunta ontológica que buscamos? Manifiestamente no. Pero veamos de dónde y cómo resulta el señalado proceso interminable de interrogaciones.

La pregunta de la forma *¿qué es . . .*?, como vimos, es aplicable únicamente a entes; supone que *aquello de que se pregunta* es un ente, porque demanda el ser y lo único que tiene ser es el ente. Ahora bien, *aquello de que se pregunta*, en la interrogación que analizamos, es justamente *el ser* y no un ente o los entes en general. El modo mismo de preguntar olvida, por así decirlo, que el ser no es ente, hace que el ser sea tratado como si fuese un ente. La pregunta misma hace que el término ser no miente propiamente el ser sino un ente, aunque, en verdad, este ente resulta ser el ser de todos los entes, pero tratado y concebido como ente.

Si yo pregunto, por ejemplo, ¿qué es el hombre?, el hombre, evidentemente, es un ente. Interrogo por su ser. Pero si este ser del hombre es descubierto, puedo volver a interrogar ¿qué es?, ¿qué es "el ser del hombre"? Esto es, concibo "el ser del hombre" como ente, y, entonces, naturalmente, fluye la pregunta por el ser de este ente. Y así sucesivamente. Este proceso de seguir interrogando es aceptable y justificable, porque nos movemos en esferas cada vez más amplias de entes y no hemos alcanzado el ser de todos los entes. Como al contestar la pregunta por el ser de una especie de entes, recurrimos a la especie inmediatamente superior, podemos interrogar por el ser de esta nueva especie, y así sucesivamente. El ser propiamente tal va retrocediendo. En su lugar se instala otra especie, otro ente. La respuesta "el animal racional", descubierta de acuerdo a la regla de la definición de la lógica clásica, que decimos nos da el ser del hombre, en verdad sólo nos coloca frente a otro ente: "el animal". La especie hombre es incluída en una especie más amplia: "animal". Y cabe, entonces, volver a preguntar, y el animal ¿qué es? No sólo la pregunta trata a *aquello de que se pregunta* como ente, sino que la respuesta misma no hace otra cosa que exhibir un nuevo ente. La pregunta, realmente, pide el ser, pero, y a pesar de ello, la respuesta mienta justamente un ente. El ser retrocede, se oculta, queda olvidado.

Al preguntar ¿qué es el ser?, pensamos que hemos dado con el límite último, y que el proceso de seguir preguntando ha terminado. Y así debería ser, en efecto, porque interrogamos por el ser, pura y simplemente, y no por el ser de un ente determinado o de una especie de entes. *Pero preguntar necesariamente es preguntar algo de algo.* Del ser preguntamos ¿qué es? Preguntamos por . . . *su ser.* El ser, que creíamos lo último, resulta albergar *otro ser*, por así decirlo. Podríamos decir: *el primer ser* resulta no haber sido ser sino ente. El llamado *otro ser* sucumbirá también en la misma transformación, y así sucesivamente. Pero esto es inadmisible. El ser es uno solo. No hay varios "*seres.*" La ontología mantiene y tiene que mantener esta afirmación fundamental si no quiere perder su objeto y disolverlo en la nada. Al decir *ser*, pura y simplemente, pensamos justamente en este ser único, *absolutamente en el ser de todos los entes*. La multiplicidad de los entes no puedo ser tan abarcadora y amplia que *fuera de* ella ya no hubiese nada más. No hay sólo entes. Hay también, necesariamente, el ser, que hace precisamente que los entes, todos, sean entes. El ser está en todos y en cada uno de los entes, en el más asombroso y grandioso y en el más humilde e insignificante. No es menester salir *fuera de* la multiplicidad total de los entes para dar con el ser. No ha de aparecer sólo al final de ese proceso de pasar de especie en especie, que implican las interrogaciones sucesivas por las diferentes especies de entes. Ya está presente, plena e íntegramente, en el primer ente que consideramos, y, por tanto, en toda la serie de entes. Y, sin embargo, el ser es radicalmente distinto de los entes. Reside en un plano totalmente diferente. Está, si se quiere, *fuera de* la multiplicidad total de entes, pero este *fuera*, naturalmente, sólo significa que el ser no puede ser colocado al lado de los entes, como una especie nueva y coordinada con las especies de entes. El ser no es una especie o un género de entes. Buscarlo como si fuese es convertirlo justamente en lo que no es.

¿Qué es el ser? Hay aquí una incogruencia radical entre el modo de plantear la pregunta y aquello de que se pregunta. El ser no es ente, y, por tanto, resulta incoherente demandar ¿qué es? Y si no es posible la pregunta, si la pregunta tiene un sentido contradictorio, entonces tampoco es posible la respuesta. La pregunta en sí es radicalmente contradictoria, tan contradictoria como la interrogación ¿qué hierro es la madera? Implica la enunciación: *el ser es ente*. Es como preguntar *¿qué ente es el ser? El ser no es . . .* , en el sentido que decimos *es . . . esto* o *aquello*. El ser no es esto o aquello, porque todo "esto o aquello" distinto del ser es ente. El lenguaje parece hecho de tal manera que el "es" sólo sirve para hablar de los entes, para conectar entes. El "es", que mienta justamente el ser, nos sirve para hablar de los entes y no del ser en cuanto tal. Por ejemplo, la definición "el hombre es el animal racional", que pretende ser la definición del ser del hombre, no hace sino ocultar este ser. El ser del hombre está, propiamente, en el "es" y no en "animal racional". "Animal racional", que constituye la definición propiamente dicha, no hace sino mentar lo mismo que "hombre", aunque de un modo más desarrollado. Parece como si no fuese posible hacer progresar la indagación ontológica hablando del ser con el mismo término ser. Naturalmente, podemos decir de manera inconmovible: "el ser *es* el ser", pero con esta afirmación no aclaramos nada, el ser queda ser. Cuando a veces declaramos *el ser es,* sólo queremos decir el ser existe, está, se da en los entes.

El ser es indefinible. Sólo se definen entes. Ahora bien, la pregunta ¿qué es el ser? pide precisamente la definición del ser. No sólo propone una tarea inalcanzable, a la que el hombre podría, sin embargo, acercarse cada vez más, y en este sentido racional, posible, sino en sí misma irracional, imposible. La pregunta encierra una contradicción interna, demanda la definición de lo que en sí es indefinible. Es como preguntar *¿qué círculo es el cuadrado?*, en que se supone que el cuadrado es un círculo. Interrogar ¿qué es el ser? es como demandar *¿qué ente es el ser?*

El análisis anterior nos permite hacer la siguiente enunciación fundamental: la pregunta ¿qué es el hombre?, en realidad, equivale a esta: *¿qué ente es el hombre?* Y, en general, todas las preguntas del tipo *¿qué es . . .?*, son susceptibles de la misma transformación: *¿qué ente es . . .?*

Entonces, la pregunta ¿qué es el ente? se convierte en *¿qué ente es el ente?*, y la demanda ¿qué es el ser? en *¿qué ente es el ser?* Ahora bien, ¿pueden constituir estas preguntas anodinas los problemas fundamentales que debe resolver o tratar de resolver la ontología?

La ontología tiene que preguntar por el ser. Pero ¿qué? *Todo preguntar es preguntar algo de algo.* ¿Qué pregunta la ontología del ser? Y además ¿cómo? La ontología no puede proponerse la tarea de decir *qué es el ser.* No hay problema de ¿qué es el ser? Y, sin embargo, *hay un problema del ser.* Pero entonces ¿cuál es propiamente este problema del ser? ¿Cuál es la pregunta ontológica fundamental?

VIII. ONTOLOGIA Y EINAILOGIA

Confrontemos ahora y pongamos en relación nuestras dos preguntas fundamentales: ¿qué es el ente? y ¿qué es el ser?

Hemos mostrado que la primera, en el fondo, tiene o puede tener dos sentidos, y puede, por consiguiente, desdoblarse en dos preguntas distintas: *¿qué son los entes?* y *¿qué es "el ente"?* La segunda, en la que "el ente" equivale a "el ser", se identifica con ¿qué es el ser?, y se agrava con todas las dificultades de esta interrogación. La pregunta ¿qué es el ente?, en su primer sentido, pide *el ser de los entes* y es distinta de la pregunta ¿qué es el ente? en su segundo sentido que demanda *el ser del ser.* Para mejor distinguirlas vamos a designarlas como *primera pregunta* o *pregunta ¿qué es el ente*? y *segunda pregunta* o *pregunta ¿qué es "el ente"*? En el primer caso la expresión ente sin comillas y en el segundo, con comillas.

Ahora bien, admitimos que *hay un problema del ente y un problema del ser*. Pero todo problema es formulable en forma de pregunta o no es tal. Entonces ¿cuáles son propiamente las preguntas en que consisten estos dos problemas?

Para caracterizar y distinguir terminológicamente ambos problemas, vamos a introducir un término nuevo en el vocabulario filosófico. Al problema del ente, utilizando el término tradicional, vamos a llamarlo *problema ontológico*, y, en cambio, al problema del ser, *problema einailógico*. (Del griego *ὄν, ὄντος* (*on, ontos*) = *ente* y *εἶναι* (*einai*) = *ser*. De manera que ontología significa *ciencia o teoría del ente* y *einailogía, ciencia o teoría del ser*.)

Nuestra aporía se desdobla, entonces, en dos cuestiones: por una parte, ¿cuál es propiamente *la pregunta ontológica*?, y, por otra, ¿cuál es en verdad *la pregunta einailógica*? En general, ¿hay verdaderamente preguntas que merezcan este nombre como cuestiones fundamentales de la indagación filosófica?

Con arreglo a la distinción terminológica propuesta, *ontológico* quiere decir relativo a todos los entes y *einailógico*, relativo al ser en general o al ser de todos los entes. El término *óntico* podría reservarse para lo relativo a entes individuales o especies de entes. La pregunta ¿qué es el hombre?, por ejemplo, resultaría ser, por tanto, una *interrogación óntica*.

El análisis realizado hasta aquí, ha puesto en evidencia que las interrogaciones ¿qué es el ente? y ¿qué es el ser?, no pueden ser las preguntas que buscamos. Las designaciones de *pregunta ontológica* y *pregunta einailogica*, no mientan una realidad ya descubierta, no corresponden, verdaderamente, a preguntas ya formuladas como preguntas. Proponen, más bien, una tarea a realizar, una tarea que la filosofía debe cumplir. Sólo una investigación especial y a fondo puede decidir si esta aspiración es posible o no.

La llamada primera pregunta ¿qué es el ente? no es la pregunta ontológica. Como interroga por el ser de los entes, podemos llamarla pregunta *einai-ontológica*, pero como interroga por los entes en general como si se tratase de una especie de entes, podemos llamarla también *pregunta óntico-ontológica*. A su vez, la interrogación ¿qué es el ser? no es la *pregunta einailógica*, pero como equivale a *¿qué ente es el ser*?

o *¿qué especie de ente es el ser?*, podemos llamarla *pregunta óntico-einailógica.*

Estas designaciones, que suenan tal vez arbitrarias, curiosas y anodinas, muestran, sin embargo, la confusión de *ente* y *ser*, en que caen ambas preguntas. La distinción entre *ente* y *ser* es fundamental, y debe guiar rigurosamente la formulación misma de los problemas. Pero ¿es posible convertir esta distinción en separación rigurosa, y preguntar por el uno sin que entre necesariamente el otro en la interrogación? ¿La esencia misma del preguntar no torna imposible la pretensión de encontrar las interrogaciones por el ente y por el ser? Nuestra posición, por el momento, tiene que ser cautelosa y prudente. Otro trabajo, en preparación, intentará la elaboración sistemática y más detenida de esta aporía. Pues, sólo un examen ontológico y einailógico, radical y llevado a cabo sobre bases nuevas y completas, apoyado en una rigorosa distinción final. Pues, sólo un examen ontológico y einaliógico, radical y llevado a cabo sobre bases nuevas y completas, apoyado en una rigurosa distinción terminológica de *lo óntico, ontológico* y *einaliógico,* podrá dar la decisión final y tal vez determinar la posible reducción de las dos preguntas en una sola fundamental. La formulación adecuada de la pregunta por el ser puede representar quizá la mitad de la respuesta.

Al querer plantear la pregunta fundamental por el ser, parece que no sabemos qué es lo que no sabemos, parece que ni siquiera podemos precisar qué es lo que no sabemos del ser. "*El que*" de la pregunta parece diluirse como una gota en el piélago, y con él el mismo "*de que*", puesto que aquel sólo emerge cuando éste es posible. "*El que*" de una pregunta es siempre *lo no sabido* y "*el de que*", *lo sabido,* pero lo no sabido es siempre lo no sabido de algo sabido. Del saber emerge inevitablemente el no saber, como un horizonte; no viceversa. Yo no llego a saber porque no sé; más bien, advierto mi no saber sólo porque sé. No hay saber porque hay no saber, sino, más bien, sólo hay no saber porque hay saber. De otro modo, la piedra un buen día despertaría hombre.

La aporía es grave, y resuena en el mismo ser del hombre. El hombre mismo, en cuanto esencialmente es el ente que interroga, parece levantarse como una pregunta, pero como una pregunta sin respuesta, esto es, como una pregunta que el hombre no puede contestar. El hombre no sabe qué es, y en la medida en que sabe que no sabe qué es, toma conciencia de su situación radical.

¿Qué soy yo? No sé qué soy. Como entre los entes me levanto interrogando, soy tal vez una interrogación. Yo soy el que pregunta. Soy tal vez eso: ¿qué soy?

El hombre, en tanto tenga una pregunta por responder, será un enigma. La hondura del enigma humano echa sus raíces justamente en "la pregunta fundamental" que, al parecer, ni siquiera puede ser balcucida en lenguaje humano.

Universidad Mayor de San Andrés
Academia Nacional de Ciencias de Bolivia
La Paz, Bolivia

INDEX OF NAMES

Analecta Husserliana

The Yearbook of Phenomenological Research

Editor-in-Chief

Anna-Teresa Tymieniecka

The World Institute for Advanced Phenomenological Research and Learning, Belmont, Massachusetts, U.S.A.

1. Tymieniecka, A-T. (ed.), *Volume 1 of Analecta Husserliana*. 1971
ISBN 90-277-0171-7
2. Tymieniecka, A-T. (ed.), *The Later Husserl and the Idea of Phenomenology*. Idealism – Realism, Historicity and Nature. 1972 ISBN 90-277-0223-3
3. Tymieniecka, A-T. (ed.), *The Phenomenological Realism of the Possible Worlds*. The 'A Priori', Activity and Passivity of Consciousness, Phenomenology and Nature. 1974 ISBN 90-277-0426-0
4. Tymieniecka, A-T. (ed.), *Ingardeniana*. A Spectrum of Specialised Studies Establishing the Field of Research. 1976 ISBN 90-277-0628-X
5. Tymieniecka, A-T. (ed.), *The Crisis of Culture*. Steps to Reopen the Phenomenological Investigation of Man. 1976 ISBN 90-277-0632-8
6. Tymieniecka, A-T. (ed.), *The Self and the Other*. The Irreducible Element in Man, Part I. 1977 ISBN 90-277-0759-6
7. Tymieniecka, A-T. (ed.), *The Human Being in Action*. The Irreducible Element in Man, Part II. 1978 ISBN 90-277-0884-3
8. Nitta, Y. and Hirotaka Tatematsu (eds.), *Japanese Phenomenology*. Phenomenology as the Trans-cultural Philosophical Approach. 1979
ISBN 90-277-0924-6
9. Tymieniecka, A-T. (ed.), *The Teleologies in Husserlian Phenomenology*. The Irreducible Element in Man, Part III. 1979 ISBN 90-277-0981-5
10. Wojtyła, K., *The Acting Person*. Translated from Polish by A. Potocki. 1979
ISBN Hb 90-277-0969-6; Pb 90-277-0985-8
11. Ales Bello, A. (ed.), *The Great Chain of Being* and *Italian Phenomenology*. 1981 ISBN 90-277-1071-6
12. Tymieniecka, A-T. (ed.), *The Philosophical Reflection of Man in Literature*. Selected Papers from Several Conferences held by the International Society for Phenomenology and Literature in Cambridge, Massachusetts. Includes the essay by A-T. Tymieniecka, *Poetica Nova*. 1982 ISBN 90-277-1312-X
13. Kaelin, E. F., *The Unhappy Consciousness*. The Poetic Plight of Samuel Beckett. An Inquiry at the Intersection of Phenomenology and literature. 1981
ISBN 90-277-1313-8
14. Tymieniecka, A-T. (ed.), *The Phenomenology of Man and of the Human Condition*. Individualisation of Nature and the Human Being. (Part I:) Plotting

the Territory for Interdisciplinary Communication. 1983
Part II see below under Volume 21. ISBN 90-277-1447-9

15. Tymieniecka, A-T. and Calvin O. Schrag (eds.), *Foundations of Morality, Human Rights, and the Human Sciences.* Phenomenology in a Foundational Dialogue with Human Sciences. 1983 ISBN 90-277-1453-3
16. Tymieniecka, A-T. (ed.), *Soul and Body in Husserlian Phenomenology.* Man and Nature. 1983 ISBN 90-277-1518-1
17. Tymieniecka, A-T. (ed.), *Phenomenology of Life in a Dialogue Between Chinese and Occidental Philosophy.* 1984 ISBN 90-277-1620-X
18. Tymieniecka, A-T. (ed.), *The Existential Coordinates of the Human Condition: Poetic – Epic – Tragic.* The Literary Genre. 1984 ISBN 90-277-1702-8
19. Tymieniecka, A-T. (ed.), *Poetics of the Elements in the Human Condition.* (Part 1:) The Sea. From Elemental Stirrings to Symbolic Inspiration, Language, and Life-Significance in Literary Interpretation and Theory. 1985
For Part 2 and 3 *see below* under Volumes 23 and 28. ISBN 90-277-1906-3
20. Tymieniecka, A-T. (ed.), *The Moral Sense in the Communal Significance of Life.* Investigations in Phenomenological Praxeology: Psychiatric Therapeutics, Medical Ethics and Social Praxis within the Life- and Communal World. 1986
ISBN 90-277-2085-1
21. Tymieniecka, A-T. (ed.), *The Phenomenology of Man and of the Human Condition.* Part II: The Meeting Point Between Occidental and Oriental Philosophies. 1986 ISBN 90-277-2185-8
22. Tymieniecka, A-T. (ed.), *Morality within the Life- and Social World.* Interdisciplinary Phenomenology of the Authentic Life in the 'Moral Sense'. 1987
Sequel to Volumes 15 and 20. ISBN 90-277-2411-3
23. Tymieniecka, A-T. (ed.), *Poetics of the Elements in the Human Condition.* Part 2: The Airy Elements in Poetic Imagination. Breath, Breeze, Wind, Tempest, Thunder, Snow, Flame, Fire, Volcano... 1988 ISBN 90-277-2569-1
24. Tymieniecka, A-T., *Logos and Life.* Book I: Creative Experience and the Critique of Reason. 1988 ISBN Hb 90-277-2539-X; Pb 90-247-2540-3
25. Tymieniecka, A-T., *Logos and Life.* Book II: The Three Movements of the Soul. 1988 ISBN Hb 90-277-2556-X; Pb 90-247-2557-8
26. Kaelin, E. F. and Calvin O. Schrag (eds.), *American Phenomenology.* Origins and Developments. 1989 ISBN 90-277-2690-6
27. Tymieniecka, A-T. (ed.), *Man within his Life-World.* Contributions to Phenomenology by Scholars from East-Central Europe. 1989
ISBN 90-277-2767-8
28. Tymieniecka, A-T. (ed.), *The Elemental Passions of the Soul.* Poetics of the Elements in the Human Condition, Part 3. 1990 ISBN 0-7923-0180-3

Analecta Husserliana

29. Tymieniecka, A-T. (ed.), *Man's Self-Interpretation-in-Existence*. Phenomenology and Philosophy of Life. – Introducing the Spanish Perspective. 1990
ISBN 0-7923-0324-5
30. Rudnick, H. H. (ed.), *Ingardeniana II*. New Studies in the Philosophy of Roman Ingarden. With a New International Ingarden Bibliography. 1990
ISBN 0-7923-0627-9
31. Tymieniecka, A-T. (ed.), *The Moral Sense and Its Foundational Significance: Self, Person, Historicity, Community*. Phenomenological Praxeology and Psychiatry. 1990 ISBN 0-7923-0678-3
32. Kronegger, M. (ed.), *Phenomenology and Aesthetics*. Approaches to Comparative Literature and Other Arts. Homages to A-T. Tymieniecka. 1991
ISBN 0-7923-0738-0
33. Tymieniecka, A-T. (ed.), *Ingardeniana III*. Roman Ingarden's Aesthetics in a New Key and the Independent Approaches of Others: The Performing Arts, the Fine Arts, and Literature. 1991
Sequel to Volumes 4 and 30 ISBN 0-7923-1014-4
34. Tymieniecka, A-T. (ed.), *The Turning Points of the New Phenomenological Era*. Husserl Research – Drawing upon the Full Extent of His Development. 1991 ISBN 0-7923-1134-5
35. Tymieniecka, A-T. (ed.), *Husserlian Phenomenology in a New Key*. Intersubjectivity, Ethos, the Societal Sphere, Human Encounter, Pathos. 1991
ISBN 0-7923-1146-9
36. Tymieniecka, A-T. (ed.), *Husserl's Legacy in Phenomenological Philosophies*. New Approaches to Reason, Language, Hermeneutics, the Human Condition. 1991 ISBN 0-7923-1178-7
37. Tymieniecka, A-T. (ed.), *New Queries in Aesthetics and Metaphysics*. Time, Historicity, Art, Culture, Metaphysics, the Transnatural. 1991
ISBN 0-7923-1195-7

Kluwer Academic Publishers – Dordrecht / Boston / London

Made in the USA
Las Vegas, NV
14 November 2024

11639802R00267